T0259413

Human Friendly Mechatronics

Human Friendly Mechatronics

Selected Papers of the International
Conference on Machine Automation
ICMA2000
September 27 - 29, 2000, Osaka, Japan

Co-Sponsored by
Japan Society for Precision Engineering
International Federation for Theory of Machines and Mechanisms

IFToMM

Edited by

Eiji Arai
Osaka University

Tatsuo Arai
Osaka University

Masaharu Takano
Kansai University

2001
Elsevier
Amsterdam - London - New York - Oxford - Paris - Shannon - Tokyo

ELSEVIER SCIENCE B.V.
Sara Burgerhartstraat 25
P.O. Box 211, 1000 AE Amsterdam, The Netherlands

First edition 2001
Transferred to digital printing 2006

Library of Congress Cataloging in Publication Data
A catalog record from the Library of Congress has been applied for.

ISBN: 0 444 50649 7

♾ The paper used in this publication meets the requirements of ANSI/NISO Z39.48-1992 (Permanence of Paper).
Printed and bound by Antony Rowe Ltd, Eastbourne

Contents

ICMA and NTF Lectures

Human Interface and Communication

Human Support Technology

Actuator and Control

Vision and Sensing

Robotics

Design and Manufacturing System

Preface

ICMA (International Conference on Machine Automation) was born in Finland, who is well known to be one of the most advanced countries in the field of science and technology. ICMA2000 is the third Conference, but is the first in Japan. In the past Conference held in Finland many people from various countries including Japan participated it, discussing on machine automation, mechatronics and robotics. It is our great pleasure that Finnish ICMA Members gave us the chance to hold it in Japan.

"Mechatronics" is the technology or engineering which has been rapidly developed in the last 2 or 3 decades. Today we can easily find "mechatronics products" in industries and in our daily life. The mechatronics products are listed as follows: Intelligent engine and cruise control systems for automobiles; Intelligent household electric appliances such as computer controlled sewing machine, washing machine, refrigerator, VTR; Auto-focusing camera; Computer peripheral equipment such as printer, magnetic/optical disk drive; Information-processing equipment such as digital communication system, portable telephone; NC machine tools such as wire-cut electric discharge machine, NC milling machine; environment control system including air conditioner, automatic door; Semiconductor manufacturing facilities; Automatic ZIP code reading machine; Sorting machine for agricultural or marine products; etc.

The word "mechatronics" is first defined as the combination of engineering in mechanics and electronics. However, the meaning of this word gradually changes to be taken widely as "intellectualization of machines and systems" in these days. Namely, machines and systems, which control their motion autonomously and appropriately based on the external information, generate a common technology of mechatronics. According to this definition, the automation for an industrial factory line, such as machining line, assembly line, strip mill line is based on the mechatronics technology. Control system for train operation and intelligent transportation system (ITS) for automobiles are also realized by the mechatronics technology. Considering these circumstances, it is evident that mechatronics has achieved the high productivity and has enriched our daily life by improving the performance of machines and systems. It is easy to understand that this trend will continue to the next century in accordance with the advances of computer technology.

However, a paradigm shift is desired and emphasized toward the 21st century in every aspect of our society, and even mechatronics is not an exception. The paradigm shift of mechatronics is that from the pursuit of the productivity or efficiency to the preference of the humanity. We will easily realize that it is still difficult for a beginner to use a personal computer skillfully, although its performance is improved year by year. For another example, much less than half of the housewives can probably master the operation of a VTR. We have to say these are never friendly to human beings. However, we find some little changes in mechatronics. One example is that the worker friendly U-shaped layout begins to be adopted in a factory production line. As for robotics, a welfare robot will be developed rapidly toward the practical utilization in the coming aging society. And it is the time when a new concept of a human friendly robot must be proposed, in order that a robot is welcomed by many human users.

In order to realize human friendly mechatronics, it is inevitable to design a product from a new point of view. It is also necessary to consider what a human friendly product concretely means, how the humanity is compatible with the technological performance. To answer these questions, psychological and physiological aspects of the human beings must be taken into account. It is important subject toward the beginning of the 21st century to point out clearly the problems existing in the system which includes both human beings and machines, and to find out the elemental technology which solves these problems.

Taking on the mission mentioned above, ICMA2000 provides a forum to discuss "Human Friendly Mechatronics" for the first time in the world. Many presentations and discussions are expected on topics such as, machine systems which interact with human beings, psychological, physiological, and physical behaviors of human being itself, robotics, human-mimetic mechanical systems, commercial application examples and so on. I hope that this conference will be a rich source of effective informations about human friendly mechatronics and send to the world the importance of this new problem.

Honorary Chair   General Chair

Masaharu Takano,   **Eiji Arai**

ICMA2000 Conference Committees

Honorary Chair:
 Takano, M. (Prof. Emeritus of Univ. of Tokyo, Japan)

General Chair:
 Arai, E. (Prof. of Osaka Univ., Japan)

International Advisory Committee:
 Itao, K. (Prof. of Univ. of Tokyo, Japan)
 Kivikoski, M. (Prof. of Tampere Univ. of Tech., Finland)

Steering Committee:

Dohi, T. (Japan)	Fujie, M. (Japan)	Fukuda, T. (Japan)
Hirose, S. (Japan)	Kajitani, M. (Japan)	Nakano, E. (Japan)
Owa, T. (Japan)	Sato, T. (Japan)	Tanie, K. (Japan)

Local Organizing Committee:

Aoyagi, S. (Japan)	Hara, S. (Japan)	Hosaka, H. (Japan)
Inoue, K. (Japan)	Inoue, T. (Japan)	Kamejima, K. (Japan)
Kamiya, Y. (Japan)	Kawato, M. (Japan)	Kume, Y. (Japan)
Morimoto, Y. (Japan)	Ohnishi, K. (Japan)	Onosato, M. (Japan)
Osumi, H. (Japan)	Sasaki, K. (Japan)	Seki, H. (Japan)
Shirase, K. (Japan)	Tadokoro, S. (Japan)	Taura, T. (Japan)
Tejima, N. (Japan)	Umeda, K. (Japan)	

Technical Program Committee:

Chair: Arai, T. (Prof. of Osaka Univ., Japan)

Airila, M. (Finland)	Angeles, J. (Canada)	Asada, M. (Japan)
Asama, H. (Japan)	Drew, P. (Germany)	Espiau, B. (France)
Fournier, R. (France)	Guinot, J.C. (France)	Halme, A. (Finland)
Higuchi, T. (Japan)	Hiller, M. (Germany)	Horowitz, R. (USA)
Kawamura, S. (Japan)	Kecskemethy, A. (Austria)	
Kerr, D.R. (UK)	Komeda, T. (Japan)	Kurosaki, Y. (Japan)
Lee, M.H. (UK)	Matsuhira, N. (Japan)	Mitsuishi, M. (Japan)
Mizugaki, Y. (Japan)	Morecki, A. (Poland)	Saito, Y. (Japan)
Schiehlen, W. (Germany)	Shirai, Y. (Japan)	Sioda, Y. (Japan)
Takamori, T. (Japan)	Tani, K. (Japan)	Tarn, T.J. (USA)
Tomizuka, M. (USA)	Trevelyan, J.P. (Australia)	Vaha, P. (Finland)
Virvalo, T. (Finland)	Vukobratovic, M. (Yugoslavia)	
Wikander, J. (Sweden)	Yamafuji, K. (Japan)	

Editorial Note

The ICMA Program Committee received 127 abstracts from 19 countries including Japan and Finland. Every abstract was surveyed and evaluated by two referees of the committees, and 118 were accepted for the presentation at the conference. Finally, 103 of the accepted papers were presented in 29 technical sessions including special organized sessions, "Wearable Information Systems" and "Medical Mechatronics & CAS". The Executive Committee and the Program Committee had an agreement on the publication of a book of selected papers, approximately 60, from the presented papers. Every presented paper was reviewed by two referees at its presentation and evaluated with respect to the following items: originality, contribution, paper quality, and presentation. According to the evaluation results, 61 technical papers were finally selected through the severe screening at the Selection Board. The selected papers with one invited paper are reorganized into 6 categories: "Human Interface & Communication", "Human Support Technology", "Actuator & Control", "Vision & Sensing", "Robotics", and "Design & Manufacturing System". The invited papers on ICMA & NTF Lectures at the conference are also included.

It would be our great pleasure to publish a volume of high quality papers dealing with the up-to-date mechatronics technologies. We hope that this book will bring advanced knowledge and valuable information to the industries as well as to the academies and will contribute to the further development in mechatronics and its related fields.

Program Committee Chair, Prof. Tatsuo ARAI

List of Referees for Paper Selection

Aiyama, Y. (Japan)	Ansorge, F. (Germany)	Aoyagi, S. (Japan)
Cotsaftis, M. (France)	Dohi, T. (Japan)	Egawa, S. (Japan)
Hayward, V. (Canada)	Inoue, K. (Japan)	Kamiya, Y. (Japan)
Kawamura, S. (Japan)	Kivikoski, M. (Finland)	Komeda, T. (Japan)
Kotoku, T. (Japan)	Kotosaka, S. (Japan)	Kurata, J. (Japan)
Mae, Y. (Japan)	Maties, V. (Romania)	Matsuhira, N. (Japan)
Morimoto, Y. (Japan)	Morimoto, Y. (Japan)	Nilsson, M. (Sweden)
Onosato, M. (Japan)	Osumi, H. (Japan)	Park, C. S. (Japan)
Riekki, J. (Finland)	Saito, Y. (Japan)	Salminen, V. (USA)
Sasaki, K. (Japan)	Seki, H. (Japan)	Shirase, K. (Japan)
Tadokoro, S. (Japan)	Takamori, T. (Japan)	Takano, M. (Japan)
Tani, K. (Japan)	Tejima, N. (Japan)	Umeda, K. (Japan)
Viitanen, J. (Finland)		

ICMA and NTF Lectures

Human Friendly Mechatronics (ICMA 2000)
E. Arai, T. Arai and M. Takano (Editors)

Mobility, Wearability, and Virtual Reality; the Elements of User Friendly Human interfaces, Modern Platforms for Cooperation

Ilpo Reitmaa, Science and Technology Counsellor

Embassy of Finland, National Technology Agency Tekes

This paper is a part of the ICMA2000 discussion with views to

- *User friendliness*
- *Benefits of being wireless*
- *Potential of virtual environments*
- *Wearability*
- *and views to all this as an obvious cooperation platform.*

JAPAN THE TECHNOLOGY SUPERPOWER

Japan is a technology superpower that has achieved amazing results in the field of electronics, information technology, material technologies, production technologies, robotics, precision mechanics and in combining mechanics and electronics, just to name a few. Simultaneously we know about the Japanese high quality, high work ethics, and about devotion to work whenever a goal is set. Japan is a country of great human power and dedication. The Japanese seem to have world record in kindness. The Japanese also seem to hold a world record in getting organized and having unsurpassable loyalty to work.

WHY USER FRIENDLINESS, WHAT USER FRIENDLINESS?

In this context we are mainly talking about combining electronics to mechanics, about mechatronics, that is. In Japan, at the consumer side this can easily be seen as a multitude of mass produced personal and home devices of unsurpassable quality and cost effectiveness. However the potential of electromechanics (mechatronics, if you like) is not only limited to the perhaps most visible consumer sector. Industrial electromechanics and mechatronics penetrate in all fronts including, manufacturing technologies, robotics, microelectronic and micromechanic devices, to name just a few. So mechatronics, at business level is important both in the business to consumer and in the business to business markets.

Along with this ever spreading of application spectrum two well known phenomena occur, both in electronics and in mechanics and in combining them:

- *Size of the devices decreases, and*
- *Complexity of devices increases.*

Along with this development, there is even doubt whether profitable applications for all these new devices and systems any more really exist. That is; do capabilities and visions of electromechanics, IT, and of all our innovations exceed the needs? That is why the popular phrase about user friendliness becomes more serious issue than ever. User friendliness becomes more than a phrase.

Making our devices, systems and services understandable and compatible with human beings is the only way to go if we intend to give room for the internal complexity of our innovations to increase.

There are examples about this at the personal computer (PC) side. Increasing visible complexity combined with deliberate (?) user-unfriendliness has made the most popular personal computer operating system (Windows) to face the question: "can this anymore be the way to go?". No, complexity must not show. User friendliness is the way to go. Complexity? It must remain inside.

Certainly the user friendliness as a term does call for a definition. There are better experts to do it. To me user friendliness is very practical things like

- *Intuitive use*
- *Functions conforming to assumptions*
- *Needless manuals*
- *Forgiving algorithms*
- *Feeling of confidence*
- *Repeatability and transparency when operated.*

The last item might be the most important; transparency to me meaning "using" without any particular feeling of "using".

User friendliness is not a phrase. It is the necessity, the bottleneck to open, in order to give technology and our new innovations room to flow. The way to go to get technology to become more sophisticated internally, less complicated externally.

USER FRIENDLINESS, EACH DAY IS DIFFERENT

We might wonder why the question of user friendliness has not been solved long ago? Certainly everybody has wished to have devices, systems and services to match the human being from the very beginning, from the early days of engineering sciences?

Yes, technology evolves fast, but basic properties of human beings practically not at all. The basic senses, perception, muscle structure, brain structure, fundamentals of behavior of this two legged creature has remained and will remain basically same for tensthousands of years.

These is no Moore's law for human beings, saying that some doubling in our genetic structure would happen within so-and-so-many years. But, yes, there are valid and proven laws of increasing complexity at the technology side. In this type of circumstances, no wonder that mismatches may occur, daily.

Thus, though the genetic evolution rate of human beings is extremely slow (practically zero in our scale) there are changes in culture, changes in social environment and in human behavioral patterns as well. As an example:

-       Early viewers of the very first cinemas were upset and deeply touched by the few minutes experience. On the other hand, today we can watch TV for hours and still easily remain organized about what we see.

This means, interaction with technology changes human behavioral patterns, too.

In Finland it has been demonstrated that the time for Internet browsing has been taken from watching TV. We also can see Internet accessed while walking. Again, along with all those changes in applying technology we get new behavioral patterns.

The actual result of all this? We will get new definitions for user friendliness in each of the new circumstances.

To conclude:
-       *Rapid development of technology, and*
-       *Zero development of human genetic properties, and*
-       *Gradual developments in social structures, and*
-       *Development generated by human interaction with technology...*

...all four mean that we will be facing a complex interfacing question, with at least four elements in it, when we are striving towards user friendliness. All this also means that user friendliness today is different from user friendliness of tomorrow.

When we wake up next morning the solutions to gain user friendliness will be different from what we believe today. Thus, there cannot be any permanent everlasting solution to user friendliness. From the research and industry point of view, I would call this a benefit. There always some work, actually; a lot of work to be done.

Even more important, from the business point of view, user friendliness can be the competitive advantage for the quick implementers. Advantage for doing business and obtaining market share, that is. All the elements of user friendliness can be a very clear criteria when consumers and customers make decisions.

To conclude: User friendliness means business. User friendliness is dynamic, something today, different tomorrow. Early and up to date user friendliness means competitive advantage. Remember genetics, social situation, effect of human interaction with technology, and the technological complexity. All call for consideration.

WHY WIRELESS?

The well known tendency towards wireless can be explained in many ways. So, why wireless? At its most fundamental level, wireless means uninterruptible 24-hours-a-day of business. When not any more limited by wires, people move, get connected anywhere, at any instant --- and become invoiced for that. In the current mobile handset (cell phone and alike) boom this can be seen most clearly.

The utmost urge of human being, being part of something (the network) and on the other hand still being free (to move) becomes realized all with one single personal device and one system solution.

To summarize, in the consumer world wireless means:

- *24 hour business*
- *Being part of, but simultaneously still...*
- *Being free*

In the industrial environment being wireless brings along other types of benefits. It means faster setup and installation times for industrial systems and extended flexibility. In a production environment, for example, production lines can be configured with reasonable effort.

Wide use of wireless solutions has for long been constrained by safety and security issues because of the common communications path, in the air. However, for example, the proven achievements in mobile phone security is likely to give thrust for believing that wireless can be relied upon in the industrial environment as well. This is going to become interesting. The obvious wireless revolution at the consumer side seems to support the wireless revolution at the industrial side, and there will be still much work to do.

To summarize:

- *Wireless means flexibility*
- *Consumer wireless is safe*
- *Industrial wireless will become even safer*

VIRTUAL ENVIRONMENTS, THE UTMOST OF USER INTERFACE

As discussed earlier, human genetic basic properties remain. Thus regardless of any technical development, the human walks in an upright position, has five basic senses and so on. Therefore all our human interface problems remain the same at the most fundamental level. But is it understood fully? Are we interested enough? If we are interested in technology, are we interested in the human being? When talking about user friendly, yes, we should.

By virtual environments we mean accurate reproduction of sensory stimuli and on the other hand arrangements in which or physical responses are sensed and processed by computer. The arrangement (realistic stimulus, realistic consequences from human actions) aims to perfect illusion of being part of and inside in an artificial computer generated world. Mostly in 3 dimensional world, but in a world for other senses as well.

Though still without any total breakthrough virtual environment technologies are something to take a closer look at. The reasons:

- In these technologies you have to take the human structures into account.
- Virtual environments will call for lot of help from mechatronics.

Virtual worlds are not really so far ahead. In military simulators we already have them. On the other hand, in the consumer market Japan definitely is the world leader in computer games, Playstations and alike. Any innovations in three dimensional processing, displays, "natural" interface devices and a like are, within the limit of consumer costs, adopted into the game world.

As already mentioned, the nice thing about virtual environments is that using electronics and mechatronics for creating perfect illusions forces us to deeply understand of human sensory organs, perception and body.

All in all, virtual environments are simply the ultimate user interface. What has been said about transparency ("using" without feeling of "using") holds especially true. Nothing comes in between and interfacing is done to sensory organs in a natural manner.

To summarize:

- *Virtual environments aim to creating perfect illusions*
- *Virtual environments force to think about the human senses and behavior*
- *Virtual environments can become the ultimate user interface*
- *Virtual environments will need mechatronics*

Japan is strong in high volume "consumer virtual reality" and has great achievements in boosting the necessary computing power for example. From the researchers point of view this research area is extremely cross disciplinary, thus interesting. It also very interesting because of the numerous problems unsolved. In our case it should be interesting because of the mechanics involved. The need of mechanics will never disappear because of the mechanics of the human being. From the business point of view there are immediate opportunities in the high volume entertainment side, and existing and future possibilities, on the industrial side.

TO CARRY OR TO WEAR?

Virtual environments are something that take our thoughts to display helmets and data gloves, to things that we wear. But they can take it further. Could electronics and electromechanics be part of our daily habitus, part of our clothing? Could we be dressed up into electronics and elecromechanincs? The answer might be yes for many reasons.

Firstly, you already see it every day. Ladies with cherry blossom color i-Mode phones fitted to the color of their jacket. Walkmans "embedded" into some proper hidden pocket of suit, and so on. As a matter of fact people seem carry substantial amount of electronics, optics, electromechanic with them. Now they carry, later perhaps wear?

There are also some obvious business aspects. Textile and clothing industry is not doing very well in many countries. Clothing can be produced, but profitability is the issue. As a result, especially in the sporting clothes section we have begun to see technologies like air cushion shock absorbers in shoes, flashlights an so on, but might be just the beginning. In order to maintain profitability, added value for clothing industry has to be generated from somewhere. Electronics and electromechanics along with the utilization of pure physics and chemistry seem one way to go. The way to go towards smart clothes and wearable electronics. To wear electronics, why not?

To summarize:

- *We already wear electronics and electromechanics*
- *Making devices part of our outfit increases transparency (use without especially noticing the use)*
- *Wearability is a challenge for multitude of technologies*
- *Wearability could save clothing industry*

BENEFITS FROM COOPERATION

Cooperation between researchers and across borders means simply extension of resources, but there is another aspect to it. As the number of available technologies increases, the variations in the possibility of combining technologies of course increase even more rapidly.

There are innumerable examples about how this works, Back in Finland we are living a phase were mobile phones are combined with security technologies resulting into mobile banking, cell phone operated wending machines, etc. Or then mobile phones are combined with satellite positioning again bringing up a myriad of new applications.

Thus cooperation is good for:

- *extending resources*
- *generating innumerable new combinations of technologies*
- *for travelling towards the unknown and exiting future*

In the area of mechatronics each researcher and expert certainly knows his/her own expertise and what might complement to it. It is always good to take a look around, especially when two or more countries meet.

Human Friendly Mechatronics (ICMA 2000)
E. Arai, T. Arai and M. Takano *(Editors)*
© 2001 Elsevier Science B.V. All rights reserved.

Haptics: A Key to Fast Paced Interactivity

Vincent Hayward

Center for Intelligent Machines, McGill University, Montréal, Québec, Canada

1. INTRODUCTION

The word "haptics" is now well accepted. Hundreds of papers are published each year on the topic of haptic devices and interfaces. Haptics, as a technological niche, has become rich with opportunities and challenges. The field borrows from, and lends to, many subjects in science and technology. Among these, two are particularly relevant: "mechatronics" on one hand, and "robot-human interaction" on the other.

Haptic devices belong to the family of mechatronic devices because their fundamental function is to take advantage of *mechanical signals* to provide for communication between people and machines. It follows that haptic devices must include include transducers to convert mechanical signals to electrical signals and vice-versa used in conjunction with one or several computational or data processing systems. These transducers appeal to a variety of technologies: electromechanical devices, optoelectronics, fluids, smart materials, exploiting the possibilities that exist to build highly integrated and cost effective devices.

The popularization of haptics as an area of investigation is due to the work of such pioneers as Brooks and Iwata [1,6].

2. TAKING ADVANTAGE OF THE HAPTIC CHANNEL

It is hard to remain insensitive to the extraordinary ability of humans to mechanically experience, and act upon their surrounding world. Part of this ability is due to the sense of touch that comprises sensations of pressure, texture, puncture, thermal properties, softness, wetness, friction-induced phenomena, adhesion, micro failures, as well as local features of objects such as edges, embossings and divets. There are also vibro-tactile sensations that refer to the perception of oscillating objects in contact with the skin. In addition, proprioceptive, or kinesthetic perception, refers to the awareness of one's body state, including position, velocity and forces supplied by the muscles. Together, proprioception and tactile sensations are fundamental to manipulation and locomotion [2]. So, why are not computers and machines in general taking better advantage of these abilities? One goal of haptics is to fill this gap.

3. AN EXAMPLE

Consider a familiar sheet of paper, viewed as a display device. This sheet is intended to support information encoded in the form of structured discontinuities created by layering ink on it in order to change its reflective properties. Next, consider a computer screen with graphics capabilities. It can be programmed to display information also using structured discontinuities. Analogously, a computer mouse contains fixed mechanically-encoded information. It is not programmable. The step that was made to move from the sheet of paper to the graphics screen is analogous to the step made to move from a computer mouse to a haptic interface. Where the graphic screen can change its optical properties under computer control, a haptic device can change its mechanical properties under computer control.

The user of a conventional mouse receives almost no information from its movements. The buttons on it, for example, are considerably richer: they respond to state change of the interface. Their mechanical detent and the small acoustical noise they produce inform the user that an event has occured. Similarly, haptic interfaces attempt to make the information flow from machine to user non zero introducing an important difference, that of being programmable or computer controlled, as in the example of moving from the sheet of paper to the graphics screen. This is illustrated in Figure 1.

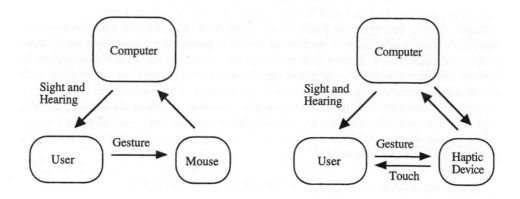

Figure 1. Left, conventional interaction loop. Right: interaction with a haptic device.

Graphical user interfaces (GUI's) have demonstrated that interactive presentation of data does not have to replicate reality, not even remotely. Being *suggestive* is what matters the most. Pull down menus and scrolling slider bars cannot be found elsewhere than on computer screens! The same holds for haptic interfaces. For example, the interaction forces we experience when moving objects generally occurs when they contact each other. With haptics, we can perfectly suggest a relationship between two distinct objects by creating a mutual interaction force, even if they are presented as being far apart.

Alternatively, some applications must recreate actual tasks. In other words, haptic interfaces should be designed to provide for a high fidelity reproduction of the phenomena which occur during actual manipulation. This is what in computer graphics is called the "quest for realism" and applies to haptic as well. The training of sensory-motor skills is one example of need for realism.

4. DEVICE EXAMPLES

The evolution of haptic technology can be appreciated appreciated by comparing three planar haptic devices, realized over a span of six years, see Figure 2.

Figure 2. Left: The Pantograph [3], custom precision machining, off-the-shelf components. Middle: The PenCat/Pro ™ by Immersion Canada Inc. Right: The Wingman™ by Logitech, a highly integrated device.

Other classes of devices with three or more degrees of freedom follow a similar evolution. For a recent survey see [5].

5. APPLICATION EXAMPLES

Bi-directionality is the single most distinguishing feature of haptic devices when compared to other machine interfaces, and this observation explains in part why they create a strong sensations of immediacy. A haptic device must be designed to "read and write" to and from the human hand (or foot, or other body parts). This combined "read and write" property may explain why the first applications of this technology involves "fast-paced interactivity" as testified by the commercial availability of haptic devices for gaming applications.

One application we have explored is the use of haptic feedback in stressful, and high paced environments, namely the operation of a GUI in absence of gravity. Haptic communication in this case could be used to reduce reliance on vision and possibly compensate for other factors caused by this type of environment [7]. Closer to everyday use, haptic devices have already been commercially introduced in certain high-end cars.

One other class of application of is interest to our group. It is the use of haptic feedback to assist operators to deal with complicated three dimensional data sets. A case presently under study is the manipulation of three dimensional angiograms which are particularly ambiguous when experienced graphically. For example, inspection of the views in Figure 3 indicate that it is difficult to realize that the shape of a vessel system could actually be highly organized. While stereopsis is helpful, the haptic experience of this system of curves should be very effective and fast [9].

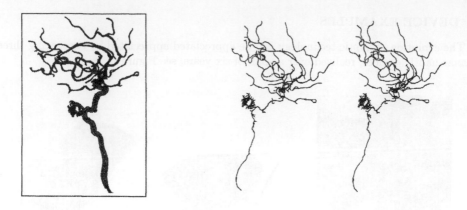

Figure 3. Left: Surface rendered vessel system. Right: a pair of stereoscopic views of its skeleton. They may be fused into a 3D image by uncrossing the eyes so that the left eye sees the left panel and the right eye sees the right panel. It is possible to realize that, despite the apparent complexity of the projection, the vessel system is actually divided into two distinct and highly organized subsystems, one in the back and one in the front.

6. HUMAN PERCEPTION

Our group is also exploring fascinating perceptual phenomena involving the haptic perception of shape, where contrary to intuition, force modulation as a function of movement seems to be a more determining factor of shape perception than geometrically induced limb movements [8]. In this study, it was shown that shapes could be "virtually experienced" when no geometrical information was available to the subject, see Figure 4.

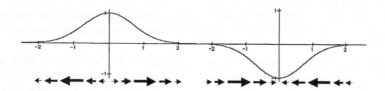

Figure 4. The arrows at the bottom show a lateral force field (LFF) that relates horizontal positions to a horizontal forces. With this stimulus, the subject receives no geometrical information, just force information, and yet what is experienced is a shape as drawn in the upper part of the figure. Note that this effect occurs over scales of centimeters and perhaps larger ones.

7. The Next Frontier

Finally, "next frontier" items were discussed. One of the them is the problem of displaying tactile information directly to the skin rather than through the manipulation of a tool. In the past, tactile displays were of one of two kinds: they were either shape displays, or relied on distributed vibrotactile stimulation. A tactile display device was described which is distinguished by the fact that it relies exclusively on lateral skin stretch stimulation. It is constructed from an array of closely packed piezoelectric actuators connected to a membrane. The deformations of this membrane cause an array of contactors to create programmable lateral stress fields in the skin of the finger pad (See Figure 5)[4]. Some preliminary observations were reported with respect to the sensations that this kind of display can produce. Quite surprisingly, one of them was shape.

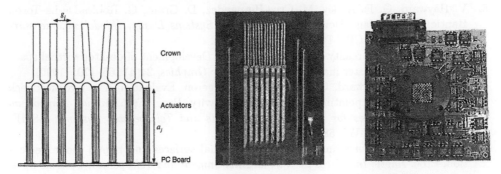

Figure 5. Left: Principle of operation. Piezoceramic actuators extend and contract, causing skin contactors to swing laterally with large mechanical movement amplification achieved through flexures. Middle: assembly detail showing and array of 64 piezoceramic actuators causing lateral swing in a array of 112 skin contactors. Right: Tactile display prototype. To give scale, the active area is 12×12 mm.

8. CONCLUSION

The field of haptics is developing rapidly. Haptic devices are mechatronic devices but with a special mission, that of facilitating communication between machines and humans. It is hoped that a fundamental benefit of this technology is enabling of fast paced interactivity.

Many challenges lay ahead of us to enable the progress of haptic interfaces connected with many areas of science and technology, as well as economic activity.

REFERENCES

1. F. P. Brooks, M. Ouh-Young, J. J. Batter, and P. J. Kilpatrick. Project GROPE: Haptic displays for scientific visualization, Computer Graphics: Proc. SIGGRAPH '90, pp. 177–185, 1990.
2. J.C. Craig, G.B. Rollman. Somesthesis. *Annual Review of Psychology*, pp. 50:305-331 (1999).
3. V. Hayward, J. Choksi, G. Lanvin, and C. Ramstein. Design and multi-objective optimization of a linkage for a haptic interface. *Advances in Robot Kinematics*, pp. 352–359. J. Lenarcic and B. Ravani (Eds.), Kluver Academic (1994).
4. V. Hayward and M. Cruz-Hernandez, Tactile Display Device Using Distributed Lateral Skin Stretch, Proc. *Haptic Interfaces for Virtual Environment and Teleoperator Systems Symposium*, ASME, IMECE-2000. DSC-Vol. 69-2, pp. 1309–1314 (2000).
5. V. Hayward, O. R. Astley, M. Cruz-Hernandez, D. Grant, G. Robles-De-La-Torre, Haptic Interfaces and Devices. In *Mechanical Systems Design (Modelling, measurement and control)*. CRC Press (2001). In Press.
6. H. Iwata. Artificial reality with force feedback: Development of desktop virtual space with a compact master manipulator. *Computer Graphics*, 24(4):165–170, (1990).
7. J. Payette, V. Hayward, C. Ramstein, D. Bergeron. Evaluation of a force-feedback (haptic) computer pointing device in zero gravity. Proc. *Fifth Annual Symposium on Haptic Interfaces for Virtual Environments and Teleoperated Systems*, ASME, IMECE-1996, DSC-Vol. 58. pp. 547-553 (1996).
8. G. Robles-De-La-Torre and V. Hayward, Virtual surfaces and haptic shape perception, Proc. *Haptic Interfaces for Virtual Environment and Teleoperator Systems Symposium*, ASME, IMECE-2000. DSC-Vol. 69-2, pp. 1081–1086 (2000).
9. D.G. Yi and V. Hayward, Skeletonization of Volumetric Angiograms for Display. In preparation.

Human Interface and Communication

Human Friendly Mechatronics (ICMA 2000)
E. Arai, T. Arai and M. Takano *(Editors)*
© 2001 Elsevier Science B.V. All rights reserved.

Micromechatronics for Wearable Information Systems

Kiyoshi ITAO

Institute of Environmental Studies, Graduate School of Frontier Sciences,
The University of Tokyo
7-3-1 Hongo, Bunkyo-ku, Tokyo 113-8656, Japan
itao@k.u-tokyo.ac.jp

The volume and mass of information machines should essentially be zero. Toward this ultimate target, efforts in electronics, machinery, physics and chemistry have all contributed to the advancement of microsystem technology. These efforts will continue going forward. At the same time, while computer development continues down the traditional path of human-operated machines, we are seeing the development of a so-called "pervasive computer" world where computers run without human operators. This paper outlines the technology related to these two trends, then introduces "nature interfacer" – a new information terminal concept that combines the movement toward smaller machines with the trend toward unmanned operation.

1. Wearable technology roots in time measurement

The first sundial is said to have been built by the Babylonians living in Mesopotamia some 3,500 years ago. Mankind grasped the concept of measuring time and used various tools to develop time-measuring devices, giving rise to the history of the clock.

The principles of the modern-day mechanical clock are rooted in the small clocks powered by spiral springs that were invented in Germany around 1500. In 1581, Italy's Galileo Galilei discovered pendular isochronism, opening the way to application of the pendulum in the measurement of time. In 1657, Dutch inventor Christian Huygens created the pendulum clock. In 1675, he invented the hair spring-balance wheel speed regulator, thereby laying the groundwork for the portable clock. Later in the 17th century, the escapement was invented in England, marking an important breakthrough in the history of time measurement. In the 18th century, the speed regulator that sits inside mechanical clocks was invented, making it possible to control the pace of the balance wheel and pendulum. As such, great advances were made in the development of the portable clock. In the 18th century, clocks were handcrafted through a division of labor primarily in London, Paris and Geneva. In the second half of the 19th century, manual production gave way to a rational manufacturing system in the United States. In this way, the mechanical clock came to have a widespread presence from the second half of the 19th century into the 20th century.

The method of using electrical power to run a clock was invented in the United States in 1804. Under this method, electromagnetic power was used to move a pendulum. This was followed by the invention of the AC electrical clock. The creation of the transistor later gave way to the transistor clock in 1954. Transistors, however, were not accurate enough.

Eventually crystal quartz clocks made their world debut in 1969. The parts consisted of a quartz pendulum, an IC and a micro step motor. Losing only 0.2 seconds a day, this was far more accurate than its mechanical predecessor.

The quartz clock technology took the miniaturization process to a new extreme through a combination of electrical parts (e.g. an LSI capable of running at low voltage, a liquid crystal display element, a long-life, small-size button-type lithium battery and a silver oxide battery) and mechanical parts (e.g. step motor, small transmission gears). The quartz clock arguably assumed a position at the very forefront of mechatronics. Such clock mechatronics technology became the basis of information machines and led to the development of new multi-function watches. Products that have already hit the market include:

(1) Medical/health devices measuring blood pressure and pulse, counting calories and performing other functions

(2) Business devices for registering telephone numbers, receiving messages (FM beepers) and handling other tasks

(3) Entertainment devices such as games with vibration boosters that produce tactile effects.

(4) Outdoor recreation devices with geomagnetic sensors, atmospheric pressure sensors, water depth sensors, temperature sensors, GPS and other functions.

The integration of communications and clock technology is giving way to an outflow of new products toward the wearable computing age.

2. Development of information microsystems

Looking back at the history of physics, Newtonian mechanics and other classical physics stand out for their tremendous impact on the industrial revolution that began in the second half of the 18th century and on the modern technology and industry that followed. Newtonian mechanics is a microcosm visible to the human eye that proved its true worth in energy innovations. In other words, it enabled the development of heavy, large-scale industries by replacing the physical labor of humans with steam engines, cars, ships, general machinery and other equipment. These industries, however, consume vast quantities of natural resources and this has resulted in destruction of the global environment.

Modern physics was basically established in the first half of the 20th century. Quantum mechanics, which is at the core of modern physics, is credited with promoting the high-tech revolution in the second half of this century and will continue to play that role in the 21st century. This formed the science of the microscopic world of molecules and atoms, and gave birth to steady stream of cutting edge technology related to information, electronics, biology, new materials and micromachines.

Miniaturization is one concrete development that occurred in the process. In order to promote miniaturization, it was necessary to deepen the technologies of integrated circuits, integrated structures and integrated data and to merge in a singular pursuit. In this way, higher performance and higher utility through miniaturization and new functions through the multiple uses of microscopic objects are becoming possible.

Figure 1 is a classic example of miniaturization showing changes in the mass and volume of information machines. Looking specifically at portable phones, Figure 2 shows that these two measurements decreased from 500g and 500cc in 1986 to 79g and 78cc in 1997. Figure

3 shows the trend among magnetic discs. As computer systems moved toward distributed processing, magnetic disc devices underwent miniaturization from the 14-inch mainframe units to 5.25-inch units, then to 3.5 inch units, 2.5-inch units and 1.8-inch units. Surface memory density increased about 10 fold every 10 years and has increased more than 50 times in the last 10 years. Around 1992, small discs actually eclipsed large discs in terms of such density. Roughly the same amount of memory capacity is now packed into a much smaller unit. While a 14-inch unit had 1.2GB of memory in 1983, a 3.5-inch unit had 1.4GB of memory in 1992 – comparable memory in a unit of 150 times less volume. The 1.8-inch units being developed in 1999 are expected to have 10 times less volume than even that.

Figure 4 borrows a diagram by R.W. Keynes showing the necessary volume and mass per bit of memory and thereby the potential capacity. As evident from the figure, memory density was 1Gbit/in2 in 1996, but now promises to enter a new range of 10Gbit/in2 in 2000 thanks to magnetic disc technology. In the future, memory of 10 to 100 times higher density will become a possibility through near-field optical memory technology and microcantilever technology applying STM technology. This is expressed in Figure 5.

3. Wearable computers take the stage

While construction moves forward on optical communications networks, digital wireless networks and other communications

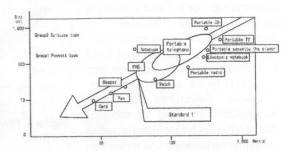

Figure 1.   Size and mass of portable machines (Tachikawa et al.)

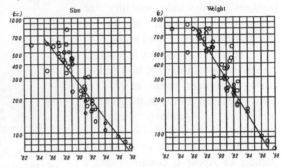

Figure 2. Size and weight reduction of the telephone

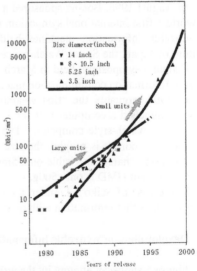

Figure 3. Increase in surface memory density of magnetic disc drives(1980-2000)

22

infrastructure, we are also witnessing an opening of networks, a diversification of equipment, services and information available, and rapid new developments in user interface equipment based on new technology. Computer downsizing has been accompanied by rapid advances in miniaturization technology (including mounting technology, LSI technology and micromachine technology) as well as revolutionary advances in the personalization and mobility of information. In short, the age of "wearable computing" has begun.

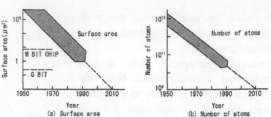

Figure 4. Physical requirements for 1 bit of memory (Keynes)

The first experiment in wearing computers is said to have involved an analog computer made at MIT in 1966 to predict results in the game of roulette. Another wearable computer was produced experimentally in 1980 by essentially putting an Apple II computer into a backpack. In 1990, a computer with a wireless network link-up was developed at Columbia University.

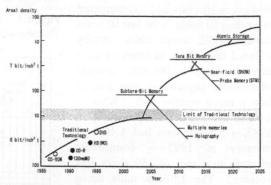

Figure 5. R/W principle and recording density of optical memories

In August 1996, Boeing sponsored a wearable computer workshop and in October 1997, the world's first international symposium on wearable computers was held. At a symposium on wearable information machines held in Japan in February 1998 by the Horological Institute of Japan, the Space Watch concept and micro information machines were exhibited. This biannual symposium, held in March and September this year, featured actual samples of portable information communication machines.

In 1998, arguably the "first wearable year," IBM Japan released a wearable PC and Compaq launched a computer that fits in the user's pocket. Seiko Instruments released its "Laputer," a watch-style computer. The IBM PC compares with popular notebook models. The main component is roughly the size of a headphone stereo and the head mount display (HMD) makes characters visible on a 5mm by 5mm PC screen when lined up with the eyes. This lightweight HMD is only 500g.

In the case of Seiko's Luputer, the user inputs the information he/she needs from a PC and exchanges that information with other Luputers via infrared transmission.

4.   Development of wearable information terminals

Figure 6 shows an attempt by the writer to classify different types of computer terminals. Computer terminals have traditionally been operated by humans who send digital input commands via a keyboard. The mass and volume of these terminals has become smaller and smaller through advances in LSI and micromachine technology. With the arrival of portable units, these terminals have become wearable. Another trend is in the technology for running computers without human operators. Under such technology, the computer senses analog information in its surroundings, converts that into a digital quantity, deciphers it based on its own knowledge (database), then sends the results of its analysis to a computer or another terminal linked to a communication circuit. On the right side of the diagram is a terminal appropriate to sensor communication. These terminals can be attached not only to humans, but also to other animals and to inanimate objects, and play an important role in monitoring health, tracking animals and measuring the deterioration of equipment and monitoring the global environment, among others. The writer proposes the use of such "nature interfacers."

	Mobile communication	Sensor-based communication
Terminal form	Portable ⇨ **Wearable** ⇨ Implant	
Operation	Human intervention	Without human intervention, automatic
Input signal	Digital	Analog
Object (interface)	Human (Human Interface)	Human, Animals, Nature, Artifacts (Nature Interface)
Operation means	Keyboard	Sensors
Role of CPU	Signal processing Output	A/D & Transfer digital signal after recognition
Connection to network	Wireless	Wireless or wire connection
KEY WORD	Portable telephone, PDA, pocket bell, note-type PC	Pervasive, Ubiquitous

Figure 6. Classification of the terminals for computer

Figure 7 shows the structure of the nature interfacer. It is a device that takes various information, converts it from an analog to a digital format and deciphers it within a watch-size computer, then makes a wireless transmission of that information. The key technology right now involves the conservation of energy through both soft and hard technology that enables the device to determine whether or not input data is meaningful, then appropriately continue to take in that information or go into sleep mode. The prospects for this technology are like that for the watch. Figure 8 shows the potential uses of such technology, e.g. tracking of migratory animals, environmental conservation through the simultaneous detection of chemical substances and other global environmental information, prevention of major accidents by measuring the deterioration of gears, axles and other moving parts. The smaller these nature interfacers become, the broader their scope of application – they will become applicable to everything ranging from airplanes to small birds. And if they can take in three-dimensional position information, fourth-dimensional time information and various chemical substance information at the same time, computers will be able to protect and enrich our lives without human operators.

24

5. Development toward bionet systems

As shown in Figure 9, the nature interfacer is the ultimate single-chip micro information network terminal integrating a traditional telephone, computer and personal information terminal.   The first step toward this ultimate picture is the adaptation of current technology into a device weighing several hundred grams. The second step will be a device weighing 100g.   In the third step, the weight will be 10g.   At that time, it stands to reason that the applicable range will undergo a shift in emphasis from quantity to quality. In other words, if a sensing terminal of less than 10g can be realized, it would be possible to attach devices to almost any information source (e.g. production sites, living environment, small birds, plants, humans) and provide windows to receive a massive inflow of information.

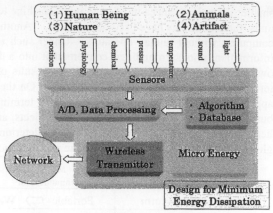

Figure 7. Composition of nature Interfacer

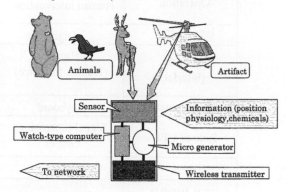

Figure 8. Application of nature Interfacer

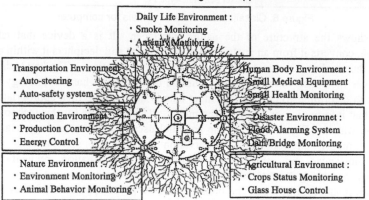

Figure 9. Nature Interfacer : Network communication system between human, artifacts and nature

Human Friendly Mechatronics (ICMA 2000)
E. Arai, T. Arai and M. Takano *(Editors)*
© 2001 Elsevier Science B.V. All rights reserved.

Monitoring of the User's Vital Functions and Environment in Reima Smart Clothing Prototype

Jaana Rantanen[a], Akseli Reho[b], Mikko Tasanen[a], Tapio Karinsalo[a], Jukka Vanhala[a]

[a]Tampere University of Technology, Institute of Electronics, P.O.Box 692, FIN-33101 Tampere, Finland

[b]Reima-Tutta, P.O.Box 26, FIN-38701 Kankaanpää, Finland

Smart clothing prototype for arctic environments has been developed. The main focus has been the creation of a truly functional concept that helps a suit's user to survive in harsh and demanding weather conditions. Communication, positioning, and several monitoring aids have been provided to the user. The monitoring system of the suit consists of temperature and acceleration sensors, a heart rate sensor, and an electric conductivity sensor measuring both the human user and the environment. According to these measurements, an automated emergency message can be sent. The suit has been tested in real conditions in Lapland during the winter 2000. The tests have shown that the suit fulfils the requirements placed for the prototype.

1. INTRODUCTION

Smaller and more powerful electronic components have made it possible to hide complex information processing systems into human clothing. The *wearable computing* destines to aid a human user to manage in different situations and environments by providing automated data processing functions and a personalised (even transparent) user interface. Wearable computers are always ready to interact with a user and similarly to their desktop counterparts, the main purpose of a wearable computer is to serve as an effective tool for information processing [1].

Smart clothing consisting of multimedia, wireless communication, and wearable computing has become a potential alternative for a wide range of personal applications, including safety and entertainment [2]. We see the smart clothing as intelligent structures based on electronics, non-electrical components, or intelligent textile materials that may enhance the functionality of ordinary clothing. Smart clothing is physically near to the user providing connection platform for several sensors needed for physiological measurements or other user's movements monitoring. Since it is worn by a human user and not explicitly used, the smart clothing tries to help the user to deal with different situations by improving or augmenting clothing's own functions, e.g. keeping the user warm, dry, and safe.

While being a usual garment covering a human user, the smart clothing may operate as a communication tool or as positioning equipment. Smart protective clothing for hazardous working environments or advanced user interfaces for different controllable devices are example application areas where smart clothing brings benefits.

Smart clothing for increasing user's safety in arctic environments has been studied in a cooperation project by Tampere University of Technology, University of Lapland, and Reima-Tutta Corporation. As one of the results, a completely functioning smart clothing prototype for arctic environments has been developed. The prototype is intended especially for experienced snowmobile users. The main reason for this target user group was that Reima-Tutta Corporation already has experience of manufacturing ordinary snowmobile suits. The application environment has obvious risks leading to danger situations that need to be prepared for, e.g. risk of getting lost or broken equipment. The target of the developed suit is to prevent accidents and to help survival in case an accident has occurred.

This paper presents the architecture of the developed suit prototype, concentrating on the sensor system used for monitoring the user's vital functions and user's environment. To our knowledge, this is the first full-scale smart clothing project that has been published. The paper is organised as follows. In Section 2, a general architecture of the prototype suit is given. Next in Section 3, the developed sensor system, the monitoring functionality utilising the sensor system, and user tests are described. Main results of the monitoring and the prototype suit development are discussed in the concluding Section 4.

2. GENERAL ARCHITECTURE

The prototype suit consists of four functional segments: communication, navigation and positioning, user and environment monitoring, and heating. This functional architecture is implemented using Global Positioning System (GPS) and electrical compass for navigation, Global System for Mobile Communication (GSM) for communication, and electrically heated Gorix panels for heating [3]. Sensor system consists of a heart rate sensor, three sensors for measuring position and movement, ten temperature sensors, an electric conductivity sensor, and two sensors for impair detection. In addition to these, the implementation requires a user interface, central processing unit (CPU), and a power source.

The final prototype consists of a two piece underwear, a supporting structure, and actual snowmobile jacket and trousers. The underwear acts as a fastening platform for the electrodes and transducer of the heart rate sensor and for heating fabric. The research of the suit's wearability was based on earlier work of Carnegie Mellon University [4]. According to the research, it was beneficial to split the whole system into small pieces. Now, it was easier to embed electrical components into the suit without disturbing the user. It was also evident that for comfortable and steady fastening, a supporting structure between jacket's lining and coating had to be built. Each main module excluding some of the sensors and the user interface is placed into the supporting vest. The supporting structure and the main components are illustrated in Figure 1.

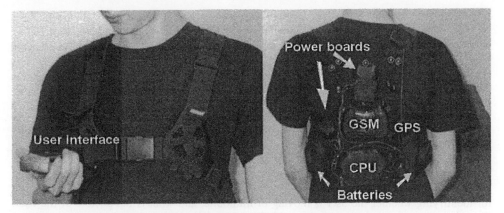

Figure 1. Photographs of the supporting structure and main components of the suit.

3. SENSOR SYSTEM AND MONITORING CONCEPT

The user and environment monitoring of the smart clothing prototype supports the survival in three basic situations: heavy impact, falling into water, and injury. If any of the three accident situations is concluded, a procedure for sending an emergency message is automatically initialised. Before message transmission, the suit informs the user with an audio signal and a warning light. If the user does not cancel the procedure, an emergency message is sent to a pre-selected destination using the GSM short message service. A single message can contain up to 160 characters. This is adequate for the emergency message that contains an emergency reason code and the current position of the user. The emergency code has currently two alternatives. The first code refers to an injury situation and the second to technical failure. The position coordinates are acquired using the GPS system. After sending this emergency message it is possible to request additional information from the suit. The prototype response to the request by sending data from the human and environment measurements. A user can also send the emergency message manually. The placements of used sensors are implemented in Figure 2.

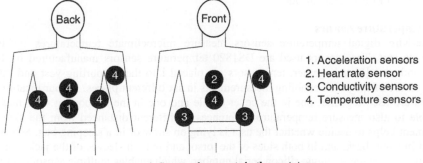

1. Acceleration sensors
2. Heart rate sensor
3. Conductivity sensors
4. Temperature sensors

Figure 2. Sensor placements in the prototype.

3.1. Acceleration sensors

Two kinds of acceleration sensors are used in the prototype suit. First, an impact situation is registered using two large-scale acceleration sensors that enable three-dimensional measurements. The used Analog Devices' ADXL150 and ADXL250 sensors can cover a range of ±50 g [5]. For greater voltage changes of the output, the sensors are adjusted to the range of ±25 g. Sensors are connected to a comparator that can generate an external interrupt to the microcontroller. The threshold voltage that can cause an interrupt is adapted to a suitable level using a potentiometer. Sensors are located on the CPU board at the small of the back. This placement assures that heavy accelerations measured are resulting only from an actual impact to a user.

Second types of acceleration sensors are used for detecting injury situations. These sensors must be more sensitive than the sensors used for impact situations. Therefore, three Analog Devices' ADXL105 sensors adjusted to the range of ±2 g are used [6]. The sensors are used for identifying whether the user is moving or immobile, and what is the posture of the user. These sensors are also placed on the CPU board, and connected to the analog-to-digital converter (ADC) of the microcontroller. In order to measure very slow movements, the sensors should be as near as possible to the place that is accelerating. In our case, the sensors are placed at user's back, which does not allow slow acceleration measurements from a limb or fingers. This compromise has been done for easier cabling connection to the rest of the system. However, the posture of the user can be measured based on gravity.

3.2. Conductivity sensor

A fall into water is detected by two conductivity electrodes that are located at the sides of the prototype suit at waistline height. Conductivity between the electrodes is measured to detect if the suit is surrounded by water. The electrodes are connected to the ADC of the microcontroller.

3.3. Heart rate sensor

In addition to acceleration sensors, heart rate information is used for detecting an injury situation. The heart rate monitoring is based on Polar Electro's wireless heart rate monitor [7] with two metal-clad-aramid-embroided electrodes. These electrodes are placed inside underwear because they need a straight contact to human skin. The electrodes are directly attached to a transmitter unit that transmits the signals from the electrodes wirelessly to a receiver on the CPU board. The heart rate receiver unit is connected directly to the counterport of the microcontroller.

3.4. Temperature sensors

One-wire digital temperature sensors measure microclimate temperatures inside the prototype suit. The sensors used are DS1820 temperature sensors manufactured by Dallas Semiconductor [8]. Altogether, ten sensors are placed into the supporting vest and into the cover textile of the jacket enabling measurements in six different places. In four locations two sensors are used together, one in the cover textile and one in the supporting vest, making it possible to also measure temperature differences at different distanced from the skin. This placement helps to decide whether the user is lying on snow or in a sleeping bag. Sensors are placed in front, back, and in both sides of the torso, and in both sleeves of the jacket. Each of the sensors contains a unique silicon serial number, which enables multiple sensors to exist in the same one-wire bus. The sensors are connected to the one-wire serial line driver [9], which is further connected to the microcontroller using serial communication port. Serial port communication makes the microcontroller software simpler than directly using the one-wire interface. The accuracy of the sensors is 0.5° C.

3.5. Measuring concept

The condition of the user is described in the CPU software using three state variables, which refers to accident situations. The values of the state variables represent the probability of an accident.

The raw data is read from four sensor groups listed above. Several signals measured need high processing capacity and therefore the signals are first processed by a feature extraction system. This also helps to avoid redundant information gathered from sensors. All features are presented using binary values, i.e. the condition is either true or false. Five features are calculated from the outputs of the acceleration sensors. These features are heavy acceleration (negative in case of an impact), the user is moving normally, the user has been immobile for a certain period, the user is in an upright position, and the user is lying. Three features are extracted from the heart rate signal: there is no measurable pulse, the heart rate is slow, and the deviation in the heart rate signal is under the normal level. Two features from the conductivity sensor are: the user is surrounded by water and the user gets off the water. The temperature sensors are used to measure the temperature in several positions inside the suit and in the same positions inside the jacket's cover textile. The feature extracted tells weather the user is lying on a cold surface.

The values of the state variables are calculated from the features by a simple inference engine. For each of the state variables there exists a corresponding rule set, which is used to calculate the probability values. If the value of any of these variables exceeds a defined threshold the system will initialize the sending of the automatic emergency message. In case there are several highly probable types of accidents, the injury condition is given the highest priority. The impact situation has the second highest priority. The block diagram of the measuring concept used in the suit is illustrated in Figure 3.

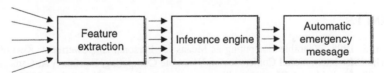

Figure 3. Measuring concept used in the suit.

3.6. User tests

The prototype suit has been tested in the actual operating environment, during the winter 2000 in Lapland, Finland. Although the operation of the suit was verified in a normal outdoor environment, the testing revealed that the extreme environment, such as the wide range of operating temperatures and changing humidity set demanding requirements for electronics. The clothing properties of the prototype have also been tested, since the added electronics and other components should not impair the wearability or washability of the garment. Falling into water detection has been tested in water, in moist air, and in moist snow environments. The conductivity measurement gives highly reliable results since the used fabric of the jacket rejects water effectively. This type of measurement is also easy and low cost to implement. The temperature sensors and the posture detecting sensors have been tested in several body positions while lying on snow and on insulating bed, walking and crawling on snow, and sitting by a camp fire. The functionality of the heart rate sensor has been verified. However, the measuring concept concluding from the heart rate has been tested only in theory.

4. CONCLUSIONS

The concept of a survival suit for snowmobile users has been described. The prototype suit has been built and its basic functionality has been tested. The monitoring concept provides intelligent aids for a user, thus increasing safety and helping in emergency situations.

The monitoring of the prototype suit has been tested to be functional. Some problems were encountered during tests. These were mainly caused by weak connections between electrical components, which have been problematic to implement reliably. Overall, the conductivity and temperature sensors give most reliable results. For cost and weight savings, the possibility to use only one-type acceleration sensors should also be researched.

Much information about smart clothing usage possibilities, advantages, and shortcomings were acquired during the project. Although the implemented prototype was designed for a special application environment, it still convinced us that because of potential advantages provided by smart clothing, the demand for various kinds of applications will appear in the near future. However, certain technical challenges concerning e.g. battery capacity, weight of components, and system costs need improvements before smart clothing can appear in common use.

5. ACKNOWLEDGEMENT

We would like to thank Polar Electro for their comments in the development of the system and their assistance with the heart rate sensor. We also like to acknowledge Suunto for their electrical compass. We thank Du Pont and Nokia for their help during the project.

REFERENCES

1. Mann, S. 1998. Wearable Computing as Means for Personal Empowerment. Keynote Address for The First International Conference on Wearable Computing, Fairfax, VA, Canada, May 12-13, 1998. Available at URL = " http://wearcam.org/icwckeynote.html ", 20.7.2000.

2. Mann, S. 1996. Smart Clothing: The Shift to Wearable Computing. Communications of the ACM. Vol. 39, no. 8, pp. 23-24.

3. Homepage of Gorix Ltd, URL=http://www.gorix.com/gorix/INDEX1.HTM, 13.3.2000.

4. Gemperle, F., Kabasach C., Stivoric, J., Bauer, M. & Martin. R. 1998. Design for Wearability. Proceedings of The Second International Symposium on Wearable Computers, Pittsburgh, PA, USA, Oct. 19-20, 1998, IEEE, pp.116 – 122.

5. Analog Devices, Norwood, Massachusetts, Data Sheets ±5 g to ±50 g, Low Noise, Low Power, Single/Dual Axis iMEMS® Accelerometers.

6. Analog Devices, Norwood, Massachusetts, Data Sheets High Accuracy ±1g to ±5g Single Axis iMEMS® Accelerometer With Analog Input.

7. Homepage of Polar Electro, URL=http://www.polar.fi/, 13.3.2000.

8. Dallas Semiconductor, Dallas, Texas, Data Sheets DS1820 1-Wire Digital Thermometer.

9. Dallas Semiconductor, Dallas, Texas, Data sheets DS2480 Serial 1-wire™ Line driver.

Human Friendly Mechatronics (ICMA 2000)
E. Arai, T. Arai and M. Takano *(Editors)*
© 2001 Elsevier Science B.V. All rights reserved.

Development of a Wearable Muscle Sound Sensing Unit and Application for Muscle Fatigue Measurement

K. Naemura[a], T. Inoue[b], K. Fujisue[b], H. Hosaka[a], K. Itao[a]

[a]Institute of Environmental Studies, Graduate School of Frontier Sciences, The University of Tokyo, 7-3-1 Hongo, Bunkyo-ku, Tokyo, Japan
[b]Department of Precision Machinery Engineering, The University of Tokyo 7-3-1 Hongo, Bunkyo-ku, Tokyo, Japan

A new wearable muscle sound sensing unit for the core device of the wearable information system for human healthcare ($WISH^2$) was developed. Muscle sound as well as electrocardiogram and body acceleration can be measured simultaneously. Fatigue of the biceps brachii after exhaustion due to isometric contraction was examined as an application for muscle sound sensing in daily life. A 1/f fluctuation from the power spectrum of muscle sound decreased after fatigue.

1. INTRODUCTION

Recent advances in micromechatronic technologies and needs for individual healthcare led the authors to develop a new wearable information system for human healthcare ($WISH^2$). This system will store the users' health information in a database. Abnormalities and emergencies will be recognized automatically using the individual database, and the system will place emergency calls though the telephone system. A small bio-sensing unit light enough to wear in daily life is one of core devices of the $WISH^2$ system, which features simultaneous, multi-channel sensing. Wireless networking, long battery life and data base for each type of sensing data will be added after the development of the bio-sensing unit. $WISH^2$ will achieve a healthcare based on chronobiology, which may differentiate the mechanisms of mental disease from rhythm disorders(Figure 2). Muscle fatigue affects daily rhythms, i.e., grater fatigue require longer periods of sleep. Though electromyogram can be used for muscle fatigue measurement, it requires surface or needle electrodes which are not suitable for $WISH^2$. Thus, the authors chose muscle sound (mechanomyogram) as a wearable muscle fatigue sensor.

Muscle sound produces a micro-vibration, which can be measured on skin surface during skeletal muscle contraction. Previous studies have reported that muscle sound is closely related to force and fatigue[1,2]. Muscle sound reflects specific aspects of muscle mechanics and can be used to follow changes in the contractile properties of muscle caused by localized muscle fatigue[3]. Recently, certain studies have demonstrated that muscle sounds can be classified into fast twitch fibers and slow twitch fibers.

Previous studies were performed as laboratory experiments, thus the equipment used for recording muscle sound was not portable or wearable. In a previous study we clarified the contact sensor (acceleration sensor, microphone) mass effect on muscle sound sensing using a laser displacement meter[5]. According to our results, a sensor mass of more than 10 grams attenuated the frequency band over 20 Hz, a band which contains significant elements. It is necessary that the sensor mass must be below at most 10 grams in order to appreciate the sound of biceps brachii by means of a contact sensor.

The purposes of the present study are to develop a wearable muscle sound sensing unit for the $WISH^2$ system and to evaluate its ability to sense muscle fatigue in daily life.

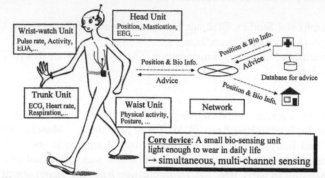

Figure 1 Wearable information system for human healthcare ($WISH^2$)

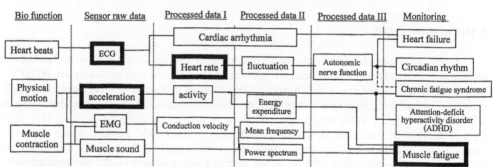

Figure 2 Relation among bio information, processed data and monitoring objectives

2. WEARABLE MUSCLE SOUND SENSING UNIT

2.1. Required specifications

The ability to record electrocardiogram, heart rate, and three-dimensional body acceleration are necessary for all wearable bio-sensing units. An electrocardiogram reflects heart function and can be used for emergency recognition (fatal arrhythmia, cardiac arrest). Heartrate variability can monitor nerve function. The long term monitoring of heart rate in daily life showed a pattern similar to simultaneously monitored rectal temperature, the golden standard for chronobiological measurement[6]. Three-dimensional body acceleration can be monitored in terms of the amount of physical activity and used for motion recognition[7]. A sensing unit, which contains an electrocardiogram amplifier and body acceleration sensor, is attached to waist. From this waist unit, an extension sensor can be added by means of a wire.

Table 1 shows a comparison of sensors for muscle fatigue measurement in terms of frequency range, consumption current, index, and sensing through clothes. Previous studies have reported that electromyograms could monitor muscle fatigue. Recently, muscle fatigue measurement has been explored using near-infrared oxygenation or muscle sound. Lower frequency range for data memory saving, lower consumption current for long period operation and sensing on clothes for easy handling are necessary characteristics for wearable sensors. A muscle sound sensor is superior to sensing through clothes, which cannot be realized by electromyogram and near-infrared oxygenation. In the present study, muscle sound was selected as the target of muscle fatigue sensors. Our sensing unit will also be used for measuring electromyogram signals and near-infrared oxygenation.

In regard to an acceleration sensor for muscle sound, the following are required: 1) a mass of below 10 grams (Figure 3)[5], 2) operation by portable batteries, 3) a frequency response from 1 to 100 Hz, 4) sensitivity 100 micrometers, 5) an installable pre-amplifier which is commercially available. A high sensitivity capacitance accelerometer (8303A2, Kistler Instrument Co.) was selected. This satisfies the weight and frequency requirements of 3 grams and a range from DC to 150 Hz, respectively.

Table 1
Comparison of sensors for muscle fatigue measurement

	Frequency range (Hz)	Consumption current (mA)	Muscle fatigue index	Sensing on clothes
Electromyogram (Electrodes)	10 ~ 5000	8	Mean frequency ↓ Muscle fiber conduction velocity ↓	NG
Near infrared oxygenation (LED & PD)	0.1 ~ 3	200	Tissue oxygenation index ↓	NG
Muscle sound (Accelerometer)	5~ 70	10	Power spectrum pattern	OK

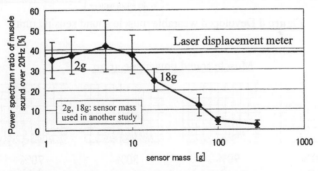

Figure 3 Relationship between acceleration sensor mass and power spectrum ratio of muscle sound over 20Hz, and comparison with non-contact sensing by laser displacement meter (n=6, biceps brachii, isometric contraction)[5]

2.2. Configuration
The prototype of a wearable bio-sensing unit is shown in Figure 4. It consists of an

34

amplifier for electrocardiogram signals (IEC-1102, Nihon Kohden Co.), a 3-axis piezo resistive acceleration (ADXL202, Analog Devices, Inc.), a 1-axis vibrating gyroscope for angular speed (ENC-03J, Murata Manufacturing Co., Ltd.), and 3 extension sensor ports. The unit can be operated for 3.9 hours with a 9 volt alkali battery. The power consumption of the sensing unit was calculated as 0.87 watts. A wearable bio-sensing unit (W 6.4 cm, H 4.7 cm and L 14 cm, 400 grams) can be attached to a waist belt along with a commercially available portable mini-computer (Libretto ff, Toshiba Co., W 22.5 cm, H 3 cm and L 13.3 cm, 900 grams).

Acceleration sensor was fixed using a flexible band around the arm. The effect of band length on muscle sound sensing was checked. From the results 80 or 90 % of upper arm circumference was adequate (Figure 5). 100 % was too loose and 70 % too tight.

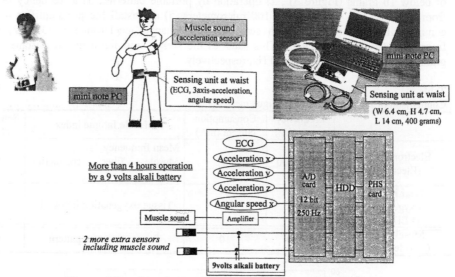

Figure 4 Developed wearable muscle sound sensing unit

Muscle sound power spectrum

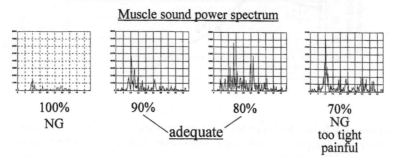

100%	90%	80%	70%
NG			NG
	adequate		too tight
			painful

Figure 5 Comparison of muscle sound power spectrum by band-length to fix the acceleration sensor normalized by upper arm circumference

3. MUSCLE FATIGUE MEASUREMENT

Muscle fatigue was evaluated as one of applications of the developed unit.   The effect of the muscle fatigue on muscle sound has not yet been determined.   As shown in Table 1, current muscle fatigue index derived from muscle sound is spectrum pattern changes.   Only calculation of some value is thought to be suitable for the $WISH^2$ system.   The authors were the first to attempt to evaluate muscle fatigue in terms of a 1/f fluctuation of muscle sound. Muscle fatigue at the biceps brachii after carrying a piece of luggage by hand was chosen as a representative daily life situation.

3.1. Methods

Two subjects (24-year-old males) joined the experiments.   They gave their written informed consent to participate in the experiment after receiving an explanation of the purpose, advantages, and risks involved.   Isometric contraction of the biceps brachii at 50% maximum voluntary contraction by pulling an iron bar was maintained for 30 minutes.   The sound of each biceps brachii was measured for 10 seconds at the beginning and after exhaustion (sampling rate 250 Hz).   Fast-Fourier transformation for two seconds was calculated.   Inclination of the power spectrum between 15 to 60 Hz was obtained as a 1/f-fluctuation parameter.   The mean and standard deviations were calculated for each subject.

3.2. Results

The inclination of 1/f fluctuation decreased after the muscle task for both subjects, which indicated that their conditions had become uncomfortable.   Changes between before and after fatigue were determined by Student's t-test to be statistically significant.

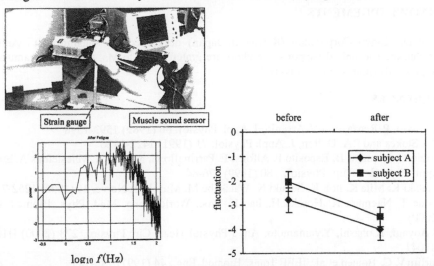

Figure 6 Experimental view of the muscle fatigue measurement (left upper side), power spectrum of muscle sound (24-year-old male, biceps brachii) after fatigue and fluctuation was calculated between 15 to 60 Hz (left lower side) and effect of muscle (biceps brachii) fatigue on 1/f fluctuation of muscle sound (right lower side)

4. DISCUSSION

Authors reported the measurement of the muscle sound in a train using the developed device as one example-exposed vibration during a daily life. Results showed the ability to sense the muscle sound not only in the experimental room, but also in outdoors. Preliminary evaluations suggest a wearable muscle fatigue sensing situation, that is people standing and carrying a heavy luggage by hand in a crowded train. Most Japanese have such experience.

Amplitude of muscle sound is too small to measure during dynamic motion like arm swinging or walking. Body acceleration becomes noise for muscle sound sensing. Subtraction of two acceleration sensors is thought to be effective for noise reduction. Authors could not obtain good results yet, for difficulty of the sensor fixation. More study is needed.

Researchers on medicine, muscle physiology, sports science and robotics can be used the developed the sensing unit. Wearable muscle fatigue sensor is not completed yet. Our current results will promote another researches on muscle sound.

5. CONCLUSION

A new wearable muscle sound sensing unit for the $WISH^2$ system was successfully developed. Muscle sound as well as electrocardiogram and body acceleration can be measured simultaneously. Fatigue of the biceps brachii after exhaustion due to isometric contraction was examined as an application for muscle sound sensing in daily life. A 1/f fluctuation from the power spectrum of muscle sound decreased after fatigue. Future study will be performed in order to construct a personal area network composed of wireless sensors that will process signals using a 1/f fluctuation of muscle sound.

ACKNOWLEDGEMENTS

Giken Kogyo Corporation (Kawasaki, Japan) developed the sensing unit Authors appreciate their technical supports. Authors are grateful also to Ms. Qunico Kawamura for her superior illustration in the Figure 1.

REFERENCES

1. C. Orizio, P. Renza, and V. Arsenio, J. Appl. Physiol. 66 (1989) 1593-1598
2. M.J. Stokes and P.A. Dalton, J. Appl. Physiol. 71 (1991) 1422-1426
3. Orizio C. Diemont B. Esposito F. Alfonsi E. Parrinello G. Moglia A. Veicsteinas A. Eur. J. Appl. Physiol. Occup. Physiol. 80 (1999) 276-84
4. Akataki K. Mita K. Itoh K. Suzuki N. Watakabe M. Muscle & Nerve. 19 (1996) 1252-7
5. Inoue T, Naemura K, Hosaka H, Itao K. Proc. World Cong. Med. Phys. Biomed. Eng. (2000)
6. N.Aoyagi, K.ohashi, Y.yamamoto. Am J Physiol Heart Circ Physiol, 278 (2000) H1035-H1041
7. Carlijn V. C. Bouten et al., IEEE Trans. Biomed. Eng., 44 (1997) 136-147

Human Friendly Mechatronics (ICMA 2000)
E. Arai, T. Arai and M. Takano (Editors)
© 2001 Elsevier Science B.V. All rights reserved.

A Novel Antenna System for Man-Machine Interface

Pekka Salonen, Mikko Keskilammi, Lauri Sydänheimo, Markku Kivikoski

Tampere University of Technology, Institute of Electronics,
Korkeakoulunkatu 3, 33720 Tampere, Finland

The development of wearable computer systems has been rapid. They are coming more and more lightweight and quite soon we will see a wide range of unobtrusive wearable and ubiquitous computing equipment integrated to into our everyday wear. This allows new possibilities for a man-machine interface. Wearable computers give the needs for controlling machines with the same equipment. If the Bluetooth system is used it is possible to connect machines and internet with the aid of wearable computers. Rapid progress in wireless communication promises to replace wired-communication networks in the near future in which antennas are in more important role. This paper presents sophisticated beam-forming patch antenna array and a dual-band antenna for wearable applications. Antenna array theory is developed and discussed with the measured results. In addition with this antenna system the transmission power could be reduced due to increased gain.

1. INTRODUCTION

During the recent years wireless communication has become in more important role. Convergence of wireless and wired networks is an essential part of this development. In near future one effective and universally available solution will be the Bluetooth system.

Bluetooth is aimed for short range connection between all kind of devices. These devices can be divided into two main groups:
- Devices connected to the wired network and fixed power supply
- Mobile devices with battery power supply

The first group contains "regular" devices like networks access points, printers, VCRs and so on. However this group will also contain completely new devices like refrigerators, ovens, coffee makers and in Finland even sauna heater i.e. bathhouse stove is going to have Bluetooth communication capabilities. The second group will contain mobile devices like PDAs, cellphones, laptop computers, headsets, health monitoring systems and so on.

In spite of the fine visions we have to bear in mind that Bluetooth technology has some limitations as well.

- One standard based limitation is that only eight Bluetooth devices can be connected to a pico cell where one is master and seven are slaves. In wired network connected access points or in case of printer it is natural that access point or printer will take the role of master and mobile units are slaves. However in the near future it is obvious that in one room there will be easily 20 Bluetooth units at the same time and effective communication requires several pico cells to be formed in this room. In this situation these devices will start to interfere each other. This will cause that the quality of service (QoS) and data throughput and reliability will reduce dramatically.
- The other limitation is dealing with mobile units battery lifetime. If the battery life time is too short the users do not feel comfortable to use this device, rather like having an virtual pet that requires continuous care taking for charging batteries.

The aim of this paper is to present a novel Bluetooth antenna system for man-machine interface. This antenna system is composed of small planar antenna to be attached to a wearable computer system carried by a human, and a beam forming array antenna for Bluetooth access point or equal device in Bluetooth neighborhood. This solution will provide actively adjustable efficient directive radiation pattern with minimal side lobes. By using this smart antenna solution in Bluetooth pico cell it will provide minimal interference for other Bluetooth devices in the same room and enabling maximum data transfer capability. This will improve QoS and reduce required connection time and reduce required transmission power. Together with this it will extend battery life time and increase usability of wearable computers. Presented beam forming antenna is also robust, low cost and easy to implement to printer or to any other networked Bluetooth device. This paper will present the antenna theory with simulations and comparison to measurements. In addition it will discuss how much power is saved using this system compared to conventional antenna systems. This shows that the same transmission range is achieved with reduced transmission power.

2. ANTENNA ARRAY FOR ACCESS-POINT-TYPE DEVICE

Antenna arrays are composed of group of similar antenna elements. The "grouping" of antennas increase the gain compared to a single element antenna. At the same the radiated power is more concentrated to one certain direction. This is due to reduced beam-width compared to the single element antenna. The radiation characteristics of an antenna array can be determined by the antenna element separation and orientation. Also the phase and amplitude difference between antenna elements control the radiation pattern and gain of the antenna array. Electronic control of these phase and amplitude differences allows the beam-forming control. This means that the radiated power can be transmitted to the direction where the receiver is. The electronic control of phase and amplitude differences allows the active tracking of mobile target. This is the main idea of apply this kind of antenna system for a man-machine interface.

The proposed beam-forming array is composed of five elements, which are equally spaced but non-uniformly excited. This section presents the array factor (AF) derivation, which is based on the isotropic point sources. Isotropic point sources can be considered as ideal antennas, which radiates the power to all direction. The array factor consists of orientation and separation information of point sources while the element factor gives the radiation characteristic of given antenna element. The antenna radiation pattern can be calculated by multiplying the array factor with the element factor of given antenna [1].

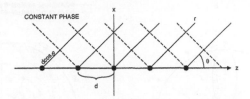

Figure 1. Incident rays are shown in solid line and constant phase plane is shown in dashed line in zx-coordinate frame.

Let the point source separation be d as shown in fig. 1. The angle between array axis z of which unit vector is z and incident ray of which unit vector is r can be calculated by the scalar product as $z \bullet r = \cos\theta$. The phase difference caused by the element spacing is thus $\beta d\cos\theta$, where β is the wave number. To obtain beam-scanning the variable phase shifter must be incorporated of which value is given by α [2]. Let A_n denote the relative amplitudes of each element where the center element relative amplitude is 1 and the phase shift $\alpha = 0°$ and the element is centered to the origin. The array factor can be evaluated.

$$AF(\theta) = \sum_{m=1}^{5} A_m \exp(j\alpha_m)\exp(j\xi_m) \tag{1}$$

$\xi_m = \beta\hat{r} \bullet \hat{r}'_m = \beta z'_m \cos\theta$

$\alpha_m = -\beta z'_m \cos\theta_0$

θ_0 = main beam pointing direction

where $\cos\theta_0$ is the angle to the direction of which maximum is wanted. The current distribution feeding the elements of the array is symmetrical in this application thus $A_1 = A_5$, $A_2 = A_4$. This expression can be simplified by collecting the equal amplitude exponential terms together. Thus the normalized array factor becomes

$$f(\theta,\theta_0) = \frac{1}{2A_1 + 2A_2 + A_3}[2A_1 \cos(2\beta d(\cos\theta - \cos\theta_0)) + 2A_2 \cos(\beta d(\cos\theta - \cos\theta_0)) + A_3] \tag{2}$$

The beam can be steered by controlling the phase shift α. The side lobe level and the half power beam-width can be controlled by relative current amplitude. The new current amplitude distribution is a compromise between binomial and Dolph-Chebyshev −30dB side lobe level distributions in which half power beam-width is identical to Dolph-Chebyshev and the side lobe level is decreased to −37.5dB. These coefficients can be calculated from normalized binomial distribution by adding 1/12 to each amplitude except the center element amplitude, thus the new coefficients are: $A_1 = A_5 = 0.25$, $A_2 = A_4 = 0.75$ and $A_3 = 1$. The resulting normalized array factor in linear scale is shown in fig. 2 a), in which point source separation d is $\lambda/2$ and in fig. 2 b) array factor multiplied with patch element factor.

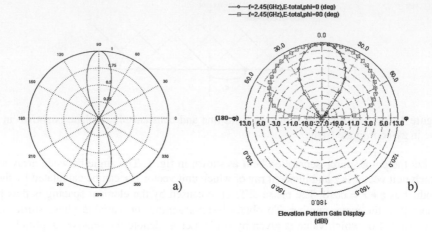

Figure 2. a) Normalized array factor in linear scale. b) Polar coordinate presentation of five element patch antenna array radiation pattern for E- and H-plane in logarithmic scale.

3. ANTENNA FOR WEARABLE COMPUTER

A wearable antenna means that it is meant to be as part of the clothing. The requirements for wearable antenna are light-weight, small, robust and it should not radiate towards human organs. In addition dual or even multi-band antenna operation at selected frequency bandwidths reduce the need of many different antenna elements in clothing. Using several antennas to cover various radio systems needs additional cabling making a wearable computer less comfortable. Dual-band antenna is an antenna, which radiates power at two different frequency bands. Dual-band planar inverted-F antenna (PIFA) is proposed as a new antenna solution for most small hand-held devices. PIFA has smallest dimensions compared to other planar antenna structures [3].

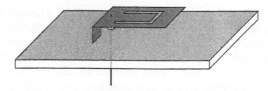

Figure 3. A standard PIFA in which a U-shaped slot is etched to form a dual-band antenna. The physical length of the antenna is remained the same as needed for the lower frequency operation.

A dual-band PIFA with a U-shaped slot has more suitable properties to operate at GSM and Bluetooth frequencies compared to other PIFAs [3], [4]. Fig. 3. shows the geometry of the U-shaped slot PIFA. The lower resonant frequency can be controlled by the outer dimensions of the antenna, and the upper resonant frequency can be controlled by the dimensions of the inner patch-shape surrounded by the U-slot. Both frequencies can be determined independently which makes the design procedure easier compared to other slot shapes.

In order to characterize the antennas a HP8722D network analyzer was used to measure the input return loss of the antennas as a function of frequency. The values for the input impedance were specified as the frequency bandwidth in which the voltage standing wave ratio (VSWR) is less than 2:1. A 2:1 VSWR is equivalent to a 10-dB return loss, the level at which 10% of the incident power is reflected back to the source. The measured and simulated return loss of the U-shaped slot PIFA is shown in Fig. 4. The bandwidth of the lower resonant frequency is 90MHz (10%) which agrees well with the simulations. However the simulated null depth is much less compared to measured results. The upper resonant frequency is slightly shifted down because of the manufacturing inaccuracies. The null depth and the bandwidth 80MHz (5%) agrees well with the simulations. This slot configuration is versatile for most wireless communication applications because of bandwidths. The lower frequency bandwidth is enough e.g. for GSM or PCS and the upper band is enough e.g. for Bluetooth, which requires only approximately 80MHz band.

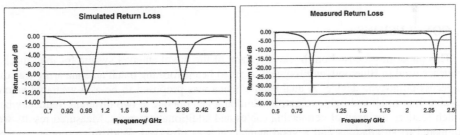

Figure 4. a) Simulated return loss as a function of frequency. A 2:1 VSWR is equivalent to a 10-dB return loss, the level at which 10% of the incident power is reflected back to the source. b) Measured return loss as a function of frequency. A good agreement with the simulations is obtained.

4. EFFECT OF ANTENNA GAIN ON TRANSMISSION RANGE

This section presents the effect of antenna gain on the distance between transmitting and receiving antenna. The results presented are based on the Friis power transmission formula, which does not include the effects of antenna impedance mismatch, polarization mismatch, or losses due to propagation effects [5].

$$P_r = \left(\frac{\lambda}{4\pi r} \right)^2 G_t G_r P_t \tag{3}$$

in which P_r is the received power, λ wavelength, r distance between receiver and transmitter, G_r gain of the receiver antenna, G_t gain of the transmitter antenna and P_t transmitted power.

To show the effect of the antenna's gain, let the sensitivity of the receiver be -70dBm and the transmitter power be 0dBm at frequency 2.45GHz. These values are typical for Bluetooth operated systems.

The calculations are based on the Friis transmission formula are theoretically maximum transmission ranges in ideal conditions. The proposed antenna system is compared to typical Bluetooth system. The access point antenna array has gain of 13dBi and PIFA 4dBi. Typical Bluetooth antenna has gain of -3dBi. For these parameters maximum transmission ranges are summarized in table 1.

Table 1.
The effect of antenna gain on transmission range.

Antenna system	Range (m)
Patch array and PIFA	220
Typical Bluetooth system	15

In practice, the maximum distance between transmitter and receiver is approximately only one third of the theoretical values. This shows that transmission power for typical Bluetooth antenna must be 10 times the power of presented antenna system. This has a strong effect on the battery life time in wearable applications.

5. CONCLUSIONS

This paper has focused on the development of the novel antenna system for man–machine interface used in Bluetooth environment. A smart beam-forming patch antenna array was used in access-point-type device. A dual-band antenna was proposed to be as a wearable antenna because it is small and light-weight. In addition it can operate at two different frequency band which reduces the need of connecting cables and additional antennas. Measured results were compared to simulations and a good agreement was observed. With this proposed antenna system the transmission range can be improved remarkably. This result shows the importance of antenna design in wireless communication applications.

REFERENCES

1. W. L. Stutzman, G. A. Thiele, Antenna Theory and Design, 2nd Edition, John Wiley & Sons, 1998.
2. S. Drabowitch, A. Papiernik, H. Griffiths, J. Encinas, B. L. Smith, Modern Antennas, Chapman & Hall, 1998.
3. P. Salonen, M. Keskilammi, M. Kivikoski, Comparison of Planar Dual-Band Antennas, AP2000, Millenium Conference on Antennas and Propagation, Davos, Switzerland, 2000.
4. P. Salonen, L. Sydänheimo, M. Keskilammi, M. Kivikoski, A Small Planar Inverted-F Antenna for wearable Applications, IEEE 3rd International Symposium on Wearable Computers, pp. 95 – 100, San Francisco, USA, 1999.
5. D. M. Pozar, Microwave Engineering, John Wiley & Sons, 1998.

Human Friendly Mechatronics (ICMA 2000)
E. Arai, T. Arai and M. Takano *(Editors)*
43

Wild animal tracking by PHS (Personal Handy phone System)

Toshiyuki Takasaki[+],　　Rie Nagato[++],　　Emiko Morishita[*]

Hiroyoshi Higuchi[*], Kobayashi Ikutarou[++], Hiroshi Hosaka[+], Kiyoshi Itao[+]

Abstract

Miniaturization and integration of portable devices have been achieved so remarkably day by day that wearable computing is being realized. It seems that vast amounts of wearable information devices will be attached not only to human beings but also to nature environments and artifacts and they will be connected with intelligent network in the future. We attached PHS location devices to urban animals such as Raccoon Dogs and Crows and tracked them. In the Crow experiments, in which lighter devices than 30g were required, we realized the Crow tracking with PHS location devices, which weighed about 28 g.

1.　　Introduction

The volume and mass of an information device must be zero ultimately. Microsystem technologies have been developed in order to achieve this ultimate goal from such various fields as electricity mechanics, physics, and chemistry and it seems the development will be continued. It has been possible to detect signals generated by an environment or living body and make a judge and process because of the rapid progresses of such technologies as microsensors, microactuators, wireless, and Internet communication these days. The category of information devices has been changing from Mobile into Wearable with such technological progresses. A cellular phone, which is always in a breast pocket, is the typical example. It is possible for wearable devices to apply not only to human environments but also to artificial environments, wildlife environments, and nature environments, by attaching the devices to artifacts and things of the nature. We proposed such an information devise attached to various things as Nature Interfacer[1].

+Institute of Environmental Studies, The University of Tokyo(7-3-1 Hongo, Bunkyo-ku, Tokyo, Japan)
*Laboratory of Biodiversity Science, School of Agriculture and Life Sciences, The University of Tokyo
(Yayoi 1-1-1, Bunkyo-ku, Tokyo)
++Department of Precision Machinery Engineering, The University of Tokyo (7-3-1 Hongo, Bunkyo-ku,
Tokyo, Japan)

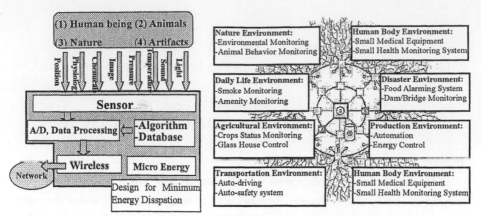

Fig.1.1 Architecture of Nature Interfacer

Fig.1.2 Concept of Wearable Information Network

Its architecture is shown in FIG1.1. First, Nature Interfacer detects such analog signals by sensors as light, sound, temperature, pressure, picture images, chemical substances, physiological information, and position, all of which are sent by human, wild lives, nature, or artifacts. Second, a CPU converts the analog signals into digital data and the data are recognized and processed based on database. Last, the data are transmitted to networks wirelessly. By using the public or wide area network, the remote collection and control of various kinds of environment information are available, which is what we call WIN[2][3] (Wearable Information Network) [Fig.1.2].

The examples of WIN application studied these days are as follows:
- Environment monitoring and Animal behavior monitoring in Nature environments.
- Smoke monitoring and Amenity monitoring in Daily life environments.
- Automation and Energy control in Production environments.
- Small medical equipment and Small health monitoring system in Human body environments.
- Auto-driving and Auto-safety system in Transportation environments.
- Crops status monitoring and Glass house control in Agricultural environments.
- Flood alarming system and Dam/bridge monitoring in Disaster environments.

2.   Wild animal tracking by PHS

Animal behavior monitoring, which part of nature environment monitoring, was done in order to find the wearable information network. Raccoon Dogs and Jungle Crows were tracked in the following two experiments.

Fig.2.1 Vinyl house damaged by raccoon dogs
(The University Farm, Tanashi City, Tokyo)

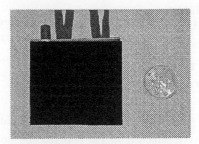

Fig.2.2 PHS location device attached to
the raccoon dogs (size[mm]:67x67x11).

2.1   Raccoon dogs tracking by PHS

2.1.1.   Introduction

Wild Raccoon Dogs, *Nyctereutes procyonoidides*, are living in such urban areas as Tanashi City, Tama City in Tokyo, Yokohama City, Kamakura City in Kanagawa prefecture. They often damage farm products, plastic greenhouses and public electrical cords. They are sometimes hit by cars and die on public roads [Fig2.1]. The number of such incidents is increasing day by day. Though they cause such problems, many aspects of their ecology, such as the daily movements and habitat use has remained unknown. There were some tracking experiments in the past by attaching a telemetry device on a raccoon and chasing its radio wave with antennae. But such experiments required a lot of time and manpower, and telemetry tracking is not a best way in an urban area where many obstacles such as buildings and so can interfere with radio waves. In this experiment, raccoon dogs were tracked by the PHS (Personal Handy-phone System), which has human position-locating service. The raccoon dogs tracked in this experiment were living wildly in the University Farm and Forest in Tanashi City, Tokyo.

2.1.2   Preliminary investigation

Before tracking the wild raccoon dogs, the location accuracy of the position tracking system was investigated by walking randomly around the University Farm where raccoons were likely to be seen. The results showed an average error range of less than 100-200 meters. This means that we can recognize by the PHS whether raccoons are inside the Farm or not, and if inside, whether they are in the center of the Farm, northern end, southern end, eastern end, or western end.

2.1.3   Raccoon dogs tracking by PHS

One male raccoon dog and one female raccoon were caught in the University

Table2.1
Results of the two raccoon dogs experiment

	Life of battery	Number of data	Number of the valid data	Frequency of the valid data
The male	20 days	676	265	39.2 %
The female	12 days	458	458	59.4 %

Forest and PHS location devices produced by Alps Electric, which weighed about 58g, were attached to the two raccoon dogs [Fig2.2]. The PHS was in a vinyl bag for waterproofing. The access frequency for getting location data from the PHS was every thirty minutes between 3:00 pm and 8:00 am, for raccoon dogs are nocturnal animals.

2.1.4   Results

The results of the tracking raccoon dogs with PHS experiment are shown in Table2.1. The number of the valid data means how many times the tracking data sent from the PHS attached to the raccoon dog was in the PHS service area (valid area). Except for the valid location, the PHS was outside the PHS service area (invalid area), and the tracking data were not sent. The frequency of the valid data means the frequency when the raccoon was in the valid area. The result showed that the male raccoon dog was more often in the invalid area than female. The number of the access to the male PHS is smaller than that of the female one, hence the life of battery for the male PHS is thought to be longer than that of the female one. It was shown that the raccoon dogs acted during the night and had its place to sleep every day and its home range was inside the Farm and the Forest.

2.2   Crow tracking by PHS

2.2.1   Introduction

Jungle Crows, *Corvus macrorhynchos*, one of urban birds in Japan, are increasing in number and causing such problems as eating garbage. And the problem is so serious that the Government of Japan has considered in the Diet[4]. Crows flying around residential areas and downtown areas are seen everyday. As to the ecology of crows in central Tokyo, their number and their active areas are roughly known. But their daily movements are not well known. It is one of the problems, which prevented us from tracking the crows, that it was very difficult to radio-track crows from the ground. In order to grasp their activities, we developed a new method using position-tracking using PHS. Network area of PHS corresponds to living area of crows and crows were tracked successfully by PHS in 1999.Sep [5]. But there remained a device's weight problem.

Table2.2

Specifications of P-doco? and remodeled device.

	P-doco?	*Remodeled device*
Size [mm]	69.1x41.4x17.4	63.1x39.0x14.0
Case [g]	19.4	8.0
Battery [g]	15.25	8.0
Others [g]	7.75	12.1
Gross weight	42.4 g	28.1 g

Fig2.3. Crow with PHS (in the circle)

In general, it is said that the weight of a device attached to birds should be designed under 4-5% of their body weight [6]. As the body weight of a Jungle Crow is about 500g-700g, the required weight is 20g-35g. Though every effort was made, the PHS device attached to the crow weighed about 44g.

2.2.2 Devices attached to Crows

The NTT DoCoMo, Inc.'s PHS location device called "P-doco? ", which originally weighed 42.4g, was remodeled to lighten. Its case was thinned down and batteries were changed to two button-shaped batteries (CR2430, 150mAh by Fujitsu). The details of the lightened device are shown in TABLE 2.2. The remodeled device was authorized by the Ministry of Posts and Telecommunications.

2.2.3 Outline of the crow experiment

Batteries perform worse under low temperature so that the experiment was done from Mid-April '00 when it's getting warmer.

Ten Jungle Crows were captured in cooperation with Ueno Zoological Park in a trapped pen and the PHS device was attached to each of them. The average weight of the crows in this experiment was 638g and so the required weight of the device (less than 5% of a crow weight) was 31.9g. The devices, which weighed 28.1g and was used in this experiment, met the demand. Toyobo's Toyobo Dyneema[R][7], which is made from polyethylene, was strong, light, and durable enough to use as the string to attach the device to the crow. Though the device was attached to crow's breast in the last experiment because the device was a little bit large, the new smaller device was attached to crow s back like a back-sack [Fig2.3]. The access frequency for getting location data from the PHS was every thirty minutes between 3:30 am and 7:30 pm in cooperation with the DoCoMo Support, Inc.

2.2.4 Results

As a result, for example, one crow went around the central parts of Tokyo, such as Ginza or Roppongi from Ueno Zoological Park and one crow went up to a suburban area in Arakawa riverbed. The place where crows act in daytime turned out to be not only one place. And as to the Ueno Zoological Park it was shown that some crows used the Park as their roost and others used as their foraging area.

3.   Conclusion

The conclusions based on the experiments of raccoon dogs and crows were as follows:

[1] One example of Animal behavior monitoring in the Wearable Information Network was shown.

[2] It was shown that PHS, originally for human use, was also available for urban animal tracking even if they are flying birds.

[3] Crow tracking with 28g PHS device was successfully achieved and the weight problem of the previous experiment was solved.

4. What's next?

We are trying to lighten the device and lengthen the battery life, and its application to other wild animal is considered. In addition to the location information of animals, we have it in view that we will construct the wearable environment information system in which various information of ecology is detected by Nature interfacers.

Great Thanks (Alphabetical order),

Alps Electric, DDI Pocket, Inc., DoCoMo Support, Inc., NTT DoCoMo, Inc., Sharp Corp., SII, Students in Laboratory of Biodiversity Science in the University of Tokyo, Toshiba Corp., The University Farm and Forest in Tanashi City, Toyobo Co., Ltd., Ueno Zoological Park, Tama Zoological Park

Reference

[1] Kiyoshi Itao, Kiyoshi Naemura,"Concept of sensor-based Wearable Communication systems", Information Processing Society of Japan SIG-MBL, pp.15-21, Japan (1999.Aug)

[2] Kiyoshi Itao, "Micromechatronics Technology for Wearable Information and Communication Equipment", Sensors and Materials Vol.10 No.6, 1998, p325-335

[3] Kiyoshi Itao, "Practical sides of wearable Information devices", THE OPTRONICS CO., LTD (1999)

[4] The second supplementary budget in National Diet: Project for promoting the measures to prevent the crow problem in urban area, Japan (1999)

[5] Rie Nagato, "Wild animal chase by PHS", Information Processing Society of Japan SIG-MBL, pp.15-21, Japan (2000.Feb)

[6] Hiroyoshi Higuchi, "Satellite tracking crane migration", Yomiuri Shinbun, Japan (1994)

[7] http://www.toyobo.co.jp/e/seihin/dn/dyneema/index.htm

Human Friendly Mechatronics (ICMA 2000)
E. Arai, T. Arai and M. Takano *(Editors)*

Active Visual Feedback Using Head Mounted Display in Human Directed Manufacturing System

N. Depaiwa[a], H. Kato[a], Yang Y.[a] and S. Ochiai[a]

[a]Dept. of Electronics and Mechanical Engineering, Faculty of Engineering, Chiba University, 1-33 Yayoi-cho Inage-ku Chiba-shi 263-8522, Japan.

In many manufacturing processes, the human operator could perform its task well and swiftly, if it could acquire sensory information helpful to the performance effectively. In this research, two kinds of head mounted display (HMD), binocular see-through HMD and monocular see-through see-around HMD, as the device for active visual feedback and the method for displaying feedback information on them have been investigated. The objective task is to make a small-diameter hole onto an aluminum workpiece with a carbide drill and the feedback information is cutting force. As the results, the monocular HMD with sector chart display method has been superior to the others in operating efficiency.

1. INTRODUCTION

In recent years, the role of human operators is still indispensable from the viewpoints of flexibility to a wide range of objective tasks and adaptability in emergency situation. In previous works, the authors investigated sensory feedback of human operators in manual machine tool tasks and developed several devices for active sensory feedback of minute cutting force [1-3]. However, the active visual feedback has not been sufficiently investigated so far. Hence, in this paper, two kinds of head mounted display (HMD) as the device for the active visual feedback and the method for displaying feedback information on them have been investigated.

2. OBJECTIVE TASK AND ACTIVE VISUAL FEEDBACK SYSTEM

The objective task is to make a small-diameter blind hole onto an aluminum workpiece with a carbide drill. The cutting conditions are shown in Table 1. This is a difficult task for even skilled operators as well as unskilled ones, because the drill flutes are apt to be choked with the cutting chip and the drill often breaks. As a machine tool, an NC milling machine in manual operation mode was adopted in order to remove the influence of force sensory feedback. That is to say, since its handwheel is just connected with a pulse generator for spindle feed, the operator cannot

Table 1
Conditions of objective task

Cutting tool	Carbide drill (ϕ1 mm)
Workpiece	Easy machinable aluminum (ϕ5 mm)
Approach distance	5 mm
Depth of hole	5 mm
Rotational speed	500 rpm
Machine tool	NC milling machine MAKINO AV III NC-85 (Manual operation mode with a handwheel)

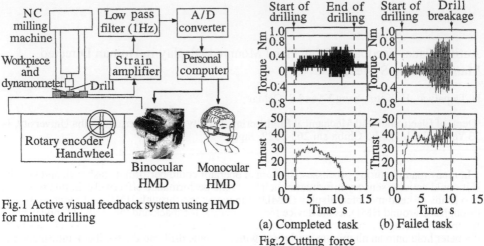

Fig.1 Active visual feedback system using HMD for minute drilling

(a) Completed task   (b) Failed task

Fig.2 Cutting force

feel cutting force through the handwheel at all. In addition, no auditory feedback exists either, because loud sound of the main spindle motor prevents the operator from hearing cutting sound. Accordingly, only the visual feedback, or view of the cutting chip flow, drill breakage, etc., exists basically in this task.

Figure 1 shows the active visual feedback system of cutting force for drilling. The force is detected with a strain gauge type dynamometer and input to a personal computer. The computer processes the force information and displays it on a screen of HMD which the operator wears. The force information is very important for the operator to complete the task safely and swiftly. And, the active visual feedback system of cutting force is constructed. As for the HMD, both a binocular one (Virtual i-Glasses, Virtual i-O Personal Display Systems) and a monocular one (Data glass, Shimazdu Corporation) were examined. The operator can see the drill and workpiece themselves through HMD because of see-through type. Data glass is a see-through see-around type and the operator can get also a view around the screen of HMD.

The cutting force is detected as thrust (axial) component and torque (rotational) component. Figure 2 shows examples of the detected force. The small pitch fluctuation is mainly caused by rotation of the drill (8.33r/s). Though a low pass filter of 1 Hz is used for the measurement, this fluctuation still appears because of dominant component in the signal. From these examples, it can be noticed that the thrust rapidly increases in the early stage of the task. In addition, the torque largely fluctuates in the cause of drill rotation and its amplitude gradually increases. Figure 2(b) shows the force signal when drill breaks during task. As can be seen, the breakage occurs at the thrust of about 40 N and the torque of about 0.8 Nm. In many cases, the breakage occurs around these values of thrust and torque. However, every breakage doesn't occur at ex-

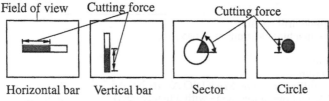

Fig.3 Chart images for active visual feedback

 Cutting force

Table 2
Subjects

Subject	Age	Skill level
A	22	Low
B	22	Low
C	35	High
M	24	Low
O	39	Middle
S	24	Low
T	26	Middle
Y	22	Low

Fig.4 Sample of HMD image

actly the same values.

On HMD, the value of either thrust or torque as feedback information is displayed by a chart image. Figure 3 shows four kinds of charts examined in this research. In both bar charts, the length of the red bar inside a fixed rectangular frame corresponds to the value of the force. In the sector chart, the angle of the red sector corresponds to it. In the circle chart, its diameter corresponds to it. The length of the frame in the bar charts is made 40 N for thrust and 0.8 Nm for torque, and the sector angle 135 degree in the sector chart is made similarly. In any chart, its size and position in the display screen can be adjusted by the operator, if it like. The sample of HMD image which the operator sees is shown in Figure 4. The updating interval of feedback information is less than 33 ms for all charts despite their position and size.

3. EVALUATION OF TASK AND SUBJECTS

In the experiment, a subject executed 20 time trials of the task one after another under the same feedback condition. For each task, the subject is requested to complete it as swiftly as possible without drill breakage. The performance of a task is evaluated by the following index;

$$I = \int_0^T |r| \cdot t\,dt \tag{1}$$

where r represents the distance from the drill point position at time t to its final position, and T represents total time consumption for the task. For measuring the distance r, a rotary encoder is attached on the handwheel as shown in Figure 1. When the drill happened to break, the task was regarded as "failed" and the index value was not evaluated. While executing a task, electromyogram (EMG) of forehead and electroencephalogram (EEG) of the operator were measured using Brain Builder (Hyper Brain Institute). These physiological data in eye closing state together with critical fricker fusion value (CFF) were also measured before all tasks and every five tasks. It took about three minutes for this measurement. In addition, the subjective symptom of the tasks as workload was evaluated using NASA-TLX [4] after the final measurement. Over ten minute recesses were taken between the experiments for a subject.

Subjects were 8 twenties- or thirties-aged male adults as shown in Table 2. Their skill level was evaluated taking into consideration their career and experiences for machine tool operation. Novices of the operation were regarded as low level skill holder and veteran technicians of Chiba University machine shop as high level skill holder. The time of completing a task was about 10 to 30 seconds for all subjects. Before the experiment, all subjects had practiced the tasks under various conditions many times in order to accustom themselves to the feedback apparatus. Especially with the monocular HMD, the subjects had to practice until overcoming disparity between left and right eye sights.

52

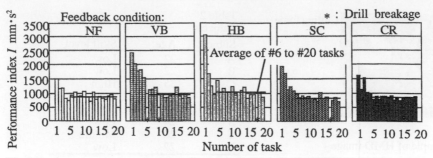

Fig.5 Performance index of subject T when using binocular HMD

4. RESULTS AND DISCUSSION

4.1 System using binocular see-through HMD

Figure 5 shows examples of the relationship between the number of task and the performance index I when using the binocular see-through HMD. Symbols NF, VB, HB, SC and CR represent the feedback conditions; "no active visual feedback", feedback with "vertical bar chart", "horizontal bar chart", "sector chart" and "circle chart" respectively. The thrust force component was displayed as the feedback information. * indicates failed tasks in which the drill was broken. As seen in this figure, the index I of the first to fifth tasks decrease apparently, but after the sixth task I goes to an almost steady value. The results of the other subjects are similar and the steady value is hardly dependent upon the order of experiments. The horizontal lines in the figure represent the average of I of the 6- to 20-th tasks. The influence of the active visual feedback upon the operating efficiency can be discussed by comparing the average values. The values and the numbers of failed task for four subjects are shown in figure 6. The feedback with the circle chart (CR) has been tried only by subject T. From the figure of average index value, it is noticed that both SC and CR improve the operating efficiency compared with NF, while both bar charts reduce the operating efficiency. Nevertheless, the effectiveness of the active visual feedback

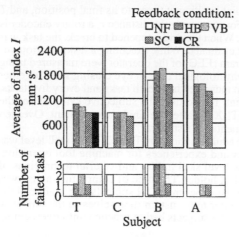

Fig.6 Average performance index and number of failed task when using binocular HMD

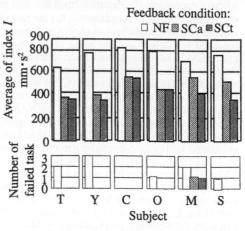

Fig.7 Average performance index and number of failed task when using monocular see-around HMD

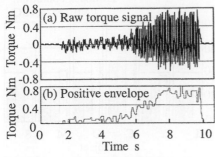

Fig.8 Envelope of torque signal

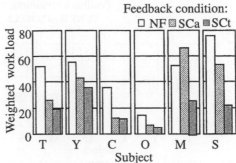

Fig.9 Subjective symptom by NASA-TLX (monocular HMD)

with the sector and circle charts is not so high as the active auditory and force sensory feedback which was described in the previous works [3, 5]. One of the reasons for it may be narrow eye sight of the operator wearing a goggles-like HMD. As for the number of failed task in Figure 6, any obvious tendency cannot be seen.

4.2 System using monocular see-through see-around HMD

Figure 7 shows the average performance index I and the number of failed tasks when using the monocular see-through see-around HMD. In the figure, only the results of feedback with the sector chart are compared with the results of no active feedback, because its relatively high effectiveness has been confirmed in the previous section. Symbols SCa and SCt represent the sector chart feedback of thrust and torque respectively. For the feedback of torque, it is impossible for human operators to recognize the raw signal through visual information because of its fast variation shown in Figure 2. Hence, in this research, the value of positive envelope of the signal is used as the feedback information. An example of the envelope processing is shown in Figure 8.

From Figure 7, it is obvious that the monocular HMD is superior to the binocular HMD in the improvement of the operating efficiency and the number of failed task. This fact must be caused by the difference of eye sight of both HMD's. Comparing the feedback of thrust and torque, it is seen that the torque is superior in the operating efficiency for most subjects, especially for un-skilled subjects Y, M and S. It is considered that this result comes from firstly the sensitivity of torque to change in feed rate and secondly the envelope processing which suppresses high frequency component from raw signal.

Figure 9 shows the average weighted workload by NASA-TLX. Low value of the workload indicates comfortability of the subject while task, and high value does difficulty of the task. As seen in the figure, the tendency of the value against the feedback condition is similar to that of the average performance index shown in Figure 8. This fact means that the feedback method which improves the operating efficiency can also reduce the workload.

Figure 10 shows the variation ratio of CFF [6]. Positive value of the ratio indicates the increase of mental fatigue by the tasks and negative value does the increase of arousal level without fatigue. From this figure, it can be noticed that suitable active visual feedback can suppress the increase of fatigue, especially for the torque feedback.

Figure 11 shows the average EMG of forehead while 20 tasks. It is known that this value is related to the magnitude of mental stress. From this figure, the same consideration of fatigue as described in CFF can be also drawn. Though the raw signal of EMG is not shown in this paper, its sudden increase appeared at the instance when the drill happened to break.

In the results of EEG, which are not shown in this paper, no obvious tendency is seen.

54

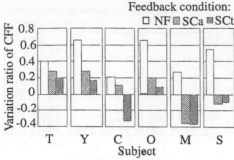

Fig.10 Variation ratio of CFF (monocular HMD)

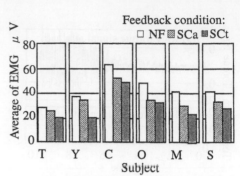

Fig.11 Comparison of average of EMG of forehead (monocular HMD)

5. CONCLUSIONS

In this study, two kinds of HMD as the active visual feedback device and the method for displaying feedback information in small hole drilling have been investigated. Main conclusions obtained are summarized as follows.

(1)The active visual feedback methods with the sector chart and circle chart are effective in improvement of operating efficiency.

(2)The monocular see-through see-around HMD is superior to the binocular see-through HMD as the visual feedback device, because the former can give wide eye sight to the operators.

(3)It is confirmed that the workload evaluated by NASA-TLX is closely related to the increase of fatigue by CFF and the stress by EMG of forehead.

(4)As the feedback information, the envelope of torque signal is superior to the thrust.

6. ACKNOWLEDGMENTS

The authors would like to thank Mr. Y. Yamamoto and Mr. Wang F. for their kind cooperation.

REFERENCES

1. H. Kato and T. Sato, Effect of Auditory feedback on Operating Efficiency and Fatigue --In the Case of Manual Cut-Off Turning with Virtual Lathe, Int. J. of Japan Soc. for Precision Eng.,30,4,(1996) 325.
2. H. Kato, K. Kobayashi and LIU S., Skill Learning in Manual Machine Tool Operation -- Development of New Virtual Lathe and Study on Minute Grooving by Using It, Int. J.of Japan Soc. for Precision Eng., 31,3,(1997) 221.
3. H. Kato and N. Taoka, Active Sensory Feedback in Manual Machine Tool Operation --Effect of Auditory and Force Sensory Feedback of Cutting Force in Fine Machining, Int. J.of Japan Soc. for Precision Eng., 64,3,(1998) 455.
4. e.g. S. Miyake and M. Kumashiro, Subjective mental workload assessment technique --An introduction to NASA-TLX and SWAT and a proposal of simple scoring methods, J. of Japan Ergonomics Soc.,29,6,(1993)399 (in Japanese).
5. N. Depaiwa et al, Active Sensory Feedback in Manual Machine Tool Operation --Effect of Feedback Information and Characteristics in Auditory Feedback of Cutting Force, Proc. of Spring Meeting of Japan Soc. for Precision Eng., (1999) 615 (in Japanese).
6. e.g. Ergonomics Handbook, Kanehara Publishing Co.,(1966) 441 (in Japanese).

Human Friendly Mechatronics (ICMA 2000)
E. Arai, T. Arai and M. Takano *(Editors)*
© 2001 Elsevier Science B.V. All rights reserved.

Dynamics of training simulator for mobile elevating platforms

E. Keskinen[1], M. Iltanen[2], T. Salonen[1], S. Launis[1], M. Cotsaftis[3] and J. Pispala[4]

[1]Tampere University of Technology, Laboratory of Machine Dynamics, P.O.Box 589, FIN-33101 Tampere, Finland

[2]Tampere University of Technology, Electronics Laboratory, P.O.Box 600, FIN-33101

[3]Ecole Centrale Electronique, Laboratoire des Techniques Mechatroniques et Electroniques 53 Rue de Grenelle, 75007 Paris, France

[4]Bronto SkyLift Ltd, P.O.Box 553, FIN-33101 Tampere, Finland

Intervention in high place such as buildings are requiring hydraulic elevating platforms commonly used for assembling outside covers, washing windows, as well as inspection and safety for repairs, rescuing persons and fire-fighting. As these platforms are aimed at providing a complex service, their operation, even when simplified by adequate design, is requiring a sophisticated action which has to be learned by operator using a concept of a man-in-the-loop simulator to be used in operator training for time-critical and accurate boom maneuvers. From various constraints the hardware consists of a boom platform mounted on a 3d Stewart platform, and virtual engineering software is used to visualize the working environment on wall screens. As a consequence for validation of the simulator, a strong limitation is coming from the needs of a simulation model satisfying contradictory properties of accuracy and real time computation which is transforming large boom movements to produce restricted motion in the Stewart mechanism.

1. INTRODUCTION

The use of mobile elevating platforms in assembly works and rescue operations in urban environment is now very common as they satisfy most criteria of mobility and adaptability to the various tasks, mainly to reach a high-rise building or other tall civil engineering construction from the outside without any attachment to the building. Needless to stress, the various boom displacements near the buildings should be very cautious because of potential risk of collision damages during the approach period. To be cost effective in construction and life saving during rescue operation, the maneuvers have to be as fast as possible, but the time needed to reach high elevations is long due to the slow approach motion though the environment is usually not unstructured which allows some predictive safety margin.

The efficiency of the process is strongly operator-dependent, because the number of manually controlled actuators is large, and the requirement to reach fast the working area while avoiding collisions is very demanding. Any improvement of speed and /or accuracy of maneuvers by means of new technology are therefore very important for end-users. Such is the Cartesian driven mode for platform motion (Ref. 1), to follow better vertical and horizontal trajectories often used in building applications.

In order to fulfill the strict requirements of accuracy, safety and speed, the operators have to follow specific training courses including different maneuvers under controlled conditions. Because of high unit price and important risk to fall in damage or in accident situations, a low-cost and more practical way to provide realistic training environment has been introduced to be used for operator training. This environment consists of the platform itself but instead of being mounted on the hydraulic boom (Fig. 1a) the platform is fixed onto a classical 6 d.o.f. Stewart platform (Fig.1b). The outside world is simulated on a large textile screen, from which the operator can see the real working environment by means of a scrolling image map composed of digital pictures taken from the actual working environment.

2. SYSTEM DESCRIPTION

Elevating platform is an hydraulic telescopic boom system consisting of two links and the platform. The first link is a chain driven expandable boom with a revolute joint at vehicle frame for turning the boom around the vertical axis and another joint for giving an inclination angle. The second link is the boom arm carrying the platform. The operator is driving the boom standing at the ground level or on the moving platform itself by controlling simultaneously four different actuators for boom turning, boom extension, boom inclination and arm tilting, a situation appearing in figure 1a. The valves governing the actuator motions are typically of proportional type. Extension of the three-stage telescopic boom is carried out using chain arrangement.

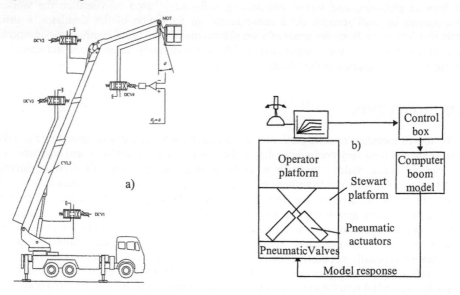

Figure 1. a) Links and actuators in the platform and b) the system diagram of the simulator.

The last joint suspending the platform works under feedback control system, which keeps the platform orientation fixed with respect to vertical direction. The main difficulty for the operator in driving the system is to manage the large number of actuator inputs for producing required Cartesian driving lines instead of curved trajectories, typically generated by actuator by actuator driving modes.

Volume limitation in indoor training stands imposes not to use the real boom structure as part of the simulator. Also, as an important simulator quality is to reproduce accelerations similar to the ones occuring with a real boom, the platform has been mounted on a classical Stewart platform as shown in the system diagram (Fig. 1b). The platform has a similar joy-stick interface and control panel as in real system. The input given by the operator with the joy-stick to drive the system manually in actuator-by-actuator mode is fed into the control electronics of the proportional valves, and valve board is producing outputs to govern the valve. As real valves and actuators are missing here, this output is transmitted to a boom computer model in order to produce the corresponding dynamic boom response in terms of the 6 d.o.f. motion of the manned platform. This motion will be transformed to the actuator positions of the Stewart platform to get the response motion within accuracy of used servo actuators. Instead of following large amplitude boom trajectories as in full lifting and turning maneuvers, the simulator platform only reaches small areas in the working space. Therefore the control of Stewart platform dynamics should be modified to produce correct driving mode with limited amplitude.

3. DYNAMIC BOOM MODEL

The boom models for in-plane and out-of-plane motions (Fig. 2) consist of rigid boom links connected by hydraulic motors or cylinders. The bodies for three-stage telescopic boom and platform arm are denoted by subscripts $i = 1,2,3,4$ while the actuators for boom inclination, arm tilting, boom extension, platform orientation and boom turning around the vertical are numbered by $j =1,2,3,4,5$ respectively. For simplicity, the platform itself has been reduced to the tip of platform arm as a concentrated mass with inertia. Another difference in the boom model is the use of motors instead of linear actuators in all revolute joints. The motion of the points in the boom system is followed in cylindrical (r,z,φ) coordinate system with origin fixed to the vertical turning joint located at offset distance e from the boom inclination joint.

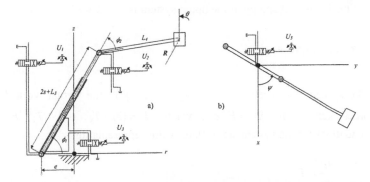

Figure 2. Rigid link boom models for a) in-plane-motion and b) out-of-plane motion.

The state variables of the boom are : relative angle ϕ_1 between boom and base, relative angle ϕ_2 between arm and boom, extension s between the first and second telescopic boom link, absolute angular position θ of the platform with respect to vertical, and the rotation angle ψ around the vertical. Recalling that the chain connects the motion of telescopic boom segments together, the Cartesian positions of the mass center of gravity of each link $i=1...4$ are

$$R_1 = \frac{L_1}{2}e_1 - ee_r \,;\, R_2 = \left(s + \frac{L_2}{2}\right)e_1 - ee_r \,;\, R_3 = \left(2s + \frac{L_3}{2}\right)e_1 - ee_r \,;\, R_4 = R_3 + \frac{L_3}{2}e_1 + \frac{L_4}{2}e_2 \quad (1)$$

while the position of the platform body at arm tip is $R = R_4 + \frac{1}{2}R_4 e_2$ with orientation θ. Unit vectors e_1, e_2, e_r correspond to boom, platform arm and boom foot directions and are functions of boom and arm orientations.

The kinetic energy of the system has contributions from linear and angular motions

$$2K = \sum_{i=1}^{4} M_i \dot{R}_i \cdot \dot{R}_i + \sum_{i=1}^{4} \omega_i \cdot I_i \omega_i + M\dot{R} \cdot \dot{R} + \omega \cdot I\omega \qquad (2)$$

where the link and platform inertia matrices are of the form

$$I_i = M_i \begin{bmatrix} L_i^2/12 & 0 & 0 \\ 0 & L_i^2/12 & 0 \\ 0 & 0 & r_i^2 \end{bmatrix} ; \, I = \begin{bmatrix} I_1 & 0 & 0 \\ 0 & I_2 & 0 \\ 0 & 0 & I_3 \end{bmatrix} \qquad (3a,b)$$

The inertia radii of links around their axis is given by r_i. The angular velocities in local body coordinate systems are for links $i=1,2,3,4$ and the platform

$$\omega_i = \begin{Bmatrix} \dot{\phi}_1 \\ \cos\phi_1\,\dot{\psi} \\ \sin\phi_1\,\dot{\psi} \end{Bmatrix} ; \, \omega_4 = \begin{Bmatrix} \dot{\phi}_1 - \dot{\phi}_2 \\ \cos(\phi_1 - \phi_2)\,\dot{\psi} \\ \sin(\phi_1 - \phi_2)\,\dot{\psi} \end{Bmatrix} ; \, \omega = \begin{Bmatrix} \dot{\theta} \\ \cos\theta\,\dot{\psi} \\ \sin\theta\,\dot{\psi} \end{Bmatrix} \qquad (4a,b,c)$$

The gravitational potential energy of the boom system is

$$V = \sum_{i=1}^{4} M_i g R_i \cdot e_z + MgR \cdot e_z \qquad (5)$$

Defining the system state vector $x = (\phi_1 \quad \phi_2 \quad s \quad \theta \quad \psi)^T$ the virtual work done by the actuator forces gets form $\delta W = F^T \delta x$, where $F = (T_1 - T_3 \quad -T_2 + T_3 \quad F \quad T_3 \quad T_4)^T$. Lagrangian equations for the system are in state vector representation

$$\frac{d}{dt}\left(\frac{\partial K}{\partial \dot{x}}\right) - \frac{\partial K}{\partial x} + \frac{\partial V}{\partial x} = F \qquad (6)$$

As the large amplitude boom motion is very slow, one can neglect higher order cross-coupling terms related to Coriolis and centripetal accelerations leading to dynamic equations

$$M(x)\ddot{x} = G(x) + F \tag{7}$$

where the equations are decoupled into in-plane, turning around the vertical and platform orientation motions. The elements of state-dependent mass matrix and gravitational load vector can be found in Ref. 4.

Hydraulic actuator dynamics is governed by two first order pressure differential equations for each actuator and valve flows are determined by two nonlinear algebraic equations for each valve. Valve inputs u_j have open loop manual control part U_j for actuators $j=1,2,3,5$ whereas classical PID control is applied for actuator $j=4$ keeping platform vertical orientation

$$u_j(t) = U_j(t) \quad ; \quad j = 1,2,3,5 \qquad u_4(t) = C_D \dot{e}_4(t) + C_P e_4(t) + C_I \int_0^t e_4(t)dt \tag{8a,b}$$

The servo error between the current and initial position of coordinate 4 is $e_4(t) = x_4(0) - x_4(t)$. Another set of PID controllers is governing the Stewart platform actuators to compensate motions going outside from its working space. The PID's have to be tuned so that they compensate fast enough the boom response input to limit the motion amplitudes compatible with the restrictions of Stewart platform. The proposed control rule means practically that the Stewart platform system always returns to its chosen initial state.

4. SIMULATION AND VIRTUAL ENVIRONMENT

The training simulator provides an environment where the platform system is driven with real-like dynamic effects combined with a view of the work object projected onto the screen and following the movements of the platform. The screen enables the operator to observe the real working environment by means of a scrolling image map composed of digital pictures (Fig. 3a). Virtual environments created by software may be used also. With combined spatial and ambient sound, a realistic and captivating 3d-audio environment is also available.

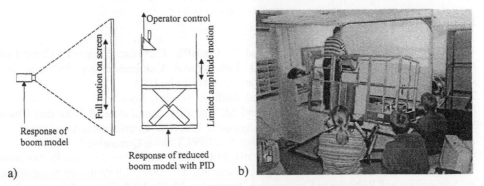

Figure 3. a) Lay-out of the training simulator and b) simulator during lifting maneuver.

As the system moves with reduced amplitudes of motion, the image controller has to obtain information from the large amplitude movement to give the operator the feeling of motion acts in full mode from the screen. This is done by scrolling the image map in synchronized way with the motion of boom tip computed from the system responses under open loop command only. This means that the open loop joy-stick output is controlling movements on the screen while the dynamic platform motions are limited by PID feedback control.

The system has been built up in laboratory environment for preliminary tests (Fig. 3b). The 3d-environment is modeled using 3D Studion MAX software tools. All buildings are textured to give better sense of immersion, and special equipment for fire-fighting such as high-pressure water shower is modeled. The simulator software runs on PC and uses common application program interfaces such as direct X and direct sound. Therefore the polygon number must be limited. The entry level system hardware is able to handle as much as 10 000 textured polygons.

The 3d-audio environment has been created using recorded sound clips. Those clips are processed and sampled separately. The target of processing was to find out the main frequency of each sample and reuse it in future.

5. CONCLUSIONS

A training simulator concept has been developed for operator training of hydraulic elevating platforms. Based on physiological operator response, the accent has been put on reproduction of acceleration peaks existing in real system. The simulator consists of the elevator platform mounted on a classical Stewart platform. Amplitudes of platform motion are reduced at actuator control level and smaller movements than in the real boom system are produced. In order to reach real-time computational speed a simplified computer model of the boom system has been used, where the flexible boom links have been modeled as rigid ones. A large screen, where image maps of the virtual working environment are scrolling and are synchronized to large amplitude platform motions, is completing the system to provide realistic touch to the work process. Preliminary tests show that the introduced simulator concept is working well within the fixed constraints of realism and real time computation on usual PC. In a next step, the computer model should be developed to represent better state-dependent flexibility of the boom.

REFERENCES

1. E.Keskinen, J. Montonen, S. Launis and M. Cotsaftis, Cartesian Trajectory Control of Hydraulic Elevating Platforms. IASTED International Conference on Robotics and Applications, October 28-30, 1999, Santa Barbara, California, USA.
2. J.J. Craig, Introduction to Robotics. 2nd ed. Addison-Wesley, 1989.
3. E. Keskinen, J. Montonen, S. Launis and M. Cotsaftis, Simulation of Wire and Chain Mechanisms in Hydraulic Driven Boom Systems. IASTED International Conference on Applied Modelling and Simulation, September 1-3, 1999, Cairns, Queensland, Australia.
4. E.Keskinen, M.Iltanen, T.Salonen, S.Launis, M.Cotsaftis and J.Pispala, Man-in-the-Loop Training Simulator for Hydraulic Elevating Platforms. 17th IFAC International Symposium on Automation and Robotics in Construction. September 18-20, 2000, Taipei, Taiwan.

Human Friendly Mechatronics (ICMA 2000)
E. Arai, T. Arai and M. Takano *(Editors)*

On Proposal of Function-Behavior-State Framework as Refinement of EID Framework of Human-Machine Interface Design

Y. Lin, W.J. Zhang, and L.G. Watson

Advanced Engineering Design Laboratory (AEDL), Department of Mechanical Engineering
University of Saskatchewan, Saskatoon, CANADA S7N 5A9

Interface design is an important issue in the human-machine systems. The principles of interface design are studied in this paper. Specifically, Ecological Interface Design (EID), which is regarded as being most powerful, is re-examined. Some problems with EID are identified. A new principle for human-machine interface design called Function Behavior State (FBS) is proposed. A comparative study for EID and FBS is performed on a simulated process plant system. The subjects operate the systems (one based on EID and the other on FBS). Measuring of subject's eye behaviors, such as the pupil diameter and eye fixation on the displays, is conducted using an eyegaze system.

1. PRINCIPLES FOR INTERFACE DESIGN

Modern complex human-machine systems, such as the control room in nuclear power plants and the autopilot or pilot-aid systems in the cockpit of aircraft, call for an extensive study on human-machine interface design. The term *machine* may be meant for an automatic controller to a machine system or a plant system. The term *plant* refers to a real-world entity, which needs to be controlled and monitored to reach one or more goals. Figure 1 shows a general model of the plant and its controller. In this figure, two interfaces related to the human operator are implied: (i) the interface between the human operator and the plant, and (ii) the interface between the human operator and the automatic controller. This study assumes that the human operator will eventually be a supervisor of both the automatic controller and the plant. If an automatic controller is viewed as a part of the plant system and thus, the human-machine interface makes sense of interactions between the human and the plant.

The main objective of developing a human-machine interface should be as follows. At a normal plant operation situation, human operator's errors should be nil, while at an unexpected plant operation situation, the human operator should be able to find a solution to incident problems in an expected time period. Principles need to be developed for designing an interface system to achieve these goals. There are two categories of principles: semantical and syntactical. Semantical principles concern information contents displayed on the media with which humans and machines can communicate; these include the audio, visual, and so on. Syntactical

62

principles concern how information contents are presented in terms of the location and appearance of them placed on a media. This study concerns the semantical principles (principles for short).

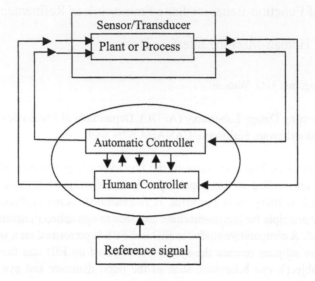

Figure 1. General configuration of a human-machine system

2. ECOLOGICAL INTERFACE DESIGN (EID)

Several principles for interface design were proposed in recent years though they are far from reaching consensus. Vicente et al. developed the Ecological Interface Design (EID) framework where a so-called "abstraction hierarchy" was described [1]. The abstraction hierarchy includes five levels of information: functional purpose, abstract function, generalized function, physical function and physical form. The EID framework has received a great attention in both academia and industries. However, the authors of this paper have identified some problems with EID, which are listed below:

(1) *The five levels of information are difficult to be comprehended.* It can hardly be possible to assume that the operator has these five layers as his or her mental model of the system.

(2) *The displaying of the five levels of information on the same front can make the operator be distracted to the information that does not contribute to solving problems at hand.* The best -known argument here would be on the need of displaying the abstract function level at the same time with the other levels of information. At this point, this study has observed an issue in developing a human-machine interface, i.e., how to trade-off between the hierarchical structure and non-hierarchical structure to display information. In fact, the EID framework takes a non-hierarchical structure in pursuit of the goal that one display of an interface system can be

used for a wide range of the operators. Philosophically, this strategy sounds like that everything is available, and it is the end-user's business to pick up the right stuff for his or her need. In reality, it may happen that an operator is simply unable to pick up what he or she needs.

(3) *In evaluating the EID framework, there is generally a lack of consideration of time constraint.* The time constraint in problem solving is very important in the case of abnormal events, and it may provide an effective evaluation means for the EID framework in response to the criticism that the efficiency of the EID framework lies in that it brings more information that other principles on the display.

(4) *In the EID framework, information may be overlapped, and furthermore, the five levels of information are displayed without concerning their syntactical issues.*

3. FUNCTION-BEHAVIOR-STATE (FBS)

A new framework is proposed to overcome these problems with the EID framework. This new framework is inspired from the general design theory called Function-Behavior-State (FBS) [2]. The FBS framework views any plant system as consisting of three levels: function, behavior and state. The *State* level is represented by entities, attributes and relationships among entities. In the case of process plants, the state shows the physical layout of a plant system, properties of each component in the plant system, and relationships among the components. The *Behavior* level is defined as changes of states. In the case of process plants, the behavior shows controlled changes of one or more states with respect to other one or more states (plant dynamics for short). Physical laws and/or physical effects govern the dynamics of a plant, and they are taken as a kind of axiom from which solutions are sought for problems. The *Function* level is a subjective description of the behavior, which describes perceived uses, by the human, of the behaviors. In the case of process plants, the vocabulary, e.g., change temperature of a tank, change flow rate in the pipe system, etc., is used to represent the functions. A prototype interface system with the FBS framework has been developed in this study for the same process plant as DURESSJ (Dual Reservoir System Simulation, Java version) which was developed based on the EID framework [1]. In the following discussion, DURESSJ associated with FBS is called FBS-DURESSJ while DURESSJ with the EID is called EID-DURESSJ.

4. EXPERIMENTAL VALIDATION

4.1. Experiment design

Validation of the finding of the problems with the EID framework, in comparison with the FBS framework, is performed in this study using EID-DURESSJ and FBS-DURESSJ programs, respectively. The subjects operate both systems. They view, supervise and control the DuressJ system for a period of time to comprehend and memorize what occurred about the plant dynamics. Two measures, eye pupil diameter and eye fixation, are collected and examined for the subjects during their operation. Pupil diameter is a kind of Autonomic Nervous Systems (ANS) measure [3]. Pupil diameter increases with the increase of task processing demands and resource investment. In another word, the pupil diameter can be used as an indicator of the mental workload. The major equipment used for recording these two measures is the Eyegaze from LC Technologies of Fairfax VA.

A 2x2x2 factorial design was taken in this experiment. The experiment has three-factors (expertise, interface and scenario) with two levels for each factor. The **expertise** factor has two levels: *expert* and *novice*, and it divides the subjects into expert and novice subjects. The **interface** factor has two levels: *FBS* and *EID*. The **scenario** factor has two levels: *Scenario 1* and *Scenario 2*, and it refers to the situations that the subjects encounter during the operation of the simulated plant. In particular, Scenario 1 is a normal supervising situation, and Scenario 2 is an abnormal problem-solving situation where a fault is introduced into the plant. The task difficulty level is increased from the normal situation to the abnormal situation. This experiment employs four subjects: two experts and two novices. Replications of each trial is 3 times. The total trial times for each subject is therefore 12 times. In total, 48 trials were performed. The duration for the eyegaze to collect data was 20 seconds. In addition to these two measures, the subjective measure (such as subject's self-assessment) was also used.

4.2. Results of the experiment
4.2.1. Pupil diameter

Analysis of variance (ANOVA) was utilized to determine effects of the subject, interface, and scenario on the pupil diameter. A Statistical Analysis System (SAS) was used to conduct the ANOVA. The results of the ANOVA is shown in Table 1. The effects of both the subject and the interface on the pupil diameter are highly significant (p=0.0001 for the subject and 0.0075 for the interface), and the effects of the scenario is not significant (p=0.329). There is no interaction among these three factors, but an interaction between the subject and the scenario (p=0.015).

A multiple comparison of means was performed using the Duncan's Multiple Range Test (α=0.05). The result is shown in Table 2. The effects of Subject 3 and 4 on the pupil diameter are not significantly different, whereas the effects of Subject 1 and 2 significantly different. The effect of the FBS interface on the pupil diameter is significantly different from the EID interface (means of the pupil diameter=3.70 mm for FBS and 3.60 mm for EID).

Two extra ANOVA were conducted for examining the assumption about the difference between the expert and novice. The results of ANOVA for the data set of the expert group show that the effects of the subject and the interface are significant (p=0.0082), whereas the effects of the scenario not significant (p=0.979). The results of the ANOVA for the data set of the novice group show that the effects of the interface and scenario are not significant (p=0.115 and 0.277 for the interface and the scenario, respectively), but the effects of the subject highly significant (p=0.0001).

4.2.2. Eye fixation

Eye fixation is another peripheral response measure. Eyegaze point and fixation are being tracked by the eyegaze system at a sampling rate of 60 Hz. Figures 2 to 5 present some of the data regarding the eye fixation. From these data, it can be seen that (i) the eye fixation is more concentrated in FBS than in EID, and (ii) the eye fixation is more concentrated in Scenario 1 than in Scenario 2. It is further observed from these data that in the case of EID and Scenario 2, the subjects seem to be more distracted.

Table 1

Analysis of Variance (ANOVA) results for the effects of the three factors (subject, interface, and scenario) on the pupil diameter (mm)

Source	DF	Sum of Squares	Mean Square	F Value	Pr>F
Model	15	5.703	0.38	26.68	0.0001*
Subject	3	5.323	1.77	124.53	0.0001*
Interface	1	0.116	0.12	8.14	0.008*
Scenario	1	0.014	0.01	0.98	0.330
Subject×Interface	3	0.065	0.02	1.51	0.230
Subject×Scenario	3	0.174	0.06	4.07	0.015*
Interface×Scenario	1	0.009	0.009	0.60	0.445
Subject×Interface×Scenario	3	0.002	0.001	0.05	0.984
Error	32	0.456	0.014		
Corrected Total	47	6.159			

* significant

Table 2

Summary of the statistical multiple comparison of means of pupil diameter for three factors (subject, interface, and scenario) (mm)

Subject	Mean* (mm)	Interface	Mean (mm)	Scenario	Mean (mm)
Sub-1	3.70^B	FBS	3.70^A	L01	3.67^A
Sub-2	4.18^A	EID	3.60^B	F01	3.63^A
Sub-3	3.35^C				
Sub-4	3.38^C				

* Means with different letter (comparison within each factor) are significantly different (Duncan's Multiple Range Test, $\alpha=0.05$).

4.2.3. Subjective measure

The self-reported measure shows that subjects prefer the FBS to EID in most of cases. The subjects can understand the configuration of the process control system more quickly and control the variables more efficiently. In the case of Scenario 2, the subjects feel that they can more easily capture the fault with FBS than they do with EID. Some subjects commented that the information in EID which shows such physical laws as energy conservation had distracted their attention. They rated that their mental workload are higher in EID than in FBS.

5. CONCLUSION

The problems with the EID framework were identified, and experimentally examined. The part which displays physics laws or effects in the EID framework appears problematic. The pupil diameter is higher in FBS than EID which implies that the subjects' mental workload is higher in FBS than EID. This is inconsistent to the subjects' self-assessment. One of the reasons for this contradictory may be that the time constraint is too demanding to make the subjects feel hopeful to get knowledge needed to solve the problem defined in Scenario 2 with the EID interface. This situation can then destroy the reliability of this measure.

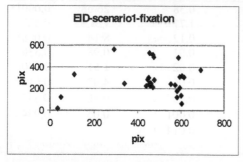

Figure 2. Eye fixation in EID (scenario 1)

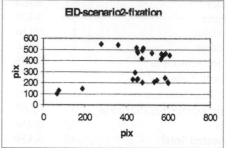

Figure 3. Eye fixation in EID (scenario 2)

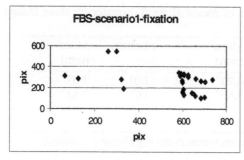

Figure 4. Eye fixation in FBS (scenario 1)

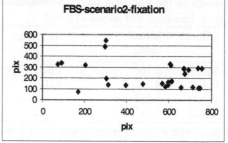

Figure 5. Eye fixation in FBS (scenario 2)

REFERENCES

1. Vicente, K. J., "Supporting Operator Problem Solving Through Ecological Interface Design", IEEE transactions on system, man and cybernetics, Vol. 25, No. 4 (1995) 529.
2. Yasushi, U., Masaki I., Masaharu Y., Yoshiki S. and Tetsuo T., "Supporting Conceptual Design Based on the Function-Behavior-State Modeler," Artificial intelligence for engineering design, analysis and manufacturing (1996) 275.
3. De Waard, D., The measurement of drivers' mental workload. Ph.D. thesis, Traffic Research Center, University of Groningen. Haren, The Netherlands (1996).

Human Friendly Mechatronics (ICMA 2000)
E. Arai, T. Arai and M. Takano *(Editors)*

A Graphical User Interface for Industrial Robot Programming in Non-Repetitive Tasks

Emilio J. González-Galván, Elizabeth Rivera-Bravo, Rubén D. González-Lizcano, Oscar A. Garay-Molina and Victor H. Miranda-Gómez [a*]

[a]Centro de Investigación y Estudios de Posgrado. Facultad de Ingeniería,
Universidad Autónoma de San Luis Potosí
Av. Dr. Manuel Nava 8. San Luis Potosí, S.L.P. 78290. México

A graphical user interface (GUI) used to facilitate the programming of an industrial robot is currently under development. In the work described herein, robot programming is understood as the process of defining a set of robot configurations that enable an accurate, three-dimensional positioning of a tool held by the manipulator over a fixed, arbitrary, smooth surface. The GUI uses a patented, vision-based method known as Camera-Space Manipulation. This paper introduces some of the basic ideas employed in the implementation of the GUI. The proposed system was tested using an industrial robot and preliminary experimental results that show the accuracy of the proposed methodology are also presented.

1. INTRODUCTION

The use of industrial robots in manufacturing plants has facilitated the massive production of goods, with high quality and low cost. However, the demands involved in the introduction of a robot system into a production line have limited their use in small and medium-size industries with a small production volume. In particular, these restrictions include the high costs associated with redesigning the manufacturing line and the addition of jigs and fixtures to hold the workpiece in a precise location and orientation. Also, the security devices required for personnel safety, the costs associated to robot programming, the hiring of employees trained in robotics and the cost of the industrial robot itself have limited the application of robots in the small and medium-size industry.

The industrial organizations for which our current development is intended share the common characteristic of requiring a system that facilitates the programming of industrial robots. This demand arises from the fact that the robot must be able to perform a given task with a small number of repetitions. As a limit, the system must be able to perform a given task only once and then it must be programmed again to perform a different one.

In general the intended tasks demand close-tolerance, three-dimensional rigid body assembly. This type of precision and versatility, without the need of special fixtures, can be achieved by using vision-based methods. Among them, the patented method

---

*Project sponsored by the National Council of Science and Technology of Mexico (CONACYT) and the Fund for Research Support (FAI) of the Universidad Autónoma de San Luis Potosí.

known as Camera-Space Manipulation (CSM) has proven its value in that a large variety of difficult, three-dimensional positioning tasks has been implemented with the method (see http://www.nd.edu/NDInfo/Research/sskaar/Home.html). These tasks have been executed reliably, robustly, and with high precision. Among them, special attention can be drawn to a technology development [1] consisting of a graphical user interface that facilitates robotic plasma application over arbitrary, curved surfaces.

The method of Camera-Space Manipulation has received considerably less attention than other vision-based methods for manipulator control such as the visual servoing (VS) methods. However [2], CSM can be more effective than VS, for instance, in situations where the uncalibrated cameras used to control the maneuver remain stationary in the time elapsed between the detection of the workpiece and task culmination, which is the case for the experiments discussed herein.

2. THEORETICAL FRAMEWORK AND EXPERIMENTAL SETUP

The general procedure for performing a task such as welding, painting, cutting, drilling, etc. over an arbitrary surface using the proposed interface is depicted in Fig. 1. This flow diagram represents the limiting case in which the work is performed in only one occasion. Once the task is finished, the surface is removed and substituted by a new workpiece, which in general has a different geometry than the previous one. The location of each smooth surface is such that it can be detected by a minimum of two control cameras, but is otherwise arbitrary.

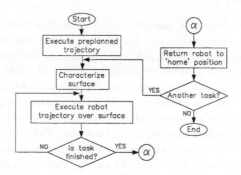

Figure 1. Flow diagram representing the execution of a given task using the GUI

In addition to the images obtained from the control cameras, the GUI makes use of an image which is used by a human operator to designate the location over the surface where the robot is going to perform the task. In order to facilitate the operation of the system, the interface was developed in a windows environment using Microsoft's Visual C++ compiler. A view of this interface is shown in Fig. 3. In the following sections, a brief description of the most important elements associated to the GUI are presented.

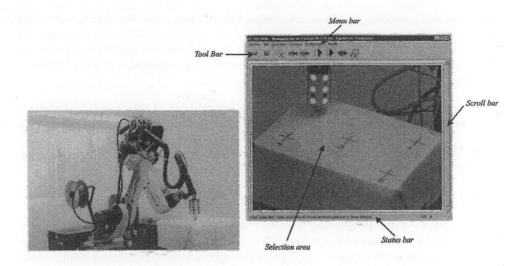

Figure 2. Fanuc ArcMate100i robot   Figure 3. A view of the GUI

2.1. Preplanned trajectory

The vision-based methodology described in this paper uses an orthographic camera model to define the nonlinear, algebraic relationship between the arm configuration and the appearance in camera-space of a number of visual features attached to the tool held by the robot. This relationship, referred to as camera-space kinematics, depends on six view parameters $\mathbf{C} = [C_1, \ldots, C_6]^T$ included in the orthographic camera model described in eq. (1):

$$
\begin{aligned}
f_x\left(x_i, y_i, z_i; \mathbf{C}\right) &\equiv b_1(\mathbf{C})\, x_i + b_2(\mathbf{C})\, y_i + b_3(\mathbf{C})\, z_i + b_4(\mathbf{C}) \\
f_y\left(x_i, y_i, z_i; \mathbf{C}\right) &\equiv b_5(\mathbf{C})\, x_i + b_6(\mathbf{C})\, y_i + b_7(\mathbf{C})\, z_i + b_8(\mathbf{C})
\end{aligned}
\tag{1}
$$

where $b_1, \ldots, b_8$ depend on the six view parameters as follows,

$$
\begin{array}{ll}
b_1(\mathbf{C}) = C_1^2 + C_2^2 - C_3^2 - C_4^2 & b_5(\mathbf{C}) = 2\left(C_2 C_3 - C_1 C_4\right) \\
b_2(\mathbf{C}) = 2\left(C_2 C_3 + C_1 C_4\right) & b_6(\mathbf{C}) = C_1^2 - C_2^2 + C_3^2 - C_4^2 \\
b_3(\mathbf{C}) = 2\left(C_2 C_4 - C_1 C_3\right) & b_7(\mathbf{C}) = 2\left(C_3 C_4 + C_1 C_2\right) \\
b_4(\mathbf{C}) = C_5 & b_8(\mathbf{C}) = C_6
\end{array}
\tag{2}
$$

In eq.(1), $f_x(\ldots)$ and $f_y(\ldots)$ describe the camera-space location of the centroid of the visual features, while (x_i, y_i, z_i) represents the physical position of the ith manipulated feature. This location, defined with respect to a robot-fixed coordinate system, is obtained by using the nominal kinematic model of the manipulator.

For the work described herein, a Fanuc ArcMate100i industrial robot was used. All algorithms were developed in a standard PC computer and the communication via ethernet between this device and the robot uses a TCP/IP protocol. Two cameras Sony SPT 304 were used and connected to a Data Translation DT3155 frame grabber, which is installed inside the computer. The visual features consist of a number of luminous cues, such as the ones depicted in Fig. 3. The cues are placed on a special tool attachment at the end of the robot, as seen in Fig. 2, with marks turned off. The advantages of using luminous cues include the following:

- The image analysis required to detect the centroid of the cues is relatively simple as they appear like white ellipses in the images obtained from the cameras.

- The luminous cues can be turned on and off using a simple computer program. This characteristic is useful for cue-identification purposes.

- The dependence on particular light conditions is reduced.

The purpose of the preplanned trajectory is to obtain samples of the camera-space location of the luminous cues, represented as (x_{ci}, y_{ci}), for the ith mark, as well as the corresponding physical location (x_i, y_i, z_i). The trajectory involves a broad displacement in both camera and robot-joint spaces. The samples are used to estimate the value of the six view parameters, according to the procedure described in [3]. This process is done independently for each control camera used in the maneuver and a set of view parameters is defined for each device. It is important to note that a minimum of two control cameras is required in order to successfully perform a 3D task. Also, it is important to emphasize that the value of the view parameters is constantly updated with samples taken during the execution of a maneuver.

2.2. Surface characterization

An important feature that share the method of Camera-Space Manipulation with other vision-based techniques is that the maneuver objectives are defined and pursued inside the images obtained from each control camera. In order to define these camera-space objectives, a way to produce non-permanent marks on the surface is desired. One alternative is the use of a low-power laser pointer, since a laser spot is in general detected by a B&W CCD camera as a bright, white spot, which can be easily detected with a relatively simple image-analysis algorithm, similar to the one that we use to detect the luminous cues on the tool attachment. Experiments have shown that the intensity of the detected spot depends on surface characteristics such as color, texture, etc.

For our experiments, a structured lighting consisting of an array of spots was used. These spots are projected over the surface as depicted in Fig. 4. The process of defining compatible, camera-space targets for each of the control cameras is described in detail in [4]. This procedure involves the use of a user-prescribed density of laser spots and a user-specified interpolation model. In general, regions of the surface with greater curvature require more laser-spots in order to "capture" better the local curvature of the workpiece.

Based on the work described in [4] a correspondence between the different laser spots detected by all control cameras can be established. This correspondence can be used in order to estimate an *approximate* physical location (t_x, t_y, t_z) of the laser spots, referred to a fixed coordinate system. This coordinate system can be considered at the base of the robot and is coincident with the reference frame used to define the physical location of the luminous cues at the tool attachment. It can be shown that the approximate 3D location of these spots can be estimated as follows,

$$
\begin{bmatrix} t_x \\ t_y \\ t_z \end{bmatrix} = \mathbf{A}^{-1}\,\mathbf{V} \;\; ; \;\; \mathbf{V} = \begin{bmatrix} \sum \left(b_1{}^i(x_l{}^i - b_4{}^i) + b_5{}^i(y_l{}^i - b_8{}^i) \right) \\ \sum \left(b_2{}^i(x_l{}^i - b_4{}^i) + b_6{}^i(y_l{}^i - b_8{}^i) \right) \\ \sum \left(b_3{}^i(x_l{}^i - b_4{}^i) + b_7{}^i(y_l{}^i - b_8{}^i) \right) \end{bmatrix}
\tag{3}
$$

and

$$
\mathbf{A} = \begin{bmatrix} \sum \left(b_1{}^{i2} + b_5{}^{i2} \right) & \sum \left(b_1{}^i b_2{}^i + b_5{}^i b_6{}^i \right) & \sum \left(b_1{}^i b_3{}^i + b_5{}^i b_7{}^i \right) \\ & \sum \left(b_2{}^{i2} + b_6{}^{i2} \right) & \sum \left(b_2{}^i b_3{}^i + b_6{}^i b_7{}^i \right) \\ Symmetric & & \sum \left(b_3{}^{i2} + b_7{}^{i2} \right) \end{bmatrix}
\tag{4}
$$

In the previous equations, the summation is performed from 1 to n_c control cameras. The value of b_1 through b_8, associated to each camera, depend on the current value of the view parameters as appear in eq. (2). The image-plane location of each laser spot is defined as $(x_l{}^i, y_l{}^i)$ for the ith camera.

Once the approximate location of n laser spots is defined, a matrix of moments $\mathbf{M_T}$ can be defined as follows,

$$
\mathbf{M_T} = \begin{bmatrix} \sum t_{xj}{}^2 & \sum t_{xj} t_{yj} & \sum t_{xj} t_{zj} & \sum t_{xj} \\ & \sum t_{yj}{}^2 & \sum t_{yj} t_{zj} & \sum t_{yj} \\ & & \sum t_{zj}{}^2 & \sum t_{zj} \\ Symmetric & & & n \end{bmatrix}
\tag{5}
$$

where the summation is performed from 1 to n detected laser spots. This compact representation combined with the user-selected image-plane location of the place where the robot has to perform and the interpolation model described in [4], enable the definition of compatible camera-space destinations in each of the control cameras.

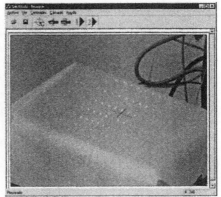

Figure 4. Array of laser dots as seen by the two control cameras

2.3. Determination of arm configuration

Once compatible camera-space destinations have been established, the strategy for determining the internal arm configuration is similar to the procedure described in [5]. This process allows the determination of the six angular joints of the robot in a single estimation procedure. In contrast with this idea, previous work [1] required two different algorithms for determining separately two partitions of the joint coordinates of the manipulator.

All previous ideas were integrated in the GUI depicted in Fig. 3. For the initial tests, the task consisted on defining a single spot that the robot has to reach with the tip of the tool. This spot was marked with ink over a cardboard surface and the camera-space coordinates of this mark were defined by clicking with a mouse over the image used to prescribe the task. Then, the system automatically generates compatible image plane destinations in each of the control cameras, and the arm configuration required to reach the destination is calculated. A large number of tests were conducted and the error was consistently under 2 mm. The error was determined by measuring the final distance between the mark on the cardboard and the tip of robot's tool.

3. CONCLUSIONS

The experimental results obtained so far have shown the viability of the proposed methodology to produce accurate and robust positioning of a tool held by an industrial robot over an arbitrary surface.

It is expected that the use of the GUI will enable a reduction in the time required for robot programming, facilitating the application of industrial robots in factories with a relatively small production volume. The interface will permit that a worker, with no particular training in robotics, will be able to define the task that the manipulator must perform. Since the computer used to define the task can be placed in a remote location, it is also expected that the use of the GUI will not increase the costs associated with personnel safety, but above all, the package will provide an user with a friendly environment adequate for simple and robust task designation.

REFERENCES

1. Seelinger,M., González-Galván,E., Robinson,M., Skaar,S.B. 1998, "Towards a Robotic Plasma Spraying Operation Using Vision". *IEEE Rob. and Aut. Mag.* 5(4):33–36.
2. Seelinger, M., Skaar, S.B., Robinson, M. 1998, "An Alternative Approach for Image-Plane Control of Robots". In *The Confluence of Vision and Control*, D.J. Kriegman, G.D. Hager and A.S. Morse (Eds.). Springer-Verlag London Limited.
3. Gonzalez-Galvan, E.J., Skaar, S.B., Korde, U.A., and Chen, W.Z. 1997. 'Application of a Precision Enhancing Measure in 3-D Rigid-Body Positioning Using Camera-Space Manipulation'. *The Intl. J. of Rob. Res.* 16(2):240-257.
4. Robinson,M.L., Skaar,S.B., Seelinger,M.J., 1998, 'Multiple Laser Spots for Camera-Space Objectives', *Proc. Sensor Fusion and Decentralized Control in Robotic Systems.* SPIE OE/Tech. Conf., Boston.
5. González-Galván E.J., Skaar, S.B., Seelinger, M.J. 1999, "Efficient Camera-Space Target Disposition in a Matrix of Moments Structure Using Camera-Space Manipulation". *The Intl. J. of Rob. Res..* 18(8):809-818.

Human Support Technology

Human Friendly Mechatronics (ICMA 2000)
E. Arai, T. Arai and M. Takano (Editors)

Development of a micro manipulator for minimally invasive neurosurgery

K. Harada[a], K. Masamune[a], I. Sakuma[b], N. Yahagi[b], T. Dohi[b], H. Iseki[c], K. Takakura[c]

[a]Department of Precision Machinery Engineering, Graduate School of Engineering, The University of Tokyo, 7-3-1, Hongo, Bunkyo-ku, Tokyo 113-8656, Japan

[b]Institute of Environment Studies, Graduate School of Frontier Sciences, The University of Tokyo, 7-3-1, Hongo, Bunkyo-ku, Tokyo 113-8656, Japan

[c]Department of Neurosurgery, Tokyo Women's Medical University, 8-1, Kawatacho, Shinjyuku-ku, Tokyo 162-8666, Japan

Surgical robots are useful for minimally invasive surgery, since it enables precise manipulation of surgical instruments beyond human ability in a small operation space. In this approach, we are developing a micro manipulator for minimally invasive neurosurgery. The micro manipulator consists of two micro grasping manipulators, a rigid neuroendoscope, a suction tube, and a perfusion tube. This paper reports on the micro grasping manipulator. It has two D.O.F for bending and one D.O.F for grasping. This prototype is 3.1mm in diameter and can bend 30 degrees in any direction. Stainless steel wire was used to actuate the manipulator.

Keywords: minimally invasive, neurosurgery, medical robot, manipulator, sterilization

1. INTRODUCTION

Endoscopic surgery, laparoscopic surgery, and other kinds of minimally invasive surgery are currently a hot topic because it can reduce a patient's post-operative discomfort and length of rehabilitation period, thus improving the patient's Quality of Life and reducing medical costs. Minimally invasive surgery has many advantages to the patient; on the contrary, the surgery is difficult for surgeons to perform. Precise manipulation of surgical forceps is indispensable for safe surgery, however, in minimally invasive neurosurgery, the precise manipulation of forceps through a small burr hole is particularly difficult. This has restricted the number of applicable clinical cases.

To achieve precise manipulation of forceps, surgical robot is useful. In this approach, various kinds of such surgical robots have been made for laparoscopic surgery [1] and some are even commercially sold. In the neurosurgery, however, the development of such surgical robots has just begun. Therefore, our aim is to develop a surgical robot for minimally invasive neurosurgery.

2. NEUROENDOSCOPIC SURGERY

In traditional neurosurgery, the patient's skull is opened widely to allow a space to see inside with a microscope (Figure 1). This kind of surgery is invasive surgery. On the other hand, neuroendoscopic surgery (Figure 2) is minimally invasive surgery. The burr hole in the patient's skull can be reduced to less than 20mm in diameter. In neuroendoscopic surgery, the surgeon inserts neuroendoscope into a hole in the patient's skull and then inserts forceps through channels situated next to endoscope's lens.

Neuroendoscopic surgery has been performed over 100 years by utilizing rigid or flexible neuroendoscope. This surgery has been applied mainly to third ventriculostomy or biopsy of tumors. Recent improvement of image diagnosis and surgical instruments lead to rapid increase in the number of applicable clinical cases. Nevertheless, due to the difficulty of this surgery, the number of clinical cases reported is small compared to that of the microscopic surgery.

One of the major difficulties in neuroendoscopic surgery is precise manipulation of forceps. The difficulty in precise manipulation of forceps can be mainly caused by following two problems. First problem is insufficient degrees of freedom (D.O.F) for positioning the tip of forceps. Different from laparoscopic surgery, there is only one hole for inserting surgical tools. To avoid damages on brain, only translation and rotation movement can be allowed. This inhibits surgeons from performing complicated task Second problem is hand trembling. Since a surgical space in a brain is packed with many thin vessels, hand trembling could cause serious damages on vessels.

To alleviate these problems, we are developing a micro manipulator for minimally invasive neurosurgery with a capability of the sterilization as reported [2]. We have focused on sterilization because it is often neglected by the researchers of surgical robot. In this paper, the micro grasping manipulator, which is one of the components of the micro manipulator, has been developed as the first step and reported. It has 5 D.O.F in total including 2 D.O.F for bending and 1 D.O.F for grasping. This prototype is 3.1mm in diameter and can bend 30 degrees in any direction. Stainless steel wire was used to actuate the manipulator and the bending motion was evaluated.

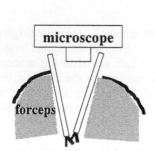

Figure 1.   Microscopic surgery

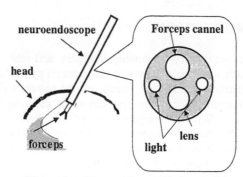

Figure 2.   Neuroendoscopic surgery

77

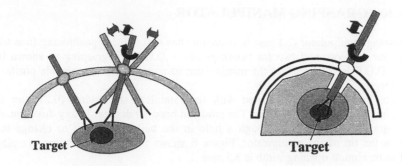

Target   Target

Figure 3.   Difference between Laparoscopic surgery and Neuroendoscopic surgery

3. CONFIGURATION OF THE MICRO MANIPULATOR

As shown in Figure 4, the micro manipulator is a unit which consists of two micro grasping manipulators (3.1 mm in diameter), a rigid neuroendoscope (4.0 mm in diameter), a suction tube (3.0 mm in diameter), and an irrigation tube (1.4 mm in diameter). The components are inserted into a rigid tube (10 mm in diameter) and are interchangeable with other surgical tools such as a surgical knife and a bipolar cautery. The micro manipulator is inserted into the burr hole with an insertion manipulator [3] that can determine the direction and position of the manipulator by reference to pre-operative images such as MRI (Magnetic Resonance Imaging).

This micro manipulator can easily be separated into sterilizable parts and nonsterilizable parts such as motors. The sterilizable parts are disposable because these parts are very complicated and 100 % sterilization cannot be guaranteed.

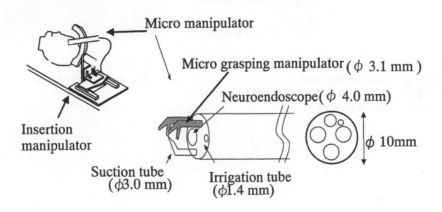

Figure 4.   The configuration of the micro manipulator

4. THE MICRO GRASPING MANIPULATOR

The micro grasping manipulator (3.1 mm in diameter) has 4 D.O.F for positioning (one for translation, one for rotation, and two for bending) and 1 D.O.F for grasping as shown in Figure 5. The 5 D.O.F in total enables the manipulator to adopt every conceivable position required in the surgery.

The manipulator has a ball-joint actuated with four stainless steel wires (0.27 mm in diameter), resulting in 2 D.O.F for bending. The joint can bend 30 degrees in any direction. A piano wire for grasping is threaded through a hole in the ball and moved to change the opening width at the tip of the manipulator. Figure 6 shows the tip of the micro grasping manipulator. The maximum opening width is 3.5 mm.

4.1. Drive unit

We fabricated a drive unit to drive the micro grasping manipulator as shown in Figure 7. The four wires for bending are driven by DC motors and optical encoders.

The drive unit for the micro grasping manipulator was designed small, because the space allowed for a drive unit is limited when manipulators are inserted with other surgical tools into one tube. Besides, the drive unit of the micro grasping manipulator was fabricated to be easily separated for the sterilization as shown in Figure 8, thus can be used in clinical cases.

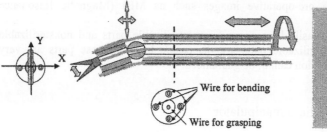

Wire for bending

Wire for grasping

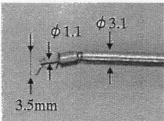

$\phi 1.1$ $\phi 3.1$

3.5mm

Figure 5
The micro grasping manipulator
In the experiment, the tip is moved in the each direction (direction X and Y)

Figure 6
The tip of the prototype

Figure 7.   Figure 8.
Prototype of the micro grasping manipulator   Separation of the drive unit

4.2. Evaluation

The bending motion (2 D.O.F) was evaluated. The following characteristics were measured in two directions shown in Figure 2.

1. Bending angle
2. Minimum change of the bending angle
    Minimum change of the tip motion (calculation)
3. Maximum bending force

Table 1 shows the result of the measurement.

The bending angle is approximately 20 degrees while the designed angle is 30 degrees; this is mainly caused by the friction at the joint and insufficient torque. The minimum movement of the tip of the micro grasping manipulator is 0.3 mm. This accuracy is small enough compared to the size of the tissue and the tip size of the manipulator.

These results will be helpful for further improvement of the micro grasping manipulators.

Table 1.   Result of the experiment   (Direction X and Y are indicated in Figure 2.)

	Direction X	Direction Y
Bending angle (degree)	$-20.2\sim+20.2$	$-18.2\sim+18.8$
Minimum change of the bending angle (degree)	1.7	1.9
Minimum change of the tip motion (mm)	0.30	0.33
Bending force (N)	0.75	0.60

5. CONCLUSION

The prototype of the micro grasping manipulator (3.1 mm in diameter) was developed for minimally invasive neurosurgery. It has 5 D.O.F and its bending motion (2 D.O.F) were evaluated. The problems such as friction should be alleviated.

After further improvement, the micro grasping manipulators, a neuroendoscope, and other surgical instruments will be installed in the micro manipulator. The micro manipulator will be applied to *in vivo* experiment.

With the micro manipulator, surgeons can precisely operate forceps and, as a result, more easily and safely achieve minimally invasive neurosurgery. As a next step, we are planning to build a human interface for this manipulator. This interface will be designed to assure easy use by surgeons.

This study was partly supported by the Research for the Future Program (JSPS-RFTF 96P00801).

REFERENCES
[1] Frank Tendick, S. Shankar Sastry, Ronald S. Fearing, Michael Cohn, Applications of Micromechatronics in Minimally Invasive Surgery, IEEE/ASME Transactions on Mechatronics, Vol.3, No,1, pp34-42, 1998
[2] S. Tadokoro et al., Development of a Micro Forceps Manipulator for Minimally Invasive Neurosurgery, Proceedings of the 13th International Symposium and Exhibition, Computer Assisted Radiology and Surgery, CAR'99, 700-705, 1999
[3] K. Masamune et al., A Newly Developed Stereotactic Robot with Detachable Drive for Neurosurgery, Medical Image Computing and Computer-Assisted Intervention-MICCAI'98, pp215-222, 1998

Human Friendly Mechatronics (ICMA 2000)
E. Arai, T. Arai and M. Takano (Editors)
© 2001 Elsevier Science B.V. All rights reserved.

81

LAN based 3D digital ultrasound imaging system for neurosurgery

Yasuyoshi Tanaka[a], Ken Masamune[a], Oliver Schorr[b], Nobuhiko Hata[c], Hiroshi Iseki[d], Yoshihiro Muragaki[d], Takeyoshi Dohi[a], Ichiro Sakuma[a]

[a]Institute of Environmental Studies Graduate School of Frontier Sciences University of Tokyo, Japan

[b]Institute For Process Control and Robotics, University of Karlsruhe, Germany

[c]Harvard Medical School Brigham and Women's Hospital, U.S.A

[d]Department of Neurosurgery, Neurological Institute Tokyo Women's Medical University, Japan

A digital three Dimensional (3D) ultrasound imaging system for neurosurgery has been developed. This system acquires 2D image data through Local Area Network (LAN) directly. It reconstructs 3D volume from a series of 2D slices by mechanical scanning of the ultrasound probe and renders 3D volume at high speed. A diagnostic ultrasound imaging-system was PC-based and communication program made it possible to transfer image datasets to other computers through LAN. To acquire 3D images, we rotated the probe and an encoder measured the accurate position of the probe. Volume data were reconstructed from sequence of 2D images acquired during the probe rotation. Another PC processed graphics, controlled the motor, and image data were transferred to the memory of it from the diagnostic ultrasound machine. It took approximately 40 seconds from the beginning of the getting the data to show the volume data. Volume rendering could be done at real time speed because of a graphic accelerator board specialized to process voxel data. The volume's opacities and colors were assigned according to their intensity and the setting could be changed easily. The system was tested in two clinical cases. It helped surgeons to understand the condition of the diseased part during the operation.

1. INTRODUCTION

In neurosurgery, intraoperative observation of tissue is important to verify whether the tissue should be cared (resected) or not. Preoperative imaging systems such as MRI and X-ray CT have become very popular and powerful tool to observe patients' pathological tissues such as tumors. There are attempts to use data from these devices for the intra-operative navigation [1–3]. But intra-operative applications of these imaging devices are not common except for ultrasound imaging system.

2D ultrasound imaging system is very convenient to use during the operation, but its resolution is inferior to that of MRI. Surgeon must have some skill to understand the image. It's not easy to increase the resolution of the ultrasound imaging system considering its principle. It will be easier for surgeons to comprehend the condition of the tissue when the image data is displayed in 3D forms.

This paper presents introduction of a 3D ultrasound imaging system we developed and reports its clinical application. By this system, surgeons can get 3D images around the part in which they are operating immediately.

2. METHODS

A diagnostic ultrasound imaging-system(Honda-Electronic, CS-300) was a PC-based (OS: Windows 95) system. Communication program made it possible to transfer image datasets to other computers with socket communication method through LAN. The probe was a convex type(75C15MF, 7.5MHz, R15mm). A stepping motor and an encoder were used to rotate the probe and to measure the rotation angle. To acquire 3D images, we rotated the probe, and an encoder measured the accurate position of the probe even if the stepping motor lost its step. Since the probe rotated around its axis, volume data could be reconstructed from sequence of 2D images obtained during the probe rotation. Because 2D images are obtained at fixed step of rotation angle, the datasets were discrete. Thus, we filled the vacant space between neighboring two images with interpolated data to make the volume cylindrical. In this study, we obtained 120 images per 180 degrees, consequently the resolution of this image is 1 mm at the edge of a circle whose radius is 38 mm, and that is enough considering the size of the operation field in neurosurgery.

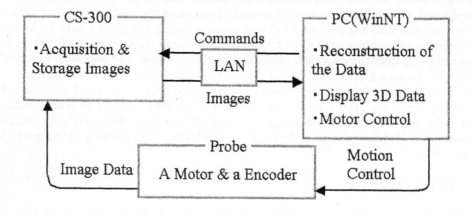

Figure 1. the structure of the system.

We used another PC (Pentium III 600 MHz dual processor, 1024 MBytes memory, a port for 100BASE-T LAN interface, a graphics card(GeForcePro), a volume rendering graphics accelerator(VolumePro)) to deal with graphics and to control the stepping motor for probe rotation. Image data were transferred to the memory of the PC from the diagnostic ultrasound machine to speed up the image processing.

Generally, the process of boundary extractions or segmentations is needed to make 3D images from medical 2D images. Especially in case of ultrasound imaging, this process is required because of its lower resolution. However, since our system displays 3D image based on volume rendering, the system does not have to do boundary extraction and segmentation. Up to the present, it has been difficult to realize real time volume rendering in a convention PC system because of vast amount of computation required. However, the volume rendering graphic accelerator enabled real time volume rendering in the PC.

After the reconstruction of 3D data, we assigned opacities and colors according to the voxel's intensity of brightness, because it could be hardly recognized the shape and the condition at grayscale.

3. CLINICAL APPLICATION

We applied the system in two clinical cases. The cases were both gliomas, which are difficult to be distinguished from normal parts to the naked eye. Thus intraoperative imaging system is highly required for this purpose. The operations were done both under craniotomies.

To fix the position of the probe and drive unit, we used a mechanical arm, point-setter (figure 2.) and made the probe touch the brain directly (figure 3.).

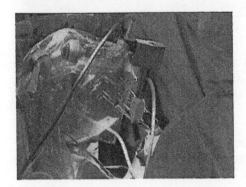

Figure 2. probe and drive unit equipped to point-setter in the operating room.

Figure 3. a situation in the clinical application.

4. RESULTS

4.1. Performance

It took 12 seconds to acquire 120 2D images and approximately 12 seconds to transport all of them (size : 512 × 450 pixel, 8 bit grayscale) with 100 base-T LAN interfaces. It

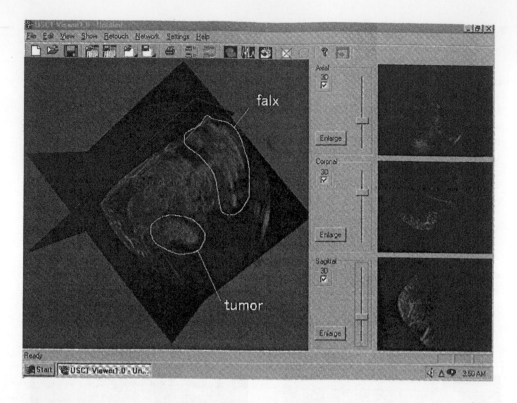

Figure 4. The user interface and the acquired data around the glioma. Three right side windows correspond to the planes in the left side large window which contains volume data. There are a tumor and a part of falx in this data.

took 13 seconds to reconstruct volume data from 2D images. Including the time of the motor turning back to the start point, the required time was about 40 seconds from the beginning of data acquisition to displaying the 3D data.

4.2. Functions

- After acquiring the data and reconstruction, the system could show the 3D data and zoom and rotate with no stress.

- Three cut planes which crossed at right angles (ex. axial, sugittal and coronal planes) could be shown arbitrarily.

- The data could be saved as both a volume data and 2D images (raw data).

- The parameters distributed according to the brightness also could be saved and be loaded afterwards. Once the display parameters were optimized in each case preoperatively, surgeons did not need to change them during the operation. In these cases, we set the parameters manually.

4.3. Limitations of the system

We acquired the phantom's data before the clinical application. The phantom data could be displayed very clearly, but the tumor was not so clear compared with the phantom. It was partly because the phantom was floating in the water, leading to better contrast of light and shade easily. But in actual brain tissue, the brightness between the ordinary part and the tumor was not so different that we have difficulties in separating them clearly, though the data include tumor, falx, etc. As a results, 2D display was sometimes more effective to distinguish the tumors. Thus the optimization of 3D display parameters was required. Registration to the pre-operative MRI was also required to realize more clear understanding of the tumor by the surgeons.

5. DISCUSSION

The processing time required in the developed system is very short compared with the other types of intra-operative 3D medical image acquisition systems. For example, when surgeons use OPEN MRI intra-operatively, they must stop their operation about 30 minute to acquire images. In the hospital where the clinical applications were conducted, most of the time required for intra-operative open MRI was consumed for preparing the environment in the operating room (moving the bed, remove instruments disturbing the imaging, etc.), rather than actual image data acquisition. The developed system was very easy to use in a short time (12 seconds). This feature of the system is one of the advantages over the other intra-operative image data acquisition systems.

RGB values and opacity, which were distributed to the voxels according to its brightness, must be optimized. We set the parameters manually. We must analyze the differences between the normal part and the diseased part about the distribution of the voxel's brightness and the recognizablirity to provide clearer distinction. The optimization of the parameters has limitation because in the tumor part and in the ordinary part, some voxels have the same brightness.

Our system should be equipped with the three-dimensional position tracer to acquire the accurate position of the probe, to increase the objectivity of the data in the aspect of 3D position. Now the attempts are in progress to connect medical imaging devices to the network and utilize its data as common properties. We use LAN-based communication method, because this system should be integrated with another imaging systems, such as MRI and CT. Ultrasound imaging system is very useful in the aspect of real time, convenience during the operation, but it cannot gain high resolution as MRI and CT. We aimed to integrate our system with pre-operation data, which have better accuracy than ultrasound's one.

6. CONCLUSION

We succeeded in obtaining 3D data within 12 seconds and show the 3D data in about 40 seconds. The system was evaluated in two clinical cases. Optimization of the parameters assigned according to each brightness of voxels and integration of the obtained information to another 3D data that is obtained from MRI or CT are needed. In the immediate future, we will develop the system that can treat pre-operative images and intraoperative 3D data at the same time.

7. ACKNOWLEGEMENT

This study is partly supported by Health Science Research Grants for Research on Advanced Medical Technology, Ministry of Health and Welfare.

REFERENCES

1. Markus H, Norbert B, Andreas F, Albrecht H. Preoperative Planning and Intraoperative Navigation: Status Quo and Perspectives. Computer Aided Surgery. Vol 3 No. 4 (1998) 153-158.
2. Kirsten S, Markus H, Albrecht H. Neuronavigation in Daily Clinical Routine of a Neurosurgical Department. Computer Aided Surgery. Vol 3 No. 4 (1998) 159-161.
3. Uwe W, Andreas F, Albrecht H. Operative Approach Due to Results of Functional Magnetic Resonance Imaging in Central Brain Tumors. Computer Aided Surgery. Vol 3 No. 4 (1998) 162-165.
4. Blass HG, Eik-Nes SH, Berg S, Hans T. In-vivo three-dimensional ultrasound reconstructions of embryos and early fetuses. The Lancet. Vol 352 October 10 (1998) 1182-1186.

Human Friendly Mechatronics (ICMA 2000)
E. Arai, T. Arai and M. Takano *(Editors)*
© 2001 Elsevier Science B.V. All rights reserved.

A Powered Wheelchair Controlled by EMG Signals from Neck Muscles

Hiroaki Seki, Takeshi Takatsu, Yoshitsugu Kamiya, Masatoshi Hikizu, [a] and
Mitsuyoshi Maekawa [b]

[a] Mechanical Systems Engineering, Kanazawa University,
2-40-20 Kodatsuno, Kanazawa, 920-8667, Japan

[b] Industrial Research Institute of Ishikawa,
1, Ro, Tomizumachi, Kanazawa, 920-0223, Japan

In the case of disabled people who cannot handle normal joysticks well, they operate special joysticks by tongue or chin, or they use some voice commands to control powered wheelchairs. However, such interfaces have much difficulties and burdens to control wheelchairs smoothly. We propose to utilize electromyogram (EMG) signals from operator's neck muscles as one of the wheelchair interfaces for them. Head motions of inclining and shaking can be detected from characteristics of these EMG signals and their motions are converted into operation commands of a wheelchair. A prototype of the powered wheelchair controlled by neck EMG interface has been developed and tested.

1. INTRODUCTION

Powered wheelchairs are used for disabled people who have disabilities to walk and drive wheels manually. They usually use joysticks to control wheelchairs and give desired direction and speed by them. However, there are many disabled who cannot handle them or hold them tilting stably. They operate special joysticks by tongue or chin, or use some voice commands [1–3]. In such cases, it is very difficult to control a wheelchair smoothly and its operation has large burden. Therefore, we propose to utilize electromyogram (EMG) signals from the operator's neck or face muscles as one of wheelchair interfaces for disabled people who cannot use normal joysticks. Since motions of operator's head or face are estimated from his/her EMG signals, the user can control a wheelchair by intended motions of his/her head or face. This EMG interface has following characteristics [4,5].

- It is relatively easy to obtain EMG signals comparing with other bioelectric signals.

- Though head or face motions can be caught by accelerometer, gyroscope, and so on, these sensors measure not only essential motion but also wheelchair vibration. On the other hand, EMG signals are generated only when a user works his/her muscles.

- This system needs the function to discriminate whether these signals (motions) involve user's intention to operate a wheelchair or not.

- EMG interface has possibilities to be applied to more serious physical handicap.

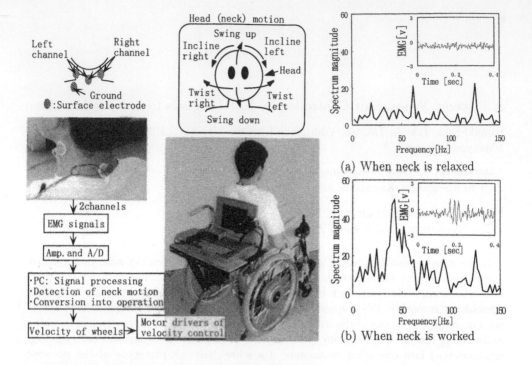

Figure 1. System of the powered wheelchair controlled by neck EMG signals

Figure 2. FFT analysis of EMG signal from neck muscle

(a) When neck is relaxed

(b) When neck is worked

The problems are how to estimate motions from EMG signals and how to control a wheelchair by their motions. A powered wheelchair controlled by EMG interface has been developed and tested in this paper.

2. SYSTEM OF EMG CONTROLLED WHEELCHAIR

A wheelchair controlled by EMG signals from user's neck is proposed. This system is shown in Figure 1. Four surface electrodes are placed at the right, left, and backside around the user's neck and EMG signals from the right and left channels are monitored. In the developed system, after EMG signals are amplified 500 times, they are filtered with a notch filter to cut commercial frequency noise (60Hz in western Japan) and they are also filtered with a band-pass filter from 3 to 140 Hz by a electric circuit. Then, they are sampled at a rate of 1 kHz by 12-bit A/D converters. In order to detect envelope patterns and amplitudes of EMG signals, simple signal processing is applied. The intended head motion of an operator is estimated from the EMG envelope signals of his/her neck. If certain head motions for operating a wheelchair are selected previously, the wheelchair can be controlled according to them. Since the motor drivers of wheels have velocity control, desired velocities of right and left wheels are given by the computer.

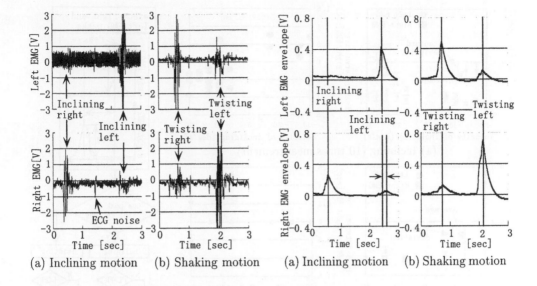

Figure 3. Neck EMG signals when head is inclined and shaken

Figure 4. EMG envelopes after signal processing and detection of their peaks

3. PROCESSING AND ANALYSIS OF NECK EMG SIGNALS

Sampled EMG are processed as follows. Figure 2 shows the examples of EMG signals and frequency spectrums when neck muscle is relaxed and worked. It can be seen that neck motion is accompanied with signals from 40 to 60Hz approximately. Based on this analysis, EMG signals are filtered with a band-pass digital filter from 40 to 60(or 80)Hz in order to eliminate ECG noise etc. Then, rectification and low-pass (1Hz) filtering extract envelopes of their waves (Figures 3, 4). This processing make EMG characteristics clear.

The relationship between the head motions and EMG signals was investigated. It is preferable that more kinds of motions can be detected from as less EMG channels as possible. The head (neck) motions are classified into three types, nodding (swinging up / down), shaking (twisting right / left), and inclining (inclining right / left). The neck EMG signals caused by these motions exhibited following characteristics (Figure 4, 5).

- In the case of nodding, EMG is small because the working muscle is deep in neck.
- In the case of inclining, the EMG envelope of the side to which head is inclined has a large peak and that of the other side has no peak or a delayed small peak.
- In the case of shaking, EMG envelopes both sides have peaks simultaneously and the amplitude of the side to which head is twisted is smaller than the other side.

Figure 5 shows the magnitude and interval of EMG peaks when the speed of each head motion is varied. Angle of motion is about 20 deg. These peaks become larger as the head moves faster. Then, four motions of shaking (twisting right / left) and inclining (inclining right / left) can be used to control a wheelchair by detecting them from EMG signals.

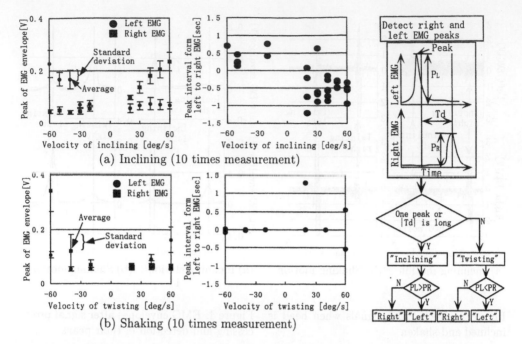

(a) Inclining (10 times measurement)

(b) Shaking (10 times measurement)

Figure 5. Relationship between EMG peaks and speed of head motion (Intervals are plotted only when both peaks exist.)

Figure 6. Algorithm to detect head motion

4. WHEELCHAIR OPERATION BY HEAD MOTIONS

Head motions can be detected by the algorithm shown in Figure 6. The shaking and inclining motions are discriminated from the time delay between the peaks of the right and left EMG envelopes. If the delay T_d is long ($0.25 < |T_d| < 0.5$ sec) or EMG of only one side has a peak, it is estimated to be inclining. If the delay is short ($|T_d| < 0.15$ sec), it is estimated to be shaking. Other cases is ignored to decrease mistake of detection. The direction (right or left) of each motion is detected by comparing the magnitudes of right and left EMG peaks. Detected head motions are converted into wheelchair operations. Since types of operations such as moving forward / backward, turning right / left, and stop are more than detectable motions, combinations and intervals of motions and different meanings by moving states should be utilized for this assignment. Information of the EMG magnitude is also used since the wheelchair velocities are continuous. Though simple and easy operation is desired, it is trade-off to safety at detection errors of motions. The conversion from head motions into wheelchair operations is proposed as shown in Figure 7.

- To prevent fatal error of detection, the special motion of inclining head right after left within 1 sec is assigned to the toggle command which starts and stops a wheelchair.

- The motions of inclining right or left are the commands to turn a wheelchair to each direction. The motions of twisting right and left accelerate and decelerate a

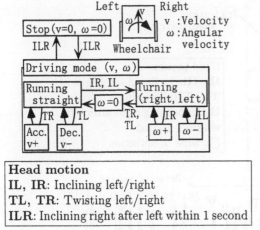

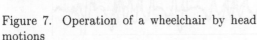

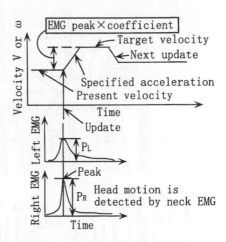

Figure 7. Operation of a wheelchair by head motions

Figure 8. Method to update velocity V / angular velocity ω from EMG peaks

wheelchair respectively. When it is turning, shaking motions also make it to run straight for easy operation.

- The velocity and the angular velocity of a wheelchair are increased / decreased by the proportional amount to the magnitude of EMG peaks from each motion (Figure 8) so that the operator can adjust them by the speed of head motions. These velocities are changed smoothly by the specified acceleration.

5. EXPERIMENT

Proposed EMG interface was implemented to a conventional wheelchair and the experiment of operation was made. Figure 9 shows an example by the user who is not disabled. EMG envelopes of user's neck, detected head motions, trajectory and velocities of the wheelchair are shown. Head motions could be detected from EMG envelope signals. After some training, a user could smoothly controlled the wheelchair by utilizing those motions. There were some lacks of detection of intended motion. However, they were not fatal for safety. The error detection of unintentional motion was few, because small EMG signal generated by slow (normal) head motion was ignored by a certain threshold.

6. CONCLUSION

A powered wheelchair controlled by EMG signals from the user's neck has been proposed for a disabled who cannot use a normal joystick. Head motions of inclining and shaking can be detected from characteristics of neck EMG signals. An operation method by using detected motion types and EMG magnitude of peaks has been also proposed to control a wheelchair smoothly. This EMG interface was tested and it was succeeded practically. Detection of smaller motions, easier and more reliable operation, control system

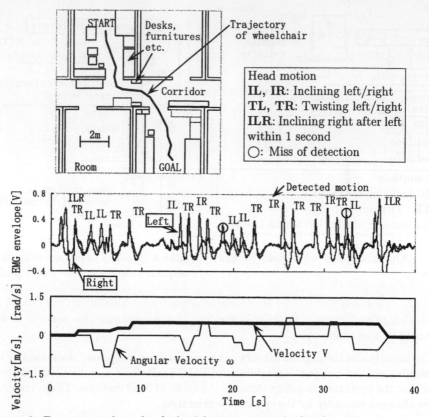

Figure 9. Experimental result of wheelchair operation by head motions

by face EMG signals are remained for our further works. A wheelchair controlled by EMG interface becomes more useful and reduces user's burden, if it has some intelligent support systems to detect/avoid obstacles and to keep running straight in spite of disturbances.

REFERENCES

1. K. Schilling, et al., "Sensor Supported Driving Aids for Disabled Wheelchair Users," Proc. IFAC Workshop on Intelligent Components for Vehicles, 267-270 (1998).
2. G. Pires, et al., "ROB CHAIR - A Semi-autonomous Wheelchair for Disabled People," Proc. 3rd IFAC Symp. on Intelligent Autonomous Vehicles, 648-652 (1998).
3. R.C. Simpson and S.P. Levine, "Adaptive Shared Control of a Smart Wheelchair Operated by Voice Control," Proc. IEEE/RSJ Int. Conf. on Intelligent Robots and Systems, 622-626 (1997).
4. T. Iberall, et al., "On the Development of EMG control for a Prosthesis Using a Robotic Hand," Proc. IEEE Int. Conf. on Robotics and Automation, 1753-1758 (1994).
5. O. Fukuda, et al., "EMG-based Human-Robot Interface for Rehabilitation Aid," Proc. 1998 IEEE Int. Conf. on Robotics and Automation, 3492-3497 (1998).

Human Friendly Mechatronics (ICMA 2000)
E. Arai, T. Arai and M. Takano *(Editors)*

93

Autonomous Mobile Robot for Carrying Food Trays to the Aged and Disabled

Tsugito Maruyama[a] and Muneshige Yamazaki[b]

[a] Peripheral-system Laboratories, FUJITSU LABORATORIES LTD.
4-1-1, Kamikodanaka, Nakahara-ku, Kawasaki, Kanagawa 211-8588, Japan

[b] Tsukuba Research Lab., YASKAWA ELECTRIC CORPORATION
5-9-10, Tohkohdai, Tsukuba, Ibaraki 300-2635, Japan

A food-carrying robot developed for the assisting the aged and disabled consists of six basic components: a compact lightweight manipulator, a moving mechanism, an environment perception unit, an information display unit, a navigator, and a remote supervisory control unit. Results from practical evaluation tests at medical and welfare facilities demonstrated that performance targets for safety, autonomy, and human friendliness were achieved. Such a robot will help resolve shortages in the number of healthcare workers and improve the overall quality of care.

1. INTRODUCTION

Japanese society is aging rapidly, a shortage in the number of nurses and care providers is feared to lead to social problems. Since the physical and mental workload on nurses and care providers is increasing in medical and welfare facilities, less nursing care and fewer services devoted to human care is already a large problem in Japan. This problem occurs especially at mealtime, when care providers have to deliver and collect food trays while feeding patients. Meanwhile, they also have to complete their regular nursing duties and provide healthcare services. They do not have enough time for face-to-face communication and have difficulties in providing sincere, tender care of their patients. If technical assistance such as autonomous mobile robots could support their work, they could have more time for face-to-face communication and personal care of patients, including helping patients eat their meals. The situation might improve substantially for both care providers and patients.

Thus far, a patrol robot for use in offices [1] and a delivery robot, HelpMate [2], for use in medical and welfare facilities have been developed. Using ultrasonic sensors and a laser range sensor, they can travel on predetermined courses and avoid certain obstacles. However, they cannot be used to deliver and collect food trays because they don't have a manipulator or any navigation cameras that are required for recognition of unfamiliar obstacles. On this background, a project for developing an autonomous mobile robot was conducted at the request of Japan's Agency of Industrial Science and Technology, which is part of the Ministry of International Trade and Industry, and the New Energy and Industrial Technology Development Organization (NEDO) from 1994 to 1998. The project was focused on building an "Autonomous Mobile Robot for Carrying Food Trays to the Aged and Disabled." To develop a robot for providing indirect care, such as the routine work and labor not involving care giving but is assigned to care providers, Yaskawa Electric Corporation and Fujitsu Ltd. conducted joint research work for the Technology Research Association of Medical and Welfare Apparatus. We developed the autonomous mobile robot with a manipulator and navigation cameras. We also performed demonstrations of food-tray delivery and collection in actual facilities.

This paper describes the configuration of the total robot system and the component technologies, and explains the results from both a technical performance test and practical evaluation test.

2. PROJECT GOAL

2.1 Purpose

Our goal is to reduce the workload on nurses and care providers at mealtime by developing an autonomous mobile robot that carries food trays to the aged and disabled. The following is the task scenario of the robot system operation:

- The robot takes four food trays (i.e., four trays for a room with four beds) from a food wagon set at a defined service area, carries them down a corridor, enters a specified room, goes to each patient's bedside, and places a tray on each patient's overbed table.
- After meals are eaten, the robot picks up the four trays from the patients' tables, carries them to the wagon, and sets them into it.
- During the task cycle, the robot runs autonomously. A remote supervisory control unit monitors the status of the robot in real time.

2.2 Performance targets

Although industrial robots are always used in separate areas outside the areas of people, the autonomous mobile robot for carrying food trays is usually operating in areas where people are residing. Therefore, we established three key words that defined our goal of making the robot adaptable to real-life environments: "Safety," "Autonomy," and "Human Friendliness."

Safety: The robot system must not harm any person or damage facilities in any way. For example, the manipulator and moving mechanism must not make any physical contact with patients.

Autonomy: The robot should operate as autonomously as possible so that staff in a facility is not disturbed during robot system operation, such as autonomous obstacle avoidance.

Human Friendliness: The robot should not give any impression of being dangerous. Instead, it should first convey a feeling of friendliness from a patient's point of view (using human-like communication and facial expressions) and from the points of view of staff at the facilities (providing a people-friendly interface and ease-of-operation).

Over the first three years, we have developed component technologies for the subsystems in the robot. In the last two years, we totally integrated the component technologies while operating and adjusting them in a mock facilities (an imitation of a real facility, it was also used as a demonstration site). In the final two months, we demonstrated and evaluated the operation of the robot system in three actual facilities.

Fig. 1 Food-carrying robot (named FUKU-chan)

3. TOTAL ROBOT SYSTEM

Figure 1 shows an overview of the food-carrying robot (named FUKU-chan) and Table 1 lists its basic specifications. Figure 2 shows the subsystems installed on the robot and Fig. 3 shows the hardware configuration of the robot system. The robot consists of six basic subsystems: a compact lightweight manipulator, a moving mechanism, an environment perception unit, an information display, a navigator, and a remote supervisory control unit. Each subsystem has its own CPU that controls its components, and all subsystems are connected via Ethernet to send and receive commands and exchange data. The navigator is shared on VMEbus with the environment perception unit functioning as the master CPU. The moving mechanism is composed of a local navigation controller and a wheel driver. The remote supervisory control unit uses infrared rays to communicate CCD camera images to a terminal at a remote site.

Table 1 Basic specifications

Outline dimensions	600×1250×1350 mm (Up to the head 1490mm)
Total weight	320 kg
Moving speed	600 [mm/s] (Rating), 1020 [mm/s] (Maximum)
Manipulator payload	3 kg
Food tray holding capacity	4 meals per 1 room
Maximum operating time between charges	3.5 hours (110AH)

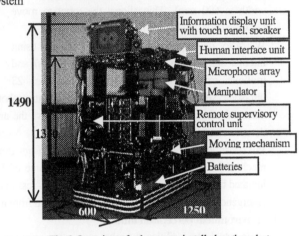

Fig. 2 Overview of subsystems installed on the robot

4. COMPONENT TECHNOLOGIES OF SUBSYSTEMS

4.1 Compact lightweight manipulator

The manipulator must place food trays on overbed tables and pick them up without spilling any food. It must also have safety factors, such as a provision against upsetting or harming patients if it were to ever make contact with them. To meet these requirements, we developed a quad-axis (three horizontal axes and one vertical axis) manipulator that includes the following safety functions.

· To absorb the impact in a collision, the manipulator is covered with a soft, pliable material. Wires and a root-pulley drive the horizontal axes. The wires are not fixed because a mechanical torque limiter is based on slipping of the wire: the limiter prevents

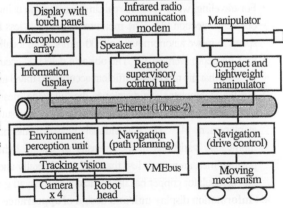

Fig. 3. Robot subsystems

excessive reaction force at contact. We confirmed that the maximum reaction force is 80.5 N during a collision, less than the human pain tolerance threshold of 250 N.

· A parameter variable impedance control based on the value of torque sensors, which are arranged along each horizontal axis, reduces the steady reaction force in a collision.

4.2 Moving mechanism and navigator

The robot must move safely and autonomously using a map to reach its destination while avoiding obstacles.

· We developed an all-direction mechanism that combines a steering mechanism with a rotation method using a differential drive, thereby enabling the robot to proceed within a narrow space of 900 mm (its body width is 600 mm). This allows the robot to move between beds without changing its orientation.

· We also developed a navigation technique using internal sensors (encoders and a gyrocompass) and multiple external sensors (4 CCD cameras, 23 ultrasonic sensors, and a laser sensor) so that the robot avoids obstacles, moves safely, and stops automatically before a collision. Figure 4 shows the arrangement of the external sensors and their detection ranges.

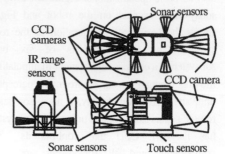

Fig. 4 Detection range of sonars, range sensor, CCD cameras

· We constructed a bumper with point-contact sensors arranged around the lower frame of the moving mechanism. The bumper detects a reaction force of 1 kg or more. In experimental results, stops initiated by the bumper had a braking distance of less than 12 mm from a velocity of 200 mm/s. The emergency stop had a braking distance of 130 mm from a velocity of 600 mm/s.

4.3 Environment perception unit

In actual care facilities, the robot must move safely and autonomously while avoiding obstacles. In addition, its manipulator must handle food trays placed irregularly on tables. Therefore, the environment perception unit is important for realizing autonomous operation by the robot.

· For traveling, we developed real-time localization using features already existing in care facilities and obstacle detection [3]. The measurement time is less than 133 ms. the robot has enough processing speed to move autonomously at 600 mm/s. The localization accuracy is ±80 mm laterally and ±100 mm for depth (vertically). The obstacle detection accuracy is less than 10% for depth. These results indicate that the robot has sufficient accuracy to travel around facilities autonomously.

· For the manipulator, we developed the following measurements: table position measurements at tray placement and at tray collection, a tray position measurement at tray collection, and obstacle detection at placement [4]. These positional measurements have the robustness to overcome the effect of momentary variations in the intensity of natural light (100~6000 lx). The accuracy of the tray position measurement is less than ±12 mm laterally and less than ±10 mm for depth. This shows that the manipulator gripper has the sufficient accuracy to grasp the tray.

4.4 Information display unit and design of appearance

The robot must behave like a friendly person by facing the source of a voice command, provide a simple people-friendly interface, and be acceptable to most care facilities. We developed the information display unit that consists of a head rotation mechanism, sound source detector, LCD display with a touch panel, speaker, and microphone array.

- We programmed the robot with a capability to carry on simple conversations, so it can respond to a voice command by turning its head, changing its eyes, and raising its voice. This makes it more people-friendly in general appearance, on its screen, and in its voice.
- For the overall appearance and design of the robot, we made a cover incorporating different colors and shapes, which is meant to evoke impressions such as friendliness, brightness, and amusement.
- For its on-screen display, images representing robot expressions (Fig. 5) and operational displays were designed to be understandable, descriptive, and easy to use.

Fig. 5 Examples of expressions on robot face
(e.g., smiling, surprised, and crying)

4.5 Remote supervisory control unit

The robot must provide easy remote control so that its operation can be supervised, its environmental situation can be monitored during unforeseen events, recovery can be made immediately, and its safe operation assured.

- We developed a highly reliable transmission method using infrared radio signals that does not cause a shutdown during communication, does not affect the operation of medical devices, and can transmit large amounts of image data from the robot to a nurses' station. We confirmed the transmission rate (19.2 kbps) is adequate for monitoring the environment state at 160 x 120 pixels, 256 levels, 0.92 frame/s, and a delay of 1.2 s.
- We also developed an intuitive, simple supervisory interface. Using a stick on the remote control unit, an operator can move the robot to desired directions and can also rotate the head with the head rotation mechanism and select the cameras (three fronts, two wrists, and one rear) for image display.

5. EVALUATION TEST OF TOTAL ROBOT SYSTEM

5.1 Technical performance test

We evaluated the total robot system in a technical performance test at our mock facility. In the safety test, the position at which the robot begins to avoid obstacles, is 2,900 mm before the obstacles, and the safe avoidance distance is 630 mm from the side of the robot (at a velocity of 600mm/s). We confirmed that the robot avoided obstacles without making any contact. The robot positioning accuracy from the center of a doorway of a bedroom, where the highest positioning accuracy is required, is ±50 mm and the space between one side of the robot and the doorway frame is ±150 mm. This indicates the sufficient accuracy for passing through doorways.

5.2 Practical evaluation test

We performed the practical evaluation test in a hospital and two care facilities to evaluate the operation of the robot by care providers. Figure 6 shows scenes of the demonstration. We performed two kinds of delivery-and-collection demonstrations using four food trays: at ordinary speed and at high speed. At an ordinary speed,

Fig. 6 Practical evaluation test
(food tray placement, obstacle avoidance)

we showed the adaptability of the robot system in an emergency, during an unforeseen event when the

98

manipulator made contact with a patient, and when the robot detected obstacles on the table and in its pathway. Then, to check safety and friendliness against moving at high speed, we operated the manipulator and the moving mechanism at double the speed of the ordinary mode.

Afterward, interviews were conducted with 100 facility employees on 23 items according to safety, autonomy, and human friendliness. Figure 7 shows a summary of their answers to 4 items: general impression, safety, response to a voice command, and autonomous avoidance of collisions. Over half of the respondents at care facilities rated the robot as "satisfactory" for safety, autonomy, and friendliness. We recognized that our goal to develop a people-friendly robot for use in areas where people reside was accomplished.

For practical use, the robot should be further developed to enable highly efficient operation and to perform consecutive actions, comparable to human behavior. In addition, many kinds of user-interfaces (e.g., voice response, infrared remote control) for autonomous interaction with the aged should be provided for simple and quick recovery in any unforeseen event. Finally, production costs should be lowered. The robot preferred by employees is not just a labor-saving device: their preferred robot not only sets and clears food trays but is also cooperative and people-friendly, delivers and collects linen, patrols the facilities at night, and provides help in cleaning or medical examinations.

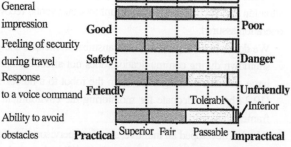

Fig.7 Questionnaires results (scale of 1 to 5)

6. CONCLUSION

We have developed an autonomous mobile robot for carrying food trays. After conducting practical evaluation tests in actual care facilities where difficult problems related to the surrounding conditions remain to be solved, we demonstrated that introducing the use of robot to care facilities is realizable. We also found where problems remain and must be solved before actual use of the robot system operation. We are planning to market the robot in the near future by improving its quality and performance based on the practical evaluation tests. We thank everyone at the Agency of Industrial Science and Technology, the Ministry of International Trade and Industry, NEDO, the Technology Research Association of Medical and Welfare Apparatus, and the Development Committee who contributed their valuable assistance.

REFERENCES

1. K.Hakamada, et al. "Improvement of Autonomous Mobile Robot to Security System," Proc. 14th Annual Conference of RSJ, Vol.1, pp143-144, (1996)

2. Kouno, T. and Kanda, S., "An Autonomous Mobile Robot for Carrying Food Trays to the Aged and Disabled," Journal of the Robotics Society of Japan, Vol. 16, No. 3, pp. 317-320, 1998.

3. Hashima, M. et al., "Localization and Obstacle Detection for a Robot for Carrying Food Trays," Proc. IEEE/RSJ International Conference on Intelligent Robots and Systems , Vol. 1, pp. 345-351, 1997.

4. Hasegawa, F. et al., "A Stereo Vision System for Position Measurement and Recognition in an Autonomous Robotic System for Carrying Food Trays," Proc. IEEE International Conference on Systems, Man, and Cybernetics, Vol. 3, 1999.

Human Friendly Mechatronics (ICMA 2000)
E. Arai, T. Arai and M. Takano (Editors)
© 2001 Elsevier Science B.V. All rights reserved.   99

A New Force Limitation Mechanism for Risk Reduction in Human / Robot Contact

Noriyuki Tejima

Department of Robotics, Ritsumeikan University,
1-1-1 Noji-Higashi, Kusatsu, Shiga, 525-8577, Japan

In this paper, a new mechanism to reduce the risk of service robots negatively contacting with their human users is proposed. The device is composed of a force limitation mechanism, a spring and a damper with anisotropic viscosity. A prototype was developed, and its basic features were experimentally evaluated. The new mechanism could avoid applying an external force stronger than the predetermined threshold. The anisotropic damper realized a quick response to excessive forces and allowed for slow restoration for safe movement. It could not respond to impulsive forces produced by collisions with metallic objects, however, it could work effectively for any contacts with a human.

1. INTRODUCTION

More then three hundred people with disabilities use rehabilitation robots for helping with their practical activities of daily living, such as feeding themselves, shaving, brushing their teeth, and scratching. The rehabilitation robot now becomes a practical tool for people with disabilities [1]. The MANUS is one of the more famous rehabilitation robots for such multipurpose applications, which has been used by many people with disabilities for several years. It was reported that many MANUS users thought that the MANUS was useful but not useful enough for totally independent lifestyles [2]; it moves too slowly, and it can handle only light goods, and its arm is not long enough. It should be easy to develop a faster, more powerful and bigger robot than the MANUS to meet these requests. However, such a high performance robot is too dangerous for everyday use, because we have no technology in place to ensure safe contact is maintained between the operator and their robot. If and when a new risk reduction strategy for safe contact between humans and robots can be formulated, it will be helpful not only for rehabilitation robots, but also for other service robots. The goal of this basic study is to establish a safety protocol for when robots are used in a human / robot coexistence environment.

2. SAFETY TECHNICS

Several fatalities and many accidents involving industrial robots have been reported [3]. The basic strategy for preventing industrial robots from harming their human operators is to completely isolate the robots from humans. However, this approach is not applicable to service robots, which need to operate near humans.

I think that designing a service robot that does not contact a human by any means will be impossible, because the devices employed to detect any contact, such as ultrasonic sensors or beam sensors [4,5], are unreliable. Therefore, a strategy for safety even if the robot contacts a human should be considered.

According to ISO/IEC Guide 51 [6], it is necessary to reduce the probability of occurrence of harm and to reduce the severity of that harm in realizing a safe robot. In this study, I considered various ways to reduce the severity of the harm in a human / robot coexistence environment. It is not clear what should be chosen as an index of the severity of the harm when a robotic device contacts a human. In this study, the force is used as an index. If the force a robot can apply to a human, is certified as below a specific threshold, it is presumed safe.

For detecting forces, force sensors, torque sensors and a monitoring of the current of the actuators are often used. However, the control method using these sensors has a low reliability caused by the intolerability of electronic devices against electromagnetic noises [7]. They can be used to reduce the risk in a human / robot coexistence environment additionally, but not essentially. The delay time in control is also a problem when a robot moves fast. Soft structures, such as soft arms, soft joints or soft covers, are effective to reduce the peak of impulsive contact force [8], however, they are not safe. When a soft structure is resonated, it may become uncontrollable and generate a strong force. Besides, a soft structure cannot reduce static forces, so it cannot prevent accidents where a human is jammed or pinned by a robot.

Low power actuators are practical solutions for a simple structure and most of the rehabilitation robots that have been introduced are so equipped [9]. However, low power actuators limit not only the force but also the moving speed of the robot. Besides, this method is not effective for a mobile base with a manipulator. If low power actuators are adopted in the mobile base, it does not move well.

Force or torque limitation mechanisms are similar to low power actuators [10]. A force limitation mechanism is rigid against forces weaker than the set threshold, but it dodges stronger forces. It can limit a force independently of its speed. It works more reliably than control circuits. However, for an articulated robot it is difficult to practically decide the threshold value because of the complex relationship between the torques of each actuator and an external force. This is illustrated by the example of a 2-dimensional, 2-degrees of freedom model as shown in Figure 1. The joint 1 is equipped with a torque limitation mechanism that slips when the torque becomes larger than threshold torque, T_{th1}, and where joint 2 is equipped with a torque limitation mechanism at the threshold torque, T_{th2}. When an external force F is applied at an angle φ to the end-effector, the force permitted is dependent upon the joint angle θ as follows:

$$T_{th1} > Fl_1|\sin\varphi| \quad \text{and} \quad T_{th2} > F|l_1\sin\varphi - l_2\sin(\theta - \varphi)|$$

Consequently, when attempting to restrict an external force below the threshold in the worst-case scenario, the threshold torques, T_{th1} and T_{th2}, must be extremely small. As a result, the robot's performance is restricted.

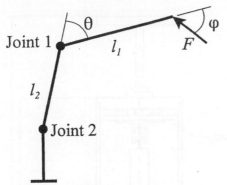 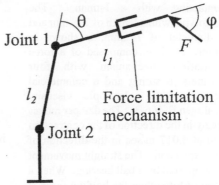

Figure 1. A 2-dimensional model with torque limitaton mechanisms on joints.

Figure 2. A 2-dimensional model with a new force limitation mechanism.

3. DESIGN RATIONALE

In order to overcome this problem, a straight-movement type force limitation mechanism was installed in the middle of the link, as shown in Figure 2. When the threshold force of the mechanism is F_{th}, the force is independent of the joint angle θ as follows:

$$T_{th1} > Fl_1|\sin\varphi| \quad \text{and} \quad F_{th} > F|\cos\varphi|$$

Consequently, the threshold torque T_{th1} and the threshold force F_{th} can be determined appropriately, so that an external force affected to the end-effector will be restricted sufficiently to avoid an accident.

A problem remains, where the original position cannot be restored, even after unloading the force, once the force limitation mechanism has been exposed to excessive force. This is because no actuator was installed for the force limitation mechanism, although the torque limitation mechanism installed on the joint 1 can be restored by the actuators that drive the joint. Then a new force limitation mechanism was proposed, which can be centrally installed in the middle of the link. The simplest actuator for restoration is a spring. However, if only a spring is used, the mechanism springs back quickly when unloaded. This quick restoration would be dangerous. Using a damper with anisotropic viscosity solved this problem. When the viscosity was set low, quick responses for safety would be available when the mechanism was activated against excessive forces, while a high viscosity would generate slow and safe restoration.

4. MATERIALS AND METHODS

I have developed a new mechanism to reduce the risk of articulated robots forcefully contacting with a human. The prototype, with a length of 238 mm and a diameter of 74 mm, is shown in Figure 3. It is composed of a force limitation mechanism with four magnets, a spring and a commercial damper with anisotropic viscosity (Takigen B-466-3). The damper moves freely in the direction of the contraction and at 0.017 m/sec in the direction of the expansion. The straight movement is supported by a ball bearing. When a weaker force than the holding force is applied to magnets stuck to a steel plate, it will be fixed rigidly. However, when a stronger force is applied, it will part from the steel plate and the force will be lessened. Force limitation is realized by this theory. It is not clear just how much force should be used as a threshold for safety. It is difficult to perform experiments to find the relationship between the force applied and the harm caused. Yamada et al, pointed out that the threshold of human pain tolerance was approximately 50 N [11]. Pain is subjective and changes easily according to psychological conditions, however, I think that a value of 50 N is a reasonable figure to use as

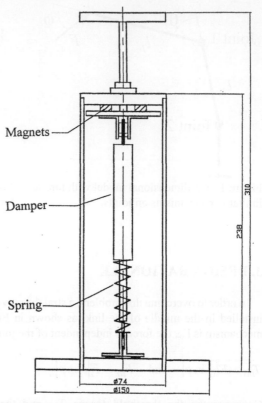

Figure 3. A prototype of a new force limitation mechanism with a spring and an anisotropic damper.

a threshold. So I used this value in the design of the prototype. Magnets, each of which had an ideal holding force of 9.8 N with steel, and a spring with a stiffness of 0.98 N/mm were used. The spring was fixed with a pre-load of 14.7 N. Since the threshold could be set as the sum of the magnetic force of the force limitation mechanism and the pre-load of the spring, the prototype has a theoretical threshold of force of 53.9 N.

Its basic features were experimentally evaluated. The prototype was rigid against weaker static forces than the threshold, but stronger forces to move out of the way activated it. The threshold of a static force against the prototype measured 52.6±2.3 N. The travel of the mechanism was also recorded with a laser displacement sensor (Keyence LK-2500) whenever a static force was loaded or unloaded. A typical example of the results obtained is shown in Figure 4. The results obtained agreed approximately with those expected. When an excessive force was applied, the mechanism was activated immediately and it traveled 40 mm within 0.15 seconds. On the other hand, the mechanism was restored slowly after the force was unloaded. Time constants of the restoration were 2.9 seconds, which would be long enough to avoid accidents. On the last one or two millimeters of restoration, the mechanism was quickly moved by the magnetic force. The distance of the quick movement depended on the force of the spring and the magnets.

The dynamic response of the mechanism was different from the static one. When the mechanism was thumped with a metal object, it was not activated, even if the peak of the impulsive force was greater than the static threshold. When the impulsive force was faded out before activation, it did not work. For example, when the peak of force was 1500 N and the force continued only for 0.1 msec, it was not activated. On the other hand, when it was thumped with a soft material such as a human limb, the impulsive force continued longer than 10 msec, and it worked the same as it did against a static force.

5. DISCUSSIONS

As expected the mechanism worked against a static force, however, it did not work against an impulsive force. This suggests that the mechanism would be effective to reduce the severity of the harm at the time of contacting against a human, but not against a metal object. Namely, it would not be effective to avoid damages to a robot arm caused by collisions with hard objects. For this purpose, soft covers should be used with the mechanism to extend the contact period and to effectively lower the peak force.

The traveling distance of the mechanism is finite. If a force were given continuously even after the activation, the mechanism would reach the end of its stroke, where finally, an excessive force would be applied to it. To avoid this, cutting off its power supply should stop the robot system when the mechanism is activated. Human movement may affect an excessive force between a robot and a human even if the robot stops. However, I suggest that in this case the responsibility should lie with the user and it is not necessary to do anything except stop the robot.

When the mechanism is mounted on an articulated robot, the payload and the weight of the end-effector influence the threshold force. When the mechanism points upward, an external

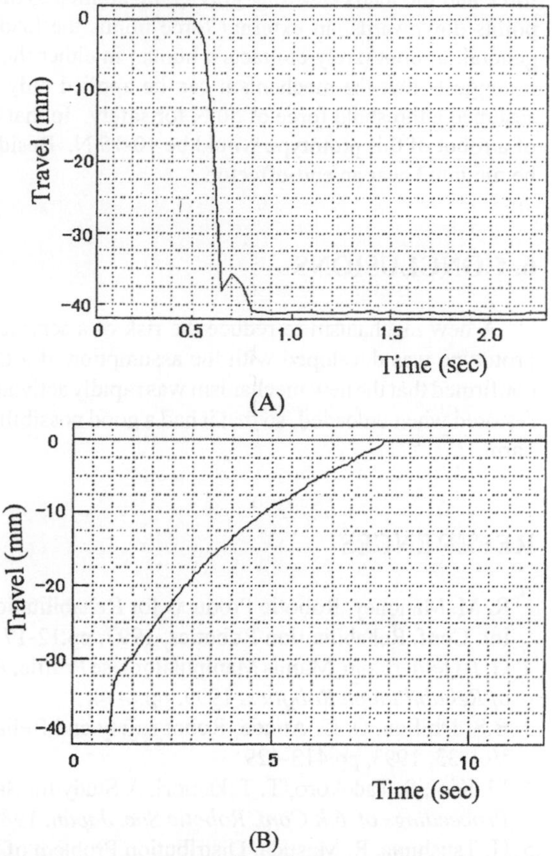

Figure 4. An example of the results of travel measurement. (A) When a force was applied, the steel plate moved within 0.15 seconds. (B) After the force was unloaded, the steel plate was restored slowly.

force and the load caused by gravity are applied to the mechanism. On the other hand, when it points downwards, an external force minus the load caused the gravity are applied to it. In general, a robot rarely contacts a human in either the upward or downward posture, however, it suggests that the mechanism can be applied only to a light robot with a small payload. I assumed a threshold force of 50 N for safety. In that case I suppose that the maximum payload of a robot of this prototype would be 10-15 N. Besides, it suggests that the mechanism should be mounted near an end-effector.

6. CONCLUSIONS

A new mechanism to reduce the risk of a service robot hitting a human was proposed. A prototype was developed with the assumption of a threshold force of 50 N for safety. It was confirmed that the new mechanism was rapidly activated against excessive forces and was slowly restored when unloaded, so that it had a good possibility of reducing the severity of any potential harm.

REFERENCES

1 R. M. Mahoney, Robotic Products for Rehabilitation: Status and Strategy, Proceedings of 5th Int. Conf. Rehabilitation Robotics, 1997, pp.12-17
2 G. J. Gelderblom, Manus Manipulator Use Profile, *Proceedings of AAATE '99 SIG1 Workshop of Rehabilitation Robotics*, 1999, pp.8-14
3 B.S. Dhillon, O. C. Anude, Robot safety and Reliability: A Review, *Microelectron Reliab.*, Vol. 33, 1993, pp.413-429
4 M. Kioi, S. Tadokoro, T. Takamori, A Study for Safety of Robot Environment (in Japanese), *Proceedings of 6th Conf. Robotic Soc. Japan*, 1988, pp.393-394
5 H. Tsushima, R. Masuda, Distribution Problem of Proximity Sensors for Obstacle Detection (in Japanese), *Proceedings of 10th Conf. Robotic Soc. Japan*, 1992, pp.1021-1022
6 ISO/IEC GUIDE 51 : Safety aspects - Guidelines for their inclusion in standards, 1997
7 K. Suita, Y. Yamada, N. Tsuchida, K. Imai, H. Ikeda, N. Sugimoto, A Failure-to-Safety "Kyozon" System with Simple Contact Detection and Stop Capabilities for Safe Human-Autonomous Robot Coexistence, *Proceedings of 1995 IEEE Int. Conf. Robotics and Automation*, 1996, pp.3089-3096
8 T. Morita, N. Honda, S. Sugano, Safety Method to Achieve Human-Robot Cooperation by 7-D.O.F. MIA ARM - Utilization of Safety Cover and Motion Control - (in Japanese), *Proceedings of 14th Conf. Robotic Soc. Japan*, 1996, pp.227-228
9 H. H. Kwee, Rehabilitation Robotics - Softening the Hardware, IEEE Engineering in Medicine and Biology, 1995, pp.330-335
10 T. Saito, N. Sugimoto, Basic Requirements and Construction for Safe Robots (in Japanese), *Proceedings of 1995 JSME Conf. Robotics and Mechatronics*, 1995, pp.287-290
11 Y. Yamada, Y. Hirasawa, S. Y. Huang, Y. Umetani, Fail-Safe Human/Robot Contact in the Safety Space, *Proceedings of 5th IEEE Int. Workshop on Robot and Human Communication*, 1996, pp.59-64

Human Friendly Mechatronics (ICMA 2000)
E. Arai, T. Arai and M. Takano *(Editors)*

Development and Motion Control of a Wearable Human Assisting Robot for Nursing Use

Takeshi KOYAMA[a], Maria Q. Feng[b] and Takayuki TANAKA[a]

[a]Department of Mechanical and Control-Engineering, University of Electro-Communications, 1-5-1 Chofugaoka, Chofu-City, Tokyo 182-8585, Japan
[b]Department of Civil and Environmental Engineering, University of California Irvine, CA 92697-2175

In order to provide quality nursing care for bed-bound and disabled people without causing physical and mental stress to both patients and nurses, we have developed a prototype of a human-assisting robotic system, referred to as HARO (Human-Assisting RObot), which is worn by an operator to amplify his or her power. This paper focuses on the realization of certain nursing related motions by the human-HARO system that are too heavy for a nurse to perform. These motions were designed based on the working experience of one of the authors at a senior nursing care facility. The safety and reliability of HARO was enhanced by significantly improving its maneuverability in such ways as limiting motor velocities, determining control parameters from mechanical characteristics, and filtering interfering forces detected by force sensors in HARO. As a result, the human-HARO system successfully performed a series of designed motions on a real patient including lifting, holding, pulling and pushing. The feasibility of using HARO for nursing care assistance was demonstrated through extensive experiments.

1. Introduction

Many developed countries including Japan and the United States are rapidly becoming more aged societies [1]. In order to meet the growing needs for nursing care of the aged and disabled, we have developed a wearable robot HARO for amplifying a nurse's power [2–4]. With the combination of human intelligence and a robot's physical strength, HARO can assist a nurse to provide quality care without causing physical and mental stress to both nurses and patients.

The objective of this study is to perform certain nursing related motions by the human-HARO system that are too heavy for a nurse. This paper will present the motions designed for HARO, the control algorithms developed in this study for improving the maneuverability of HARO, and the experimental results showing that HARO successfully maneuver a real patient.

2. HARO system and motion

2.1. HARO system

HARO is designed to be worn on the arms to amplify arm power. Its control system is illustrated in Figure 1 and its multi-link model in Figure 2. As shown, HARO is equipped with 15 AC servo motors. 4 Human Force Sensors (HFSs) are installed on the upper arms and forearms to detect the force of the operator. Two computers are used, one for processing the operator's behavior and the other for controlling HARO. Being connected by a dual port RAM, they can communicate with each other. When the operator moves his or her arms and HFSs detect the forces, HARO will move smoothly under the impedance control.

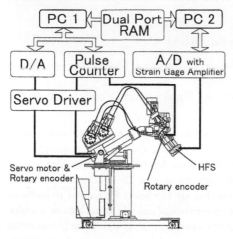

Figure 1. Configuration of the control system

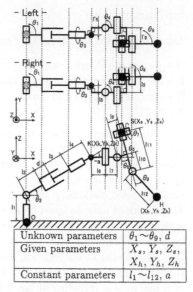

Unknown parameters	$\theta_1 \sim \theta_9$, d
Given parameters	X_s, Y_s, Z_s, X_h, Y_h, Z_h
Constant parameters	$l_1 \sim l_{12}$, a

Figure 2. 3D link model of HARO

2.2. HARO motion

In order to understand the demands of routine nursing care, one of the authors worked at a senior nursing care facility for one year as a volunteer. He observed how nursing care was actually provided to the patients and investigated what kind of assistance to the nurses was needed. Based on his experience, the patients often need assistance transferring between bed, stretcher and chair, using the toilet, bathing, moving around, and communicating from a distance. Most of these motions demand the strong physical power of a nurse. Therefore, as the first step of this study, the motions to be performed by HARO were designed to be lifting, holding, pulling and pushing. They are not only performed very frequently but also demand strong power.

3. Improvement of maneuverability

3.1. Saturation of actuators

The HARO system must guarantee the safety of the patient and the operator. In this system, the minimum safety of the operator is achieved by the fail-safe functions added in the hardware, electric circuitry and software. On the other hand, the safety of the patient depends on the operator's nursing skills and HARO's maneuverability. The operator's nursing skills can be obtained through proper training. This study, therefore, focuses on the improvement of HARO's maneuverability in order to improve its safety and reliability for patients.

3.2. Problems affecting maneuverability

It is difficult to quantitatively evaluate HARO's motions because they are complex. So in this study the maneuverability was improved by solving problems that adversely affected the maneuverability, instead of evaluating HARO's motions. Such problems were discovered as follows:

(a) Limited movement range of the structure due to saturation of the motors,
(b) Way of determining values of control parameters, and
(c) Interference in forces detected by HFSs.

In order to solve these problems, the following approaches were proposed and studies.

3.3. Control with limited angular velocity

For HARO that is a multi-link system, the control system should focus on the maximum angular velocity of each link in order to guarantee the safety for the operator and to improve the maneuverability. The maximum response angular velocity of each link under a step input detected by the HFS is regarded as the maximum maneuverable angular velocity of that link, ω_{max}. In this study, it was proposed to saturate the motor torque when the link controlled by the motor reaches ω_{max}. By doing so, improved maneuverability of HARO was confirmed through a series of experiments at the sagittal plane.

However, HARO showed poor response while moving in the three dimensions (3-D). In order to improve that, impedance at each joint was adjusted by the control system. For achieving optimal impedance, both feed-forward and feed-back type control of the angular velocities were examined. However, neither was able to achieve satisfactory response performance of HARO. Therefore, a control algorithm combining the feed-forward with the feedback control was proposed. The 3-D experimental results using this algorithm are shown in Figure 3. The effectiveness of the proposed control algorithm in improving the response of HARO is demonstrated.

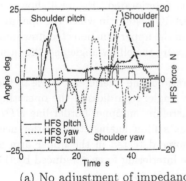

(a) No adjustment of impedance

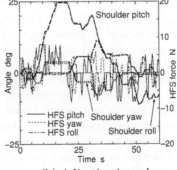

(b) Adjusting impedance

Figure 3. Improvement in HARO response

3.4. Approach for determining control parameters

The impedance control algorithm used in this study to control the angular displacement of a link of HARO is shown in Eq.1.

$$\delta\theta_r = \frac{\alpha F_{HFS}}{l\,k} \tag{1}$$

Where $\delta\theta_r$ is the control commend for the angular displacement during a time interval of δT, F_{HFS} is the operator's force detected by HFS, α is the force assisting rate defined as the ratio of the output force of the link to F_{HFS} , k is the equivalent stiffness and l is the length of the link. In other words, $\delta\theta_r$ is determined based on α, F_{HFS}, l, and k. While F_{HFS} is measured at each time instant, k and α are the control parameters determined as follows. On the other hand, $\delta\theta_r$ leads Eq.2.

$$\delta\theta_r = \omega \times \delta T \tag{2}$$

From Eqs.1 and 2, Eq.3 can be derived for calculating k.

$$k = \frac{\alpha F_{HFS}}{l\omega\delta T} \tag{3}$$

In Eq.3, F_{HFS} and ω are limited by the capacity of the force sensors and motors and l is constant. Equivalent stiffness k was determined by substituting maximum ω and F_{HFS} (determined by the capacities of the sensors and motors), $\alpha = 1.0$ (meaning no force amplification by HARO), and the optimal δT (investigated through systematic experiments) into Eq.3. Setting the value of k in such a way can lead to effective use of the capacities of the sensors and motors.

3.5. Interference in HFS

In order to reduce interference in HFS, an interference filter was developed and added to the control algorithm. Interference means that HFS detects an unwilling force of the operator and as a result the performance of HARO is deteriorated. Figure 4 shows an example from experiments carried out on the human-HARO system, in which the operator intended to bend and expand his elbow, but the HFSs in the upper arm detected unwilling forces from the shoulder. The key parameters studied were the elbow angle and the ratio of the interference force to the operating force.

From extensive experiments, it was found that the ratio of the interference force to the operating force is almost constant for the same elbow angle, regardless of the operating force and the elbow velocity, as shown in Figure 4. Therefore, an upper border line BDR of the ratio-angle plots in Figure 5 was established as a filter against the interference. When the ratio is below line BDR, the interfering force detected in the upper arm is treated as zero. Once the ratio exceeds line BDR, the interfering force is reduced by the force achieved to line BDR.

The experimental results are shown in Figure 5. Using the proposed filter, the interfering force was effectively reduced. As a result, the maneuverability of HARO was significantly improved.

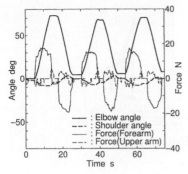

Figure 4. Behavior of the angles and the forces

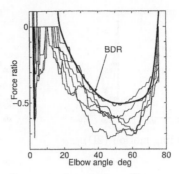

Figure 5. Interference filter

4. Experiment

In order to verify the effectiveness of the proposed approaches in improving the maneuverability of HARO, experiments were conducted on HARO worn by an operator. At first, lifting and holding was tested at the sagittal plane using a 14-kg dummy, as shown Figure 6(a). The operator wearing HARO could lift up the dummy. However, the motion was not stable because the operator could not make placket.

Next, 3-dimensional motions including pulling, lifting, holding, and pushing were performed on the dummy (Figure 6(b)–(d)). All these motions were performed safely. The dummy was pulled 150mm horizontally. For a longer pulling distance, it is safer to pull the patient several times, each achieving a short distance, rather than to pull at one time. The dummy was lifted 60mm which is high enough to transfer a patient. In addition, the dummy was pushed 100mm horizontally.

Finally, lifting and holding were successfully performed by the human-HARO system on a real patient, as shown in a photo in Figure 7. This patient weighs 59.4kg and obviously it is very difficult, if not impossible, for a nurse to lift him up. As a qualitative evaluation of these motions performed by the human-HARO system, the operator did feel being assisted by HARO and the patient felt comfortable and safe.

5. Concluding remarks

A wearable robot HARO has been developed in this study. In particular, this study demonstrated that, with a HARO worn by an operator, this integrated human-HARO system can increase the operator's physical strength for performing pulling, holding and pushing motions, which are often required in the routine of nursing care and demand an unrealistically large strength of a nurse. In order to improve the maneuverability of HARO, control algorithms were developed which take into consideration the saturation in the motors and the interference in the force sensors. In addition, control parameters were determined from mechanical characteristics of HARO. As a result, the human-HARO system successfully and safely performed the designed motions on a real patient. The experimental results demonstrated the effectiveness of the proposed control algorithms.

110

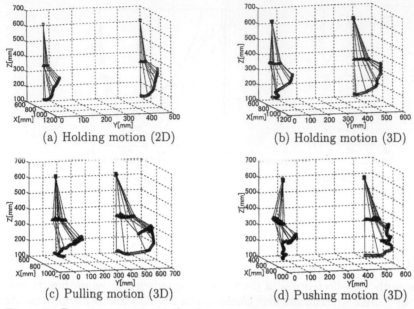

(a) Holding motion (2D)

(b) Holding motion (3D)

(c) Pulling motion (3D)

(d) Pushing motion (3D)

Figure 6. Care motions againt the dummy

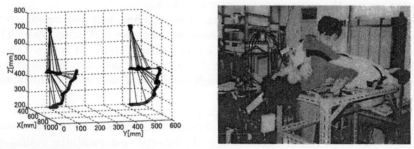

Figure 7. Holding motion against the patient

REFERENCES

1. Social Security and Population Research Institute, Ministry of Health and Welfare, "Prediction of Population in Japan" (1998).
2. T. Koyama, et al., "Human-Assist System for Nursing Use (8th Report, Transferring and nursing motion by wearing the human-assist system developed)", JSME Ann. Conf. ROBOMEC'99, 1P1-09-022, 1-2 (1999).
3. T. Koyama, et al., "Human-Assisting System for Nursing Use (1st Report, Conepts, Design of System and Development of a Prototype)", Trans. of JSME (Accepted).
4. S. Morishita, et al.,"Improvement of Maneuverability of Man-Machine System for Wearable Nursing Robots", Journal of Robotics and Mechatronics, Vol.11, No.6, 461-467 (1999).

Human Friendly Mechatronics (ICMA 2000)
E. Arai, T. Arai and M. Takano *(Editors)*
© 2001 Elsevier Science B.V. All rights reserved.

Development of an active orthosis for knee motion by using pneumatic actuators

S.Kawamura[a], T.Yonezawa[a], K.Fujimoto[a], Y.Hayakawa[b], T.Isaka[a], and S.R.Pandian[a]

[a]Department of Robotics, Ritsumeikan University
Kusatsu 525–8577, Japan

[b]Department of Control Engineering, Nara National College of Technology
Nara 639–1058, Japan

In this paper, we develop an active orthosis to support standing up motion, by using pneumatic rubber actuators. Orthosis flexibility is realized by making use of the compressibility of air. The pattern of load acting on the knee part in standing up movement is investigated by biomedical measurements. Further, we consider the relation between bending angle of knee and maximum torque which is exerted on the knee. The effectiveness of the proposed orthosis is illustrated by experimental results.

1. INTRODUCTION

In recent years, developed countries such as Japan have been undergoing a demographic transition, wherein increased longevity and declining birth rates result in the emergence of the so-called *aging society*. This trend presents several challenging research problems for the robotics and automation community, as the increased number of elderly people with limited sensorymotor capabilities need to be assisted by the development of inexpensive, safe, and human-friendly rehabilitation and orthosis systems. Motivated by this need, in this paper we consider the development of a *soft mechatronic* system, to support knee motion by using pneumatic actuators.

So far, active orthosis systems have usually been employed by patients whose lower limb is completely paralyzed. However, in the case of support of elderly people, an active orthosis which is effective for muscle force degradation and reduction of joint pain is needed. In this study, we propose a new type of mechanical orthosis to decrease the peak value of knee torque in order to reduce load acting on the knee in the standing up movement. In other words, the orthosis does not supply the total knee torque but uses available human muscle force effectively. In this way, the adaptability of the mechanism with humans can be realized when the orthosis is used in contact with humans.

The main requirements of such an orthosis are (1) light weight and miniaturization, (2) safety, (3) flexibility, (4) simplicity, (5) softness or compliance, and (6) ease of use (to put on and take off). Use of pneumatic actuators is ideal to meet these requirements.

A major limitation of active orthosis systems relates to their portability: batteries are heavy and hazardous in the case of electrically actuated systems, and pneumatically actuated systems require large air reservoir space and have low energy efficiency. To alleviate the latter problems, in the present study we harness the human power of walking by footwear-based air compression.

Conventional cylindrical actuators have many limitations for use in active orthosis. Therefore, in this study we focus on the use of hexahedron rubber actuators (HRA) proposed in [1] for developing an active orthosis supporting the rising phase movement easily. For HRA, miniaturization and light weight of orthosis are possible by using silicon rubber for the chamber department. Orthosis flexibility is easily realized by utilizing the compressibility of air. In this paper, the load acting on the knee part in standing up movement is analyzed by biomedical measurement. Further, we investigate the relationship between bending angle of knee and maximum torque which is exerted on the knee.

From these results, we design an active orthosis to support standing up movement. The rubber actuator comes into contact with a plate in sitting position condition. Hence, the orthosis assists the human with a force necessary for the standing up movement. As the standing up movement progresses, the supporting force of the actuator decreases, and the actuator separates from the plate. Therefore, the orthosis does not interfere with the walking and enables normal walking.

The rubber actuator in the orthosis is driven by air which is compressed by a bellows type compressor attached to the heel part of shoes or by a compressed CO_2 tank. Furthermore, the rubber actuator is controlled by using a valve. The valve is controlled by two types of method. One is a manual control method in which the user switches on the valve by hand. The other is a sensor feedback control method in which the valve is switched on by the force signal of the pressure element sensor seat on the heel of the shoes. We compare the experimental results of the two control methods by measurement of myoelectricity in geniculum part. Moreover, the effectiveness of the proposed orthosis is verified by experimental results.

2. ESTIMATION OF KNEE JOINT TORQUE AND DETERMINATION OF SUPPORTING TORQUE

In order to develop an orthosis to support standing-up motion, we investigate the mechanism of standing-up motion. When we stand up from a chair, we usually bend forward. At this time, a joint torque is exerted on the knees. Now, we consider that the human body can be modeled by 3 links, viz., shank, thigh, and arm [2]. Therefore, each joint torque is estimated by Lagrangian equations using each joint (ankle, knee, hip) angle, angular velocity, and angular acceleration signals. Experimentally, the measurement points are fifth metatarsophalangeal button head, lateral malleolus point, shank point, trochanter point and acromion point. Each joint angle is measured by the position measurement device (QuickMag System). The subjects are 3 healthy male students. Their age, height and weight were in the range 22.5 ± 0.5 years, 171.5 ± 3.5 cm and 57.0 ± 1.0 kg respectively. In order to estimate the joint angle and support torque, the subjects stands up from a chair, i.e., the subject bends his knee angle by more than 90 degrees. Further, the subject stands up slowly after moving the body center of gravity to front position.

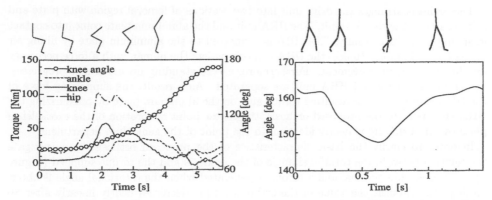

Fig. 1: Slow standing up   Fig. 2: Variation of knee angle in walking

The experimental result of the standing up motion is shown in Figure 1. From this result, it is considered that the time when the subject starts to stand up from a chair is about 1 sec. In this study, we develop the active orthosis to decrease burden of the geniculum torque. Further, when we walk in the condition that the orthosis is mounted on the lower limbs, the orthosis must not interrupt the walking motion. The knee angle in walking motion is shown in Figure 2. It is clear that the knee angle is over 140 degrees in the walking process. As a result, the specification of the orthosis is as follows. The motion range of the rotational angle of the orthosis is between 80 degrees and 120 degrees. The support torque is over 30 Nm.

3. BASIC CHARACTERISTICS OF THE ORTHOSIS

The proposed orthosis is an exoskeleton type HRA-based orthosis. The drive part using rubber actuator is illustrated in Figure 3. The orthosis configuration to support standing up motion is shown in Figure 4.

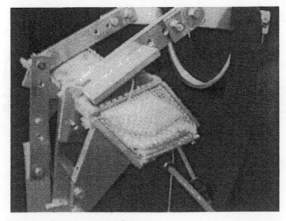

Fig. 3: Actuator mechanism

Fig. 4: Active orthosis

The orthosis arranges the drive unit into two sections of femoral region with plate and calf department with HRA unit. The HRA unit and the aluminum plate come into contact in sitting position mutually. The HRA is connected to the aluminum plate by filling air in the HRA chamber. Hence, the HRA orthosis assists the human with a force necessary for the standing up movement. Furthermore, as the standing up movement progresses, the aluminum plate and HRA unit are separated. As a result, the supporting force of the HRA decreases. The resulting torque acts on the aluminum plate by pressurizing the HRA chamber. In the proposed orthosis, the center point of rotation of the exoskeleton plate by HRA driving is nearly identical to the point of the geniculum department.

In order to clarify the basic characteristics of the proposed orthosis, we investigate the relation between the rotational angle of the orthosis and the torque supplied (Figure 5). It is clear that the maximum torque is generated when the rotational angle is about 80 degrees. Further, the value of the orthosis torque decreases nearly linearly after 80 degrees. As a result, the torque becomes zero when the rotational angle is about 117 degrees.

4. ENERGY SOURCE

In this section, we investigate methods to supply compressed air to the orthosis. Two approaches are considered, as discussed below.

4.1. Cartridge Type

Here, we used a small-size CO_2 tank. The total weight is 557 g(tank :293 g, regulator: 264 g). The CO_2 gas (97 cc) is charged into the tank and the pressure is about 10 MPa. When we use the tank gas to drive the active orthosis, the orthosis can be driven about 73 times.

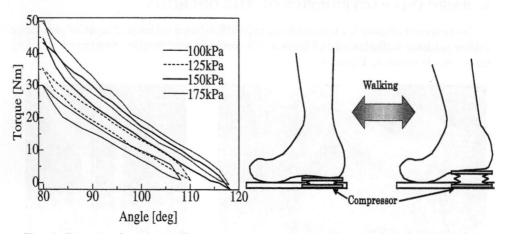

Fig. 5: Rotational angle vs. Torque   Fig. 6: Footwear-based compression

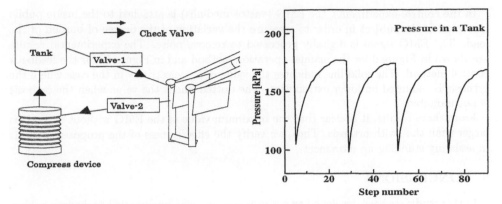

Fig. 7: Recompression system   Fig. 8: Step number vs Pressure

4.2. Compressor Type

Here, we propose a foot compressor device. The mechanism is shown in Figure 6. Further, we illustrate the recompression system in Figure 7. The compressed system uses a bellows compressor (inner diameter: 40 mm, outer diameter: 50 mm, free length: 38 mm). The compressor characteristics of inner pressure and compression cycle number are shown in Figure 8. From the experimental results, it is clear that this compressor can supply 100 kPa pneumatic pressure and about 20 steps are necessary for one standing up motion. The maximum pressure in the tank is about 180 kPa.

5. CONTROL OF THE ACTIVE ORTHOSIS

We have considered two methods for charging air into the HRA actuators. One is a manual operation method to control valves and the other is a sensor feedback method. In the manual operation method, the subject operates a valve manually to supply the driving pressurized air into the HRA actuators. However, this method is cumbersome for the user. In the sensor feedback method, a pressure sensor is attached to the calcaneal region of the subject. The signal of the sensor seat is utilized to detect the start of the standing up motion. As a result, the valves are controlled by the sensor signal.

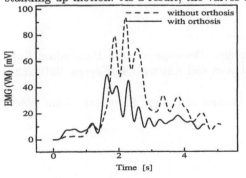

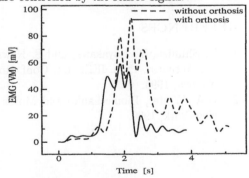

Fig. 9: Manual valve operation   Fig. 10: Control using heel pressure

In the control experiments, the EMG (vastos medialis) is attached to the inside public caro part of the subject in order to measure the variation of the degree of burden of the load. The EMG signal is digitally processed to remove noise. The experimental results are shown in Figure 9 for the manual operation method and in Figure 10 for the feedback control method. The solid line indicates the value of myoelectricity in the case where the orthosis is mounted on lower extremities. The dotted line is the value when the orthosis is not mounted.

From these results, it is clear that the maximum value of the EMG without orthosis is larger than that with orthosis. Thus, we verify the effectiveness of the proposed orthosis in assisting standing up movement.

6. CONCLUSIONS

In this study, we have developed an orthosis system using pneumatic hexahedron rubber actuators for assisting the standing up movement of humans whose muscular strength of lower extremities has declined. Experimental evaluation of the prototype orthosis systems was presented to illustrate its practical effectiveness. In future, the following problems will be investigated:

1. Quantification of joint burden,

2. Miniaturization of the actuator and the tank,

3. Simplification of the structure,

4. Enhancement of energy efficiency,

5. Detection algorithm for the standing up motion, and

6. Method for performance evaluation of the orthosis.

Acknowledgment

This study was supported by the *Future Frontiers Project* of the Japanese Society for the Promotion of Science.

REFERENCES

1. T. Shimizu, Y. Hayakawa, and S. Kawamura, Development of a Hexahedron Rubber Actuator, Proc. IEEE Int. Conf. Robotics and Automation, Nagoya, 2619-2624, Nagoya, 1995.
2. D.A. Winter, Biomechanics and Motor Control of Human Movement, John Wiley, New York, 1990.

Human Friendly Mechatronics (ICMA 2000)
E. Arai, T. Arai and M. Takano (Editors)
© 2001 Elsevier Science B.V. All rights reserved.   117

Gait Improvement by Power-Assisted Walking Support Device

Saku Egawa, Yasuhiro Nemoto, Atsushi Koseki,
Takeshi Ishii and Masakatsu G. Fujie

Mechanical Engineering Research Laboratory, Hitachi, Ltd.,
502 Kandatsu, Tsuchiura, Ibaraki 300-0013, Japan

A walker-type power-assisted walking support device has been developed and tested by individuals who can hardly walk without help. Adjusting the parameters of the control system enabled 73% of the test participants to walk by themselves. The variable dynamics capability provides high adaptability to various needs of the users.

1. Introduction

In an aged society, where elderly population is rapidly increasing, the elderly need to live a self-supported life. By keeping their independence, they can enjoy a high-quality life and also reduce their own and social health care expenses. Walking is a basic ability essential to independence. It is required not only for mobility purposes but also for maintaining physical and mental health.

Some people who have difficulty in walking use conventional walkers. A walker is an assistive device that has a supporting pad or grips, which a user can lean on or grasp, and legs, which may have wheels. Walkers are commonly seen in hospitals or nursing homes, but only people with fairly good motor function can use them. Walkers should thus be more adaptive to the various needs of the users. Several researches have been conducted on walkers with enhanced functional capabilities. Lacey [1] and MacNamara [2] developed several robotic walkers for frail visually impaired. Miyawaki [3] reported a motorized walker that keeps a constant distance from its user.

The authors have developed a power-assisted walking support device for the elderly who have difficulty in walking [4,5]. This walker-type device assists the user's gait by a power-assist control based on force input. Its dynamical characteristics can be easily adjusted by modifying the control parameters so that a wide range of users can use it. It is shown by field tests that the power-assisted device can improve gaits of elderly people with various physical needs.

2. Power-Assisted Walking Support Device

The power-assisted walking support device has four wheels and a supporting pad that holds the user as shown in Figure 1. Its rear wheels are actively driven by two independent motors. And its front casters can freely rotate and turn. It also has a lift mechanism that uses two electric linear actuators to control the height and slant angle of the supporter. A 6-axis force sensor between the supporter and the lift detects the forces and torques acting on the supporter. Specifications are listed in Table 1.

Figure 2 illustrates the power-assist control system. The system drives the two motors so that the walker moves forward at the speed in proportion to the propulsive force applied to the supporter by the user and turns in the same way according to the applied turning torque. To improve usability for hemiplegic patients, the control system is equipped with an imbalance compensator that suppresses unintentional turning motion by virtually shifting the center of the force sensor [5]. The control system also has safety and usability functions such as slope adaptation by means of an inclinometer and obstacle avoidance by means of infrared range sensors [4].

With this control system, the user can intuitively manipulate the walker at their own pace. If the user pushes or twists the supporter, the walker moves forward or turns accordingly. The user can also walk backward by pulling the supporter. By modifying the control parameters through a simple dial-type interface, the user or a therapist can easily adjust the dynamical characteristics of the walker in order to achieve a steadier gait. The variable parameters include viscous resistances and virtual inertias in the forward, backward and rotational directions and the parameter given to the imbalance compensator, that is, the amount of the virtual shift of the force sensor.

Figure 1. Photograph of the walking support device.

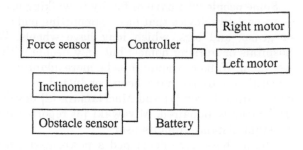

Figure 2. Control system

Table 1. Specifications

Overall size	length 880 mm, width 750 mm
Height control	630 – 1220 mm
Weight	120 kg
Battery life	2 h
Maximum speed	1.5 km/h

Table 2.   Participants of field evaluation.

(a) Age

Range	%
< 50	7
50 – 59	26
60 – 69	41
70 – 79	12
80 – 89	12
> 89	2

(b) Cause of gait disorder

Disease	%
Brain stroke	51
Nervous system disorder	30
Spinal cord injury	12
Other orthopedic injury	7

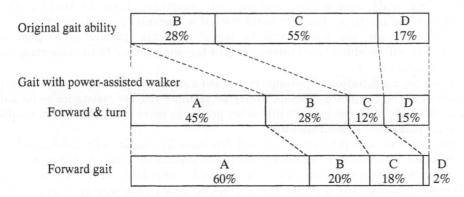

Original gait ability

B	C	D
28%	55%	17%

Gait with power-assisted walker

Forward & turn

A	B	C	D
45%	28%	12%	15%

Forward gait

A	B	C	D
60%	20%	18%	2%

(A) Self-supported   (B) Escort needed   (C) Assist needed   (D) Unable to walk

Figure 3.   Field evaluation result.

3. Field Evaluation

The power-assisted walking support device was evaluated in a physical therapy room in a hospital, and 60 individuals who could hardly walk without help participated in the test. As shown in Table 2, most of the participants were above 50 years old and the average was 62. Half of them had experienced a brain stroke and had hemiplegia. Other causes of their gait disorder were diseases in the nervous system such as parkinsonism or ataxia, paresis or paralysis by spinal cord injury, and other orthopedic injuries such as bone fracture. To achieve the best gait, the control parameters were tuned for each person. Forces and torques from the user and the speed of the walker were recorded by a data-logging system on the device. The speed was measured by rotary encoders attached to the motors that drive the wheels.

4. Results

Figure 3 shows the original gait ability of the participants and the improved one as a result of using the walking support device. Since some people who had difficulty in turning could walk forward easily, ability of forward gait is separately shown. In this chart, the gait ability of the participants is classified according to walking independence into the following categories.

(A) Self-supported: Can walk steadily and independently
(B) Escort needed: Can walk by themselves but someone should watch them for safety
(C) Assistance needed: Can walk if someone gives physical assistance
(D) Unable to walk

Among the participants, 45% were marked as (A) and 28% as (B); that is, in total, 73% could walk without any physical assistance. Only 2% could neither walk forward nor turn. The rest could walk if a therapist gave slight assistance. For these people, the walker will be useful for walk training since physical therapists will be able to concentrate on gait teaching rather than supporting the body of patients.

The best combination of parameters varied among the participants. Generally people with better motor function preferred smaller viscous resistance to be able to walk easily at faster speed, while other people in more severe condition needed heavier resistance to gain stability.

Figures 4 (a) and (b) show speed and direction fluctuation of a left-hemiplegic male's gait. Two different configurations of the imbalance compensator were used: (a) disabled and (b) the force sensor was virtually shifted 0.15 m to the right. Since his right arm was stronger than the left one, he pushed the walker mainly by the right hand. Without compensation, his gait drifted to the left although he intended to walk straight. After the compensator was adjusted, his gait was significantly stabilized.

Parkinsonian patients have difficulty in controlling gait speed. They need an adequate resistance to stabilize their gait. Figures 5 (a) and (b) show speed and force of the gait of a parkinsonian male using the walker set with two different resistance values. In this case, he could walk more stably and easily with smaller resistance of 20 Ns/m.

Figures 6 (a) and (b) show gait data of an ataxic male who has severe balance disorder and loss of joint position sense. To prevent him from falling down on his back, reverse motion of the walker was suppressed by setting a large backward resistance. In the first trial, he could walk but his gait was irregular and intermittent. To improve his gait pace, the virtual inertia of the walker was reduced in the second trial. In this trial, he could walk continuously for several seconds.

As shown in the above cases, the gait of each person could be improved by modifying the dynamical behavior of the walker. These results demonstrate that the variable dynamics capability of the power-assist system provides high adaptability to various needs of the users.

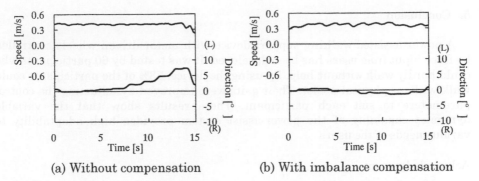

(a) Without compensation　　　(b) With imbalance compensation

Figure 4.　Speed and directional fluctuation of a hemiplegic person's gait.

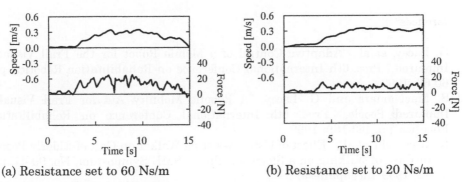

(a) Resistance set to 60 Ns/m　　　(b) Resistance set to 20 Ns/m

Figure 5.　Speed and propulsive force of a parkinsonian person's gait.

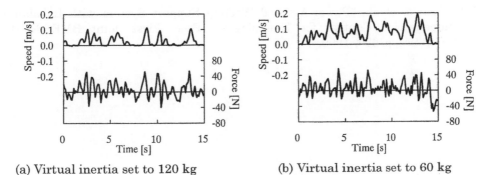

(a) Virtual inertia set to 120 kg　　　(b) Virtual inertia set to 60 kg

Figure 6.　Speed and propulsive force of an ataxic person's gait.

5. Conclusion

A power-assisted walking support device with motor driven wheels controlled by force input from users has been developed. It was tested by 60 participants who could hardly walk without help. By using the device, 73% of the participants could walk without physical assist. Their gaits were improved by adjusting the control parameters to suit each participant. These results show that the variable dynamics capability of the power-assist system provides high adaptability to various needs of the users.

Acknowledgement

This work was performed partly under entrustment by the New Energy and Industrial Technology Development Organization (NEDO).

References

1. G. Lacey, et al., "Adaptive Control of a Mobile Robot for the Frail Visually Impaired," Proc. 6th International Conference on Rehabilitation Robotics, pp. 60-66, 1999.
2. S. MacNamara and G. Lacey, "A Robotic Mobility Aid for Frail Visually Impaired People," Proc. 6th International Conference on Rehabilitation Robotics, pp. 163-169, 1999.
3. K. Miyawaki, et al., "Effect of Using Assisting Walker on Gait of Elderly People — In Case of Walking on a Slope —," Proc. JSME Symposium, No. 99-41, pp. 68-72, 1999. (in Japanese)
4. Y. Nemoto, S. Egawa and M. Fujie, "Power Assist Control for Walking Support System," Journal of Robotics and Mechatronics, Vol. 11, No. 6, pp. 473-476, 1999.
5. S. Egawa, et al., "Power-Assisted Walking Support System with Imbalance Compensation Control for Hemiplegics," Proc. First Joint BMES/EMBS Conference, p. 635, 1999.

Human Friendly Mechatronics (ICMA 2000)
E. Arai, T. Arai and M. Takano (Editors)

A Horseback Riding Therapy System using Virtual Reality and Parallel Mechanism

Hitoshi Kitano[a],Tatsuji Kawaguchi[a],Youji Urano[a] and Osamu Sekine[b]

[a]Matsushita Electric Works,Ltd. Production Engineering R&D Center
1048,Kadoma,Osaka 571-8686,Japan

[b]Matsushita Electric Works,Ltd. Advanced Technology Research Laboratory
1048,Kadoma,Osaka 571-8686,Japan

We developed a horseback riding therapy system using a mechanical horse for activating physical functions of healthy elderly people.

The horseback riding therapy system is composed of three parts, a parallel mechanism, a virtual reality system, and a control system.

The system provides also a way for practicing riding while enjoying it.

The system has proved effective for aged persons to improve the muscles and capability of maintaining their balance.

1. Introduction

It is well known that horseback riding is a good therapy for both body and soul, and spreading mainly in Europe under the name of horseback riding therapy or disabled persons riding therapy. These days it is getting popular also in Japan [1].

Horseback riding therapy is effective for improving the reflex for keeping a posture or balance during motion, improving the reaction for keeping standing, suppressing the degradation in the function of nerve centers, and reinforcing the muscles at the lower part of the back. This suggests that the therapy is effective for removing or preventing low-back pains, stoop and tumbling that are likely common among elderly people.

Another advantage of horseback riding therapy is that it provides an unintentional practice of maintaining his (her) balance as he (she) tries not to fall from the horse. Unlike other therapies, horseback riding therapy does not therefore cause the aged users to feel mental pains or reluctance [2].

However, the therapy using real horses has some problems especially in Japan. We don't have much space to feed and train horses. Moreover, using real horses cannot permit us to repeat the same quality of therapy or evaluate its results quantitatively.

The horseback riding therapy system using an artificial, mechanical horse, presented here, is a solution for these problems. Note that unlike our system the existing horseback riding simulators are not intended for therapeutic use.

Our horseback riding therapy system provides a way for practicing riding while enjoying it.

The system has functions:
· To improve the muscles around the waist using a simulated motion of a horse, and
· To give the user a joyful time by providing the entertaining elements of animation,
  background music and game so that he (she) can stick to and continue practicing.
   This paper describes the components of the system and the results of our studies
on the effect of the system on elderly persons.

2. Analysis of Horse Motions

   To design a machine that can emulate the motion of a horse, it is necessary to
analyze the motion of the horse to determine the specifications of the machine.
   Horse motions were measured using a motion capture equipment. As shown in
Figure 1, the positions (x, y and z coordinates) of four markers on a saddle on a
horse with a person thereon were measured at intervals of 1/120 second using
multiple cameras.
   Then the position (x, y and z) and posture (yaw(α), pitch(β) and roll(γ)) of the
saddle were calculated from the data of the markers positions. The position of the
saddle was defined by the marker (a) at the center of the saddle. The posture
(yawing and pitching motions) of the saddle was defined by the positions of the
markers (a) and (b). The rolling motion of the saddle was defined by the positions of
the markers (c) and (d).The marker (d) was on the opposite side of the saddle.

$$\alpha = \tan^{-1} \left((y_a - y_b) \diagup (x_a - x_b) \right)$$
$$\beta = \tan^{-1} \left((z_a - z_b) \diagup (x_a - x_b) \right)$$
$$\gamma = \tan^{-1} \left((z_c - z_d) \diagup (y_c - y_d) \right)$$

The position (x, y and z) of marker (a) : (x_a, y_a, z_a)
   Note : before preforming the above calculations, the change in the x coordinate due
to walk at constant speed was subtracted.
   The above measurements and calculations derived the position and posture of the
saddle at intervals of 1/120 second.
   Among these position and posture data describing a series of steps of the horse, a
data typical of the step was selected for the purpose of determining the specifications
of the system. The data was used also for defining the operation of the horse
machine.

Figure 1.The location of the mark

Table 1.The characteristics of the "walk"

Movable range	50 mm in each of X, Y and Z directions, $\pm 5°$ in each of the yawing, pitching and rolling directions
Operating cycle	50 mm in amplitude, 2 Hz

Among various types of gaits, "walk" is said to be the most appropriate for horseback riding therapy. We defined the "walk" as follows. The amplitude in each direction was defined as the distance between the minimum and maximum change in position. The velocity and acceleration were calculated from the measurements sampled at intervals of 1/120 second.

Table 1 summarizes the characteristics of the "walk" as defined above.

Using this table, the operating characteristics of the horse machine were defined as follows.

3. Design of Parallel Mechanism

As a driving mechanism to emulate the "walk," a Stewart platform-shaped parallel mechanism was chosen. The mechanism features:

· Higher stiffness leading to a smaller size of the horse machine.

· Smaller number of rubbing parts and smaller inertia force leading to smaller sound.

· Needing a smaller capacity of actuator, and economical due to a larger number of standard parts.

The specifications of the mechanism was determined by extensively using computer analyses so that the before mentioned specifications of the machine could be met. Figure 2 shows the model used for analyzing the parallel mechanism. The model (actually its end effecter) simulated the walk that was defined by the position (x, y and z) and posture (yaw, pitch and roll) of the saddle. During this simulation, the stroke in the motion of the links, movable range of the joints, and interference between the links were checked and measured. During loading the end effecter so that the walk can be simulated, the force generated by the links was also measured. The specifications (power and reduction gear ratio) of the driving motor were determined based on the information on the force generated by the motion of the links.

Each link operates by a driving mechanism consisting of a pair of a lead screw driven by a servo motor and a nut that moves on the screw keeping contact with the link.

Figure 2. The model used for analyzing

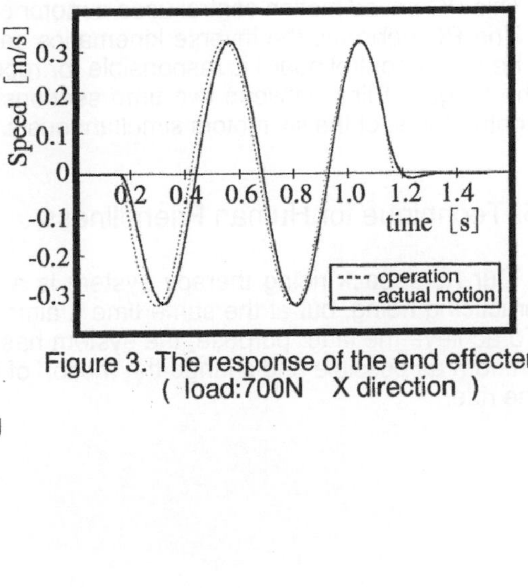

Figure 3. The response of the end effecter
(load:700N   X direction)

Table 2. Performance of the parallel mechanism

Dimension	900mm×1040mm×850mm
Motor	750W×6 units
Fidelity	max. error of 1mm throughout the operation
Acceleration	8m/s² in each of X, Y and Z directions
Noise	60 dB (at a distance of 1m)
Link stroke	200 mm

The geometry of the end effecter was determined so that a person can straddle it and the person's soles do not reach the floor during rido. The links were configured so that the legs of the rider do not touch them during the operation.

Table 2 shows the actual measurements of the parallel mechanism determined as above.

Figure 3 shows the response of the end effecter at walking direction.

4. Motion Control

To be able to control the position and posture of the end effecter of the parallel mechanism, it is necessary to determine the length of each link using the theory of inverse kinematics. We used the solutions shown in [3] to calculate the length of each link.

The motion control applied uses the following algorithm to let the end effecter emulate the motion of a saddle on a horse.

· Divide the saddle's position and posture data obtained by a motion capture equipment into time sections corresponding to the predetermined sampling interval.
· Perform the inverse kinematics calculation for each of a series of the time sections to obtain the length of each link. Calculate the difference in the length of link between two adjacent time sections.
· Operate the motor at such a constant speed that the difference in the length of link calculated above can be actually obtained during the relevant time section.

This PC based motion control uses a motor control board.

The PC performs the inverse kinematics calculation for each of the time sections. The motor control board is responsible for receiving the calculation results (change in the length of link between two time sections), and performing the continuous path control for all of the six motors simultaneously.

5. Technique for Human Friendliness

Our horseback riding therapy system is a virtual system for providing a way for practicing riding, but at the same time it aims to give the user a joyful time of riding. To achieve the latter purpose, the system has some ingenuity to move the saddle as smooth as possible and change the motion of the saddle according to the intention of the rider.

To move the saddle smooth, it is essential that the end effecter does not produce sudden acceleration or deceleration. The control of the end effecter repeatedly uses a single data representing a typical cyclic motion of a saddle set on a real horse that was obtained with a motion capture equipment. A special algorithm was employed in order to avoid sudden change in speed in the event of start, stop, continuing operation by inputting a new command or changing the type of gaits.

To continue the operation by inputting a command, the current cycle is multiplied by:

$f(t) = 0.5 \times (1 + \cos(\pi \times t / T))$, and

the next cycle is multiplied by:

$f(t) = 0.5 \times (1 - \cos(\pi \times t / T))$, and

the sum of the current and next cycles is calculated.

t: sampling interval, T: 1 cycle time.

The two sensors shown in Figure 4 are to realize the change in motion according to the intention of the rider. One of them located on the reins is a switch that turns on when the reins are tighten. The other one located on the belly is a switch that turns on when it is tapped with the knee of the rider. By tightening the reins, the horse decelerates. Tightening the reins when the rider (horse) is at a "virtual" fork on the display indicates which way to go. By tapping the belly, the rider can start or accelerate the horse.

6. Horseback Riding Therapy System

Figure 4 shows the configuration of the horseback riding therapy system.

The system consists of a parallel mechanism, virtual reality system comprising a display, speaker and operator console, and control system.

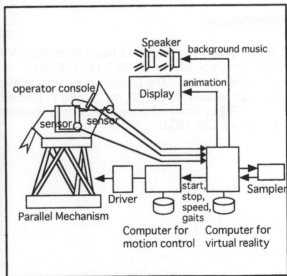

Figure 4. The horseback riding therapy system and its configuration
(○: The locations of sensors)

128

They say that "walk" is best for horseback riding therapy, though this system can simulate "trot" and "canter" as well.

We investigated the effect of our horseback riding therapy system on elderly persons.

Ten men in their sixties used this system for 15 minutes per time, two or three times a week during a three-month period. As a result, their back and abdominal muscles became stronger by 1.4 and 1.5 times on average respectively. Their capability of keeping standing on one leg closing their eyes could be improved from 8.7 seconds to 13.2 seconds.

They all said that they could enjoy practicing. All of them could complete the three-month test.

7. Conclusion

We developed a horseback riding therapy system employing an artificial, mechanical horse through gathering data on the motion of a real horse, determining the specifications of the system based on the data, and developing a parallel mechanism that can satisfy the specifications. The system can emulate several types of gaits including "walk" that is said to be the most appropriate for horseback riding therapy.

The system contains a function for moving the saddle smooth and a virtual reality system with a operation console mimicking an actual harness and reins, so that the users can enjoy the practice of riding.

The system has proved effective for improving the muscles of aged persons and capability of maintaining their balance. This suggests that the system can be used also as a training machine.

References

1.Emiko Ohta, "The Curative Effects of Horseback Riding," Stock Breeding Research, Vol. 51, No. 1, 1997, pp. 148-154.
2.Tetsuhiko Kimura, "The Development of an Intelligent Simulator of Horses," Medical Treatment, Vol. 40, No. 8, Also Published Separately, pp. 749-755.
3.Japan Robot Society Parallel Mechanism Research Special Committee Report, pp. 85-98, Oct. 1993.

Human Friendly Mechatronics (ICMA 2000)
E. Arai, T. Arai and M. Takano *(Editors)*

Development of a driving-support system for reducing ride discomfort of passenger cars

F. Wang[a], S. Chonan[a] and H. Inooka[b]

[a]Graduate School of Engineering, Tohoku University, 04 Aoba-yama, Aoba-ku, Sendai, 980-8579, Japan

[b]Graduate School of Information Sciences, Tohoku University, 01 Aoba-yama, Aoba-ku, Sendai, 980- 8579, Japan

This paper describes the development of a driving-support system for reducing ride discomfort of passenger cars. Based on results of psychophysical experiments of ride discomfort evaluation, we investigated ride discomfort caused by speed fluctuations of a passenger car and set a multiple regression model that evaluates the ride discomfort index (RDI) using longitudinal acceleration and jerk of the car as explanatory variables. Using a microcomputer, the system evaluates the RDI and informs the driver of the RDI in real time. With the help of our system, the driver can improve his driving skills and modify his/her driving method to reduce ride discomfort.

1. INTRODUCTION

In recent years, our daily life relies more and more on automobiles, thus improvement of the ride comfort of automobiles has become an important task. In the past decades, many researches have been done on evaluating ride comfort. Most of these researches are concerned with ride discomfort caused by the vibration of automobiles, which depends mainly on the mechanical characteristics of automobiles and on road paving[1-4]. Active or passive control suspension systems absorbing the vertical vibration of automobiles have been proposed by many authors for the improvement of ride comfort[5-6]. On the other hand, except for when being driven in superhighways, the motion of automobiles is always accompanied with speed fluctuations. However, the speed fluctuations, depending largely on driving skills and driving habit of drivers, while have been considered as an important factor of ride discomfort—even might cause motion sickness in extreme situations[7]— have seldom been discussed in the researches of ride discomfort.

Considering the fact that for a driver to improve the ride comfort, improving his/her driving skills and modifying his/her driving habit is more practical than changing a better car, we investigated ride discomfort caused by speed fluctuations of passenger cars by psychophysical experiments. By imitating the passengers' subjective judgment of ride discomfort, we defined a ride discomfort index (RDI) and set up a model to evaluate the RDI from longitudinal acceleration and jerk. And then we developed a driving-support system to feedback the RDI in real time to the driver. The system enables the driver to reduce ride discomfort by improving his/her driving skills and habit.

2. MODELING OF RIDE DISCOMFORT

In order to investigate the relationship between speed fluctuation and passengers' ride discomfort, we first designed and performed experiments for continuous subjects' evaluation of ride discomfort to collect data, and then by analyzing the data, we established a ride discomfort model to estimate ride discomfort index.

2.1. Experiments
Continuous category judgment method[8] was adopted in the experiments. To put it more concretely, the subjects continuously rated ride discomfort they sensed in a moving automobile on a five-point scale from no discomfort sensed to extremely discomfort sensed in the experiments.

A mini van type passenger car (TOYOTA HIACE) was used in our experiments. It was driven in a straight level lane. Each test run included random alternation of normal acceleration, normal braking, brisk acceleration, and brisk braking maneuvers. The accelerations during the test were monitored by a 3-axis linear accelerometer. Eight healthy subjects (male, 22.8 ± 1.4 years of age) participated in experiments. At each test run, four subjects took one of the following two most common sitting postures: (a) leaning on backrest, and raising head or (b) bowing heads without leaning on backrest. Using a special keyboard, they rated ride discomfort caused by the speed fluctuation of the car by pressing the corresponding keys continuously at a pace of 1 second generated by an electronic metronome during the test runs. Their rating scores were recorded in a computer together with accelerations of the car. Each test run took about 5 minutes. Eight test runs were performed.

Figure 1 shows an example of the records of longitudinal acceleration (above) and average subjects' rating score of ride discomfort by four subjects (bottom) during the test run. In the bottom figure, '0' refers to no discomfort sensed and '4' refers to extremely discomfort sensed. Therefore the higher is the score the stronger is the discomfort. From the figure we notice that there is obvious correlation between physical strength of acceleration and the ride discomfort.

2.2. Modeling of ride discomfort
To understand human judgmental indices of ride discomfort, seeking explanatory variables is of significant importance. The essential question is what a subject is sensing in a moving car and what a rating score given by a subject represents. Physiologically, a passenger can percept acceleration and jerk stimuli of a moving car[4], therefore it is natural to consider acceleration and jerk of the car as the sources of ride discomfort. By analyzing the correlation between the subjects' rating scores of ride discomfort and the longitudinal acceleration and jerk of the car, we found that rating scores have a linear correlation with root mean square (r.m.s.) and peak of acceleration and jerk, and highest correlation exists between rating scores

and acceleration/jerk within the 3-second period in advance of the rating times, with the correlation coefficients of 0.661~0.752[9]. This suggests that pooling of both acceleration and jerk is the source of ride discomfort, and it is possible to evaluate ride discomfort by using acceleration and jerk. And analyses also show that the subjects' rating scores depend not only on the amplitudes of acceleration and jerk, but also on the directions of them.

Next, we tried to establish a linear multiple regression model that imitates subjects' response to the movement of the car. By stepwise method, we chose the following parameters of plus and minus peak of longitudinal acceleration $a_p(t)$, $a_m(t)$ and r.m.s. of jerk with the sign of mean jerk $j_p(t)$, $j_m(t)$ in the 3-second period in advance of evaluating time t as explanatory variables. The definitions of explanatory variables are as follows:

For any evaluating time t, let $T \in (t-3, t]$,

$$a_m(t) = \begin{cases} 0, (|a(T)|_{max} = [a(T)]_{max}) \\ \\ -a(T)_{min}, (|a(T)|_{max} = -[a(T)]_{min}) \end{cases} \quad (1)$$

$$a_p(t) = \begin{cases} a(T)_{max}, (|a(T)|_{max} = [a(T)]_{max}) \\ \\ 0, (|a(T)|_{max} = -[a(T)]_{min}) \end{cases} \quad (2)$$

$$j_m(t) = \begin{cases} 0, (\overline{j(T)} \geq 0) \\ \\ -\sqrt{\frac{1}{3} \int_{t-3}^{t} j^2(\tau) d\tau}, (\overline{j(T)} < 0) \end{cases} \quad (3)$$

$$j_p(t) = \begin{cases} \sqrt{\frac{1}{3} \int_{t-3}^{t} j^2(\tau) d\tau}, (\overline{j(T)} \geq 0) \\ \\ 0, (\overline{j(T)} < 0) \end{cases} \quad (4)$$

Using these 4 terms, we defined a multiple regression ride discomfort model as:

$$RS(t) = \alpha_0 + \alpha_1 a_m(t) + \alpha_2 a_p(t) + \alpha_3 j_m(t) + \alpha_4 j_p(t) + \varepsilon(t) \quad (5)$$

where $RS(t)$ is the average rating score of ride discomfort by subjects and $\varepsilon(t)$ is the error that cannot be explained by the model. Imitating RS, we defined a real time ride discomfort index $RDI(t)$ as:

$$RDI(t) = \hat{\alpha}_0 + \hat{\alpha}_1 a_m(t) + \hat{\alpha}_2 a_p(t) + \hat{\alpha}_3 j_m(t) + \hat{\alpha}_4 j_p(t) \quad (6)$$

With data collected from experiments, we obtained the partial regression coefficients $\hat{\alpha}_0$, $\hat{\alpha}_1$, $\hat{\alpha}_2$, $\hat{\alpha}_3$ and $\hat{\alpha}_4$ using least square method from Equation (5). And then with these coefficients and acceleration and jerk of a car, we can estimate the RDI with Equation (6).

Figure 2 shows an example of comparison of RS of ride discomfort obtained in an experimental test run and the RDI estimated with Equation (6). From the figure we can see that RS and RDI are quite similar, with a correlation coefficient of 0.907 and the mean squared error of 0.167 between them in this case. For all 8 test runs, the correlation coefficient varies form 0.807 to 0.917, with the average mean squared error of 0.218. These results verified the efficacy of the model.

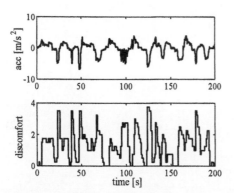

Figure 1. An example of experimental records of acceleration (above) and average rating score of ride discomfort of 4 subjects (bottom) of a test run.

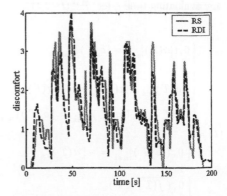

Figure 2. Comparison of rating score (RS) of ride discomfort judged by subjects and ride discomfort index (RDI) estimated by the model.

3. PROTOTYPE OF THE DRIVING-SUPPORT SYSTEM

The previous section verified that with acceleration and jerk, it is possible to evaluate ride discomfort index with an enough high accuracy. It is expected that if we inform the driver of the RDI in real time, he can adjust his driving to reduce ride discomfort. Based on the results of the previous study, a driving-support system for reducing ride discomfort was developed.

The driving support system we developed consists of three main modules, namely, a data acquisition module, an RDI-estimating module, and a driver-informing module. Schematic diagram of the system is shown in Figure 3.

As shown in figure 3, the data acquisition module consists of an accelerometer, a low-pass filter, an amplifier, and an AD converter. Acceleration is detected by the accelerometer, which is mounted on the floor of an automobile underneath the seat of the driver, and is amplified by the amplifier to fit the range of AD converter. The low-pass filter, with the cut-off frequency of 20hz, cuts off the high frequency vibration components of the acceleration caused by engine vibration and also acts as an anti-aliasing filter. With the A/D converter, acceleration of the automobile is fed into the microcomputer at a sampling rate of 100Hz.

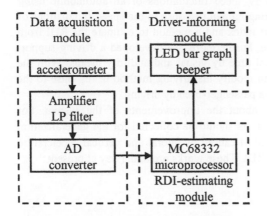

Figure 3. Schematic diagram of hardware of the driving-support system

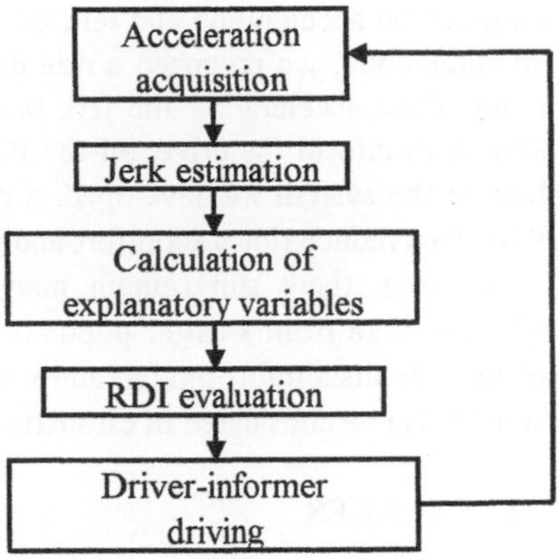

Figure 4. Flow chart of software of the system

The RDI-estimating module is realized with a MC68332 one-chip microcomputer. Flow chart of the software is shown in Figure 4. First, it gets longitudinal acceleration of the automobile from ADC. As differential of acceleration, jerk cannot be directly measured. Simple numerical differential operating is unsuitable for applications because it tends to amplify high frequency noise. Though there are some sophisticated methods of smoothing and differentiation by least square procedures, they cannot be used in real time. By analyzing acceleration data, we proposed a method to estimate jerk with a Kalman estimator that is very simple to be realized in real time. With the Kalman estimator, the computer estimates jerk from acceleration in real time. Next, it calculates explanatory variables using Equations (1)~(4). Then with Equation (6), it evaluates RDI in real time. Finally, it sends RDI to the driver-informing module to inform the driver. This procedure is repeated as a timely interrupting routine.

In realization of traffic information systems (TIS), designing of driver-informer is a vary important topic[10]. During driving a automobile, a driver receives most of information with visual perception, therefore using of visual output to inform the driver of information is most popular in TISs. However, with the evolution of TIS, second-order driving tasks during driving a automobile such as glancing at dashboard displays or watching a navigator display tend to increase visual workload of drivers, and might even bring some bad side-effects on traffic safety[11]. In order to reduce visual workload, we use a 3-color LED bar graph display mounted above the dashboard as RDI indicator because it can provide both color and space information at a glance. For further reducing of visual workload, auditory output is also used as RDI indicator in our system. In other words, a beeper that beeps at different tones and paces corresponding with the RDI is also used to inform the driver of the RDI.

4. CONCLUSIONS

In this paper, based on results of psychophysical experiments, we demonstrated that it is possible to estimate ride discomfort caused by speed fluctuations of an automobile using longitudinal acceleration and jerk of it. Imitating human response to the speed fluctuations of an automobile, we proposed a ride discomfort index and a method to estimate the RDI from longitudinal acceleration and jerk in real time. Furthermore, we developed a driving-support system to inform the driver of the RDI in real time by visual and auditory means. With the help of the system we developed, a driver can modify his driving method in response to the RDI, thus reduce ride discomfort and provide a passenger-friendly driving.

However, there still remain many topics about the improvements of the system. For instance, data from a larger population of subjects should be collected for the generalization of the ride discomfort model, and ride discomfort caused by lateral and vertical acceleration should also be considered in estimation of the RDI. These works are underway at present.

REFERENCES

1. S. Cucuz, Evaluation of ride comfort, Int. J. Vehicle Design, Nos. 3/4/5, (1994) 318-325.
2. J.D. Leatherwood, T.K. Dempsey, S.A. Clevenson, A design tool for estimating passenger ride discomfort within complex ride environments, Human Factors, No. 3, (1980) 291-312.
3. P.R. Payne, Comparative ride comfort studies, Noise & Vibration Worldwide, No. 7, (1997) 25-41.
4. C. Liu, D.C. Gazis, T.W. Kennedy, Human judgment and analytical derivation of ride quality, Transportation Science, No. 3, (1999) 290-297.
5. J.A. Tamboli, S.G. Joshi, Optimum design of a passive suspension system of a vehicle subjected to actual random road excitations, J. Sound & Vibration, No2, (1999) 193-295.
6. H.D. Taghirad, E. Esmailzadeh, Automobile passenger comfort assured through LQG/LQR active suspension, J. Vibration & Control, No. 5, (1998) 603-618.
7. H. Vogel, R. Koholhaas, R. J. von Baumgarten, Dependence of motion sickness in automobiles on the direction of linear acceleration. Eur. J. Applied Physiology, 48, (1982) 399-405.
8. S. Number, S. Kuwano, The relationship between overall noisiness and instantaneous judgment of noise and the effect of background noise level on noisiness, J. Soc. Acoustics of Japan (E), No. 2, (1980) 99-106.
9. F. Wang, K. Sagawa, H. Inooka, A study of the relationship between the longitudinal acceleration/deceleration of automobiles and ride comfort, Japanese J. Ergonomics, No. 4, (2000) 191-200, (in Japanese).
10. A. Naniopoulos, E. Bekiaris, M. Dangelmaier, TRAVELGUIDE: Towards integrated user friendly system for drivers traffic information and management provision, Proc. Intn'l. Conf. Machine Automation 2000, Osaka, Japan, (2000) 585-590.
11. A. Okuno, Driver vision and driver support technologies, J. Soc. Automobile Engineering of Japan, No. 1, (1998) 22-27, (in Japanese).

Human Friendly Mechatronics (ICMA 2000)
E. Arai, T. Arai and M. Takano (Editors)
135

Characteristic mapping of stance posture control systems from a control engineering standpoint

M. Kikuchi[a] and M. Shiraishi[b]

[a]Graduate School of Ibaraki University, Nakanarusawa 4-12-1, Hitachi, Japan

[b]Department of Systems Engineering, Ibaraki University, Nakanarusawa 4-12-1, Hitachi, Japan

Techniques for objectively measuring human movement traits and mapping the characteristics of these traits in artificial systems have a wide range of applications [1-2]. A problem associated with mapping techniques of this sort is that the characteristics are obscured by fluctuations in the movement traits, which makes it difficult to objectively evaluate and distinguish between the measured results while taking fluctuations into account.

In our current line of research, as a first evaluation method, we have already proposed a method whereby the stance characteristics of a test subject are treated as changes in a pole assignment by focusing on the human stance posture control system and measuring the swaying of center of gravity [2]. Furthermore, by applying loads to test subjects, we have shown that the characteristics of each test subject's posture control system can be objectively measured in terms of mutual differences in their pole assignments, and that it is possible to extract the temporal variation of characteristics [3]. As a second evaluation method, we have also used stance posture control to determine the attractors that give rise to swaying of center of gravity, and we have used an error-back-propagation method to identify the two-dimensional frequency components contained therein with a layered neural network. As a result, by using changes in the output values of the neural network, we have confirmed that the control characteristics of test subjects can be evaluated even more quantitatively.

1. MEASUREMENT OF SWAYING OF CENTER OF GRAVITY

In this research, the stance state measurement method hitherto used to evaluate the degree of handicap of patients was improved to allow it to be applied to able-bodied subjects, and the swaying of center of gravity of healthy test subjects was measured in the standing state. The measurement system and measurement method are shown in Figure 1. The test subjects remained standing during the swaying of center of gravity measurements, and stood on the left leg or right leg according to instructions from a display device (an alternating left/right display of a 1 cm square mark on the right side of a monitor screen). In the measurements, subjects were asked to stand on their left and right legs alternately for 5 seconds each, for a total of 20 seconds (two repetitions of right-leg and left-leg stances). This measurement is referred to as "measurement I". The swaying of center of gravity r_G along the x axis during this task was measured with a swaying of center of gravity meter, whose output was digitized

136

with an A/D converter and fed into a computer. Measurement I was repeated 15 times at 60-second intervals, and these measurements are collectively referred to as "measurement II". In the experiments, measurement II was performed at two-hour intervals from 10 a.m. to 4 p.m. Figure 2 shows some typical results for measurement I, which were obtained at 10 a.m. with a 36-year-old male test subject "A".

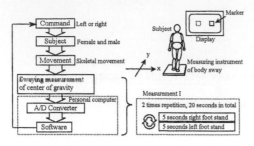

Figure 1 Measuring system

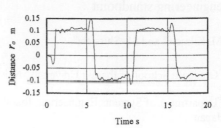

Figure 2 Measurement result of swaying of center of gravity for Subject A

2. POLE ASSIGNMENT AND TEMPORAL VARIATION OF POSTURE CONTROL SYSTEM

When using the identification method for measurement I, the posture control characteristics of the test subject at each time can be represented as a pole assignment. By varying the measurement times, the temporal variation of the characteristics can also be ascertained. We therefore decided to study the test subjects' posture control system traits in terms of the pole assignment of the discrete transfer function.

2.1 Experimental results and discussion

Figure 3(a) shows the experimental results for test subject A at 10 a.m. The poles can be broadly divided into three groups. Poles that lie on the real axis are classed as poles (1), and poles that lie above the real axis are classed as poles (2). The complex conjugate poles that lie below the real axis are symmetrical about the real axis with poles (2), and are thus omitted. As Fig. 3(a) shows, the group of poles (2) is positioned more or less within a certain region, although there is a certain amount of variation within the group. Also, in Fig. 3(a), since the distance from the origin to the most remote pole Pd is about 0.9981, all the poles in Fig. 3(a) lie within the unit circle. Since a z-transform maps the half-plane to the left of the imaginary axis to within the unit circle in the z

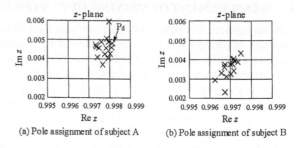

(a) Pole assignment of subject A   (b) Pole assignment of subject B

Figure 3 Pole assignment by the measurement I at 10 a.m.

plane, the pole groups of Fig. 3(a) are all positioned in the stable region.

Next, Fig. 3(b) shows the experimental results obtained at 10 a.m. with a 23-year-old female test subject "B". Basically these results exhibit the same overall trends as those of test subject A. The poles (2) in Fig. 3(b) are scattered in a similar manner to those shown in Fig. 3(a). However, the overall position is closer to the point (0,0) and to the real axis. It can thus be concluded that the stance posture control system of test subject B has better control ability at the point of stability than the system of test

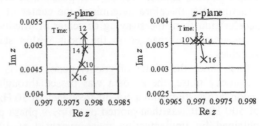

(a) Pole locus of subject A   (b) Pole locus of subject B

Figure 4   Pole locus of stance posture control from 10 a.m. to 4 p.m.

subject A. In this way, the measurement procedure and analysis method proposed here allow human posture control characteristics to be investigated with good reproducibility. They also provide an effective way of investigating the traits of a test subject's posture control state.

Figure 4(a) shows the average pole assignment of test subject A from 10 a.m. to 4 p.m. In this figure, an average pole is determined for each measurement time from the average of the real parts and the average of the imaginary parts of each pole, The symbol ✕ marks its position. From the locus of Fig. 4(a), it can be conjectured that the complex conjugate poles move so as to describe a closed loop. Figure 4(b) shows the average pole assignment of test subject B. As can be seen from this figure, this method also allows the temporal variation of posture control characteristics of a test subject to be investigated.

3.   STANCE POSTURE CONTROL SYSTEM ATTRACTORS

In general, the swaying of center of gravity during stance posture control can be regarded as one type of information that is representative of the control system characteristics. Although the stance posture control system cannot be represented as a simple mathematical model, simplifying each element makes it possible to construct a computationally feasible mathematical model using ordinary differential equations. These simplified differential equations contain no chaotic factors. Accordingly, if the form of a differential equation is temporally invariant and stable, then the locus described by the states obtained from the equation is absorbed into the characteristic point. That is, the attractors are invariant. However, since chaotic factors are thought to be included in real stance posture control systems, the attractors fluctuate around this point due to small differences in the states and initial values of the test subjects. By observing the changes in the attractor pattern of the

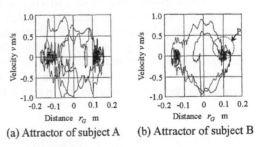

(a) Attractor of subject A   (b) Attractor of subject B

Figure 5   Attractors by the measurement I at 10 a.m.

stance posture control system, it is possible to infer temporal changes in the stance control structure. We therefore determined the center of gravity position r_G, the velocity, and the acceleration from the swaying of center of gravity data obtained in measurement I, and we investigated the changes in the locus within state space with these parameters as state variables.

First, since the center of gravity position r_G is obtained as discrete data including high-frequency noise, the r_G data was passed through a digital low-pass filter by digitizing a secondary system with a cut-off frequency of 30 Hz and a fall-off rate of 0.707 at a frequency of 500 Hz. It was then plotted on an r_G–v phase plane. Figure 5 shows the attractor pattern obtained by implementing measurement I at 10 a.m. Measurements were subsequently performed in the same way at midday, 2 p.m. and 4 p.m. As Fig. 5(b) shows, the attractor pattern of test subject B starts from close to the origin, and moves every time the instruction on the display apparatus changes between the left foot stance position close to (0.1, 0) and the right foot stance position close to (−0.14, 0). It can also be seen that the test subject moves with a slight degree of swaying close to the left/right foot stance positions. The same trend is also exhibited in Fig. 5(a), but test subjects A and B differ in terms of the overall swaying intensity and the shape of the attractors. Therefore, we propose the following method for automatically extracting the traits of test subjects from the attractor patterns.

4. ATTRACTOR TRAIT EXTRACTION

To extract the traits contained in the attractor pattern, a two-dimensional Fourier transform is performed on the attractor pattern image. It is possible to check the power spectrum of each frequency component contained in the attractor pattern. First, the attractor pattern is plotted on the r_G–v phase plane, and the differences in the shading of this plotted image (200 × 200 pixels) are acquired as pixel data $F_d(i,j)$. The resulting pixel is subjected to a two-dimensional complex-base discrete Fourier transform (DFT), and the resulting each datum is defined in the form of $F_{td}(i,j)$ as shown in Eq. (1).

$$F_{td}(i,j) = DFT_i(DFT_j(F_d(i,j))) \tag{1}$$

Where i and j are integers in the range $1 \leq i \leq 200$ and $1 \leq j \leq 200$. Formula (1) means that a one-dimensional discrete Fourier transform is calculated for each column of matrix $F_d(i,j)$, and then its transformation is performed for each row of the results.

Next, a new matrix obtained by extracting a 100 × 100 region of positive frequencies in both the rows and columns is defined as $F_{cd}(k,l)$. That is, $F_d(i,j)$ is substituted with $F_{cd}(k,l)$ so as to satisfy the relationships $k=i-100$ (where i is an integer in the range $101 \leq i \leq 200$) and $l=j-100$ (where j is an integer in the range $101 \leq j \leq 200$). The region of $F_{cd}(k,l)$ is then classified into 100 groups by partitioning $F_{cd}(k,l)$ into groups of 10 elements in the row and column directions. The average value of the power spectrum in each group is determined from Formula (2), and the results are stored in a 10 × 10 matrix $F_{ed}(m,n)$ with $m=n=10$.

$$F_{ed}(m,n) = \frac{1}{100} \sum_{k=10(m-1)}^{10m} \left\{ \sum_{l=10(n-1)}^{10n} F_{cd}(k,l) \right\} \tag{2}$$

This procedure allows the information contained in the attractors to be converted into a matrix of 100 elements.

Figure 6(a) shows the results obtained for $F_{ed}(m,n)$ from the attractor pattern in Fig. 5(a) for test subject A. This figure is a three-dimensional representation of each element of the 10×10 matrix stored in $F_{cd}(m,n)$ (which stores the power spectrum values), with the magnitude in the height direction, rows in the depth direction, and columns in the width direction. In this figure, larger column and row numbers indicate transitions to higher frequency components. Similarly, Figure 6(b) shows the results of measurement I at 10 a.m. for test subject B. Although Figs. 6(a) and 6(b) have basically similar curve shapes, the component in the vicinity of $F_{ed}(1,10)$ is almost 10 dB larger for test subject A. If a correlation exists between the characteristics of the test subjects and these curve profiles, then it is possible to objectively infer the state of a test subject's characteristics based on the curve profiles.

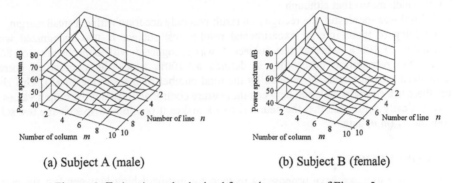

(a) Subject A (male)   (b) Subject B (female)

Figure 6 $F_{ed}(m,n)$ result obtained from the attractor of Figure 5

5. INFERRAL OF STATES WITH A NEURAL NETWORK

Neural networks are widely known to be effective at recognizing speech and images. Here, we trained a neural network with the $F_{ed}(m,n)$ data of test subjects A and B for 10 a.m., and then tried to distinguish between the test subjects from arbitrary input data,

5.1 Inferred results

Inferrals were made based on 15 sets of data obtained with test subjects A and B starting from 10 a.m. These results are shown in Figure 7. The ■ symbols indicate the output values for test subject A, and the ● symbols indicate the output values for test subject B. The average and standard deviation of the output values for test subject B are 0.832 and 0.065, respectively. The corresponding quantities for test subject A are 0.656 and 0.132, respectively. In this figure, the horizontal axis shows the number of measurements, and the vertical axis shows the values of the output units corresponding to each test subject. Since there were two test subjects, the output layer of the neural network contained two output units o_1 (for test subject B) and o_2 (for test subject A). Accordingly, when the $F_{ed}(m,n)$ data of test subject B is input as the data to be inferred, the value of output o_1 is shown on the graph. Output values for test subjects A and B closer to 1 indicate that the inferral has a greater probability of success. Also, measurements where the output values are 0.5 or less mean that an incorrect inferral is made (the shaded parts of Fig. 7). For example, the fifth measurement of test subject A produced an output value of 0.490, which means that the inferral system misinterpreted the $F_{ed}(m,n)$ data of test subject A as that of test subject B. Also, the sixth measurement of test subject A produced an output value of 0.509, which means that although

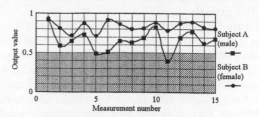

Figure 7 Estimated result by neural network at 10 a.m.
Subject A: average=0.656, standard deviation=0.132
Subject B: average=0.832, standard deviation=0.065

the inferral was successful, the recognition result was only accurate within a small margin.

According to these inferral experimental results, test subject B was recognized with a probability of 100%, whereas test subject A was recognized with a probability of 86.7% (where the recognition probability is defined as 100% times the number of correctly recognized measurements 13 divided by the total number of measurements 15). This shows that there is a strong correlation between the posture control actions and the output values. We have basically confirmed that the neural network makes it possible to specify the test subject from the $F_{ed}(m,n)$ data.

6. CONCLUSION

In this research, we have proposed a method for distinguishing between test subjects and quantitatively evaluating changes in their posture control states. This approach determines the attractors associated with the swaying of center of gravity during human stance posture control, forming the frequency components contained therein into a pattern, and inferring the posture control status with a neural network. Our future goals are to perform measurements with a large number of test subjects with different conditions and to study in more detail the temporal transitions of human posture control characteristics, thereby constructing methods for experimentally evaluating the dynamic characteristics of humans while comparing them with system identification methods.

REFERRENCES

1. M. Shiraishi and H. Watanabe: Pneumatic Assist Device for Gait Restoration, Trans. ASME, J. Dyn. Syst., Meas., 118, 1 (1996) 9.

2. M. Kikuchi and M. Shiraishi: Characteristic Mapping of Human Dynamics and Evaluation of Its Control Performance (1st Report)-Identification of Stance Posture Control System and Its Pole Assignment-, The Japan Society for Precision Engineering, 65, 6, 1999, pp. 840-844, (in Japanese).

3. M. Shiraishi, S. Sugano and S. Aoshima: High Accuracy Positioning in SCARA-type Robot by Sensor-based Decoupling Control, Trans. ASME, J. Manuf. Sci. Eng., 222, 1 (2000) to appear.

Actuator and Control

Human Friendly Mechatronics (ICMA 2000)
E. Arai, T. Arai and M. Takano *(Editors)*

143

Micro actuator with two piezoelectric elements placed at right angles

Shinya Matsuda*, Takashi Matsuo and Masayuki Ueyama

Minolta Co.,Ltd. Takatsuki Laboratory
1-2, Sakura-Machi, Takatsuki-Shi, Osaka 569-8503, Japan
*matsuda@eie.minolta.co.jp

This paper describes a micro actuator with two piezoelectric elements placed at right angles. Two driving methods at resonance frequency are proposed, phase-difference drive and single-phase drive. On and close degeneracy of the natural vibration modes are necessary for these methods. In this paper necessary conditions are described through kinetic models. These conditions can be controlled by mass of the tip. Theories are demonstrated through the finite element method and experimental studies.

1.Introduction

Piezoelectric actuators are expected to be applied to handheld or wearable equipment because their force is large compared with their volume [1-2]. But high driving voltage is necessary for desirable output, because their displacement is small. Some methods utilize resonance phenomenon for higher efficiency and lower driving voltage [3-5]. Ultrasonic motors with traveling waves are typical examples for these methods. But plural driving circuits are necessary to generate these waves and feedback mechanism is also necessary to drive at a resonance frequency. Drive circuits will be complicated and expensive.

This paper describes a micro actuator with two piezoelectric elements placed at right angles. Two driving methods at resonance frequency are proposed, phase-difference drive and single-phase drive. Torque and speed can be controlled easily by the former, while driver circuit can be simplified through the latter. Lower driving voltage and higher efficiency are also realized by both methods.

2.Design of the micro actuator

Configuration of this actuator is shown in *Fig.1*. Two piezoelectric elements are placed at right angles. The tip is placed at the crossing point of these elements, while the base holds the ends of these elements. These members are fixed with adhesives.

The tip will move in an elliptical orbit if piezoelectric elements make simple harmonic motions. It is called Lissajous' figure. The rotor driven element is forced through these motions.

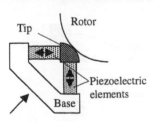

Fig.1 Configuration
of actuator

3.Phase-difference drive
3-1.Principle

The tip will move in an elliptical orbit if piezoelectric elements are driven at off-resonance frequency. But near the resonance frequency the orbit will transform from desirable one by restriction of the natural vibration.

144

Cross-phase mode

Same-phase mode

Fig.2 Natural modes
by *FEM*

There are two natural vibration modes that cause piezoelectric elements to extend: same-phase mode and cross-phase mode. Configurations of the two natural vibration modes demonstrated through the finite element method *(FEM)* are shown in *Fig.2*.

The chip crosses the symmetry axis at cross-phase mode and moves along the axis at same-phase mode.

Natural vibration modes are independent because they are generally orthogonal. Linearly combined vibrations of the two degenerated natural vibration modes will also be degenerated.

Longitudinal vibrations of piezoelectric elements cross these natural vibration modes at 45 degrees. If resonance frequencies of the two natural vibration modes are equal, or degeneracy, longitudinal vibrations will be independent. On this condition the orbit of the tip can be controlled freely by the vibrations of piezoelectric elements. The actuator is driven at high torque when the elliptical orbit extends along the symmetry axis, and at high speed across the axis.

3-2.Kinetic models

At the cross-phase mode there exist longitudinal displacements of piezoelectric elements and a translational motion of the tip. An angle between piezoelectric elements will not change by these displacements. There is no bending force at the end of these elements. Piezoelectric elements correspond to springs and cantilever beams fixed at the base. But the latter effect can be neglected because their flexural rigidities are preferably small.

At the same-phase mode there also exist longitudinal displacements of piezoelectric elements and a translational motion of the tip. However the angle between piezoelectric elements will change by these displacements. There are bending forces at the end of these elements. Piezoelectric elements correspond to springs and fixed beams. There is also a bending force at the tip. The effect can also be neglected because its displacement is preferably small.

To simplify the analysis piezoelectric elements are considered to be fixed at the base, and adhesives are also neglected. Masses are considered to be concentrated, because there is no high degree vibrations.

3-3.Cross-phase longitudinal vibration

A kinetic model of the cross-phase longitudinal vibration is shown in *Fig.3*. In this model cross-phase mode corresponds to a spring-mass system. Mass of the spring: m can be considered to an equivalent mass: $m_l = m / 3$ concentrated to the end of the spring.

Fig.3 Cross-phase mode

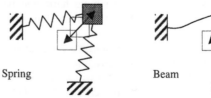

Spring   Beam

Fig.4 Same-phase mode

The spring constant is shown as $k_1 = SE/L$, where S is cross section, E is elastic modulus, L is length.

Natural angular frequency of the cross-phase mode is shown as *Formula1*.

$$\omega_1^2 = \frac{k_1}{M + m_1} = \frac{SE}{L\left(M + \frac{m}{3}\right)} \qquad M: \text{mass of the tip.} \qquad (1)$$

3-4.Same-phase longitudinal vibration

A kinetic model of the same-phase longitudinal vibration is shown in *Fig.4*. In this model same-phase mode corresponds to a spring-mass-and-beam system. One end of the beam is fully fixed while another end is fixed with its angle.

Natural frequency of the complex vibration with plural masses and springs can be estimated as individual vibrations. In this model spring factors are connected in series, while mass factors are connected in parallel. Natural frequency of this model can be calculated as the addition of masses and springs. Spring constant: k_2 and equivalent mass: m_2 of the beam are shown as $k_2 = 12EI/L^3$ and $m_2 = m/2.63$, where I is second moment of area.

Natural angular frequency of the same-phase mode is shown as *Formula2*.

$$\omega_2^2 = \frac{SE/L + 12EI/L^3}{M + m/3 + m/2.63} \qquad (2)$$

3-5.Necessary condition for Phase-difference drive.

As described before resonance frequencies of the two natural vibration modes are necessary to be equal. Necessary condition is shown as *Formula3*, where cross section is shown as $S = WH$, second moment of area is shown as $I = WH^3/12$, H is height and W is width of the cross section. When mass: m and aspect ratio: L/H of piezoelectric elements are given, mass of the tip for the degeneracy can be calculated.

$$M = \frac{L^2}{H^2} \times \frac{m}{2.63} - \frac{m}{3} \qquad (3)$$

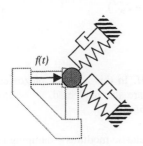

Fig.5 Natural vibrations
in a kinetic model

4.Single-phase drive
4-1.Priciple

Piezoelectric elements can be excited by the natural vibrations if their frequencies are close. The amplitudes and phases of these elements are restricted by these natural vibration modes. The fact suggests that the orbit will be elliptical by controlling these natural vibration modes. This paper proposes a new driving method using only one piezoelectric element.

4-2.Kinetic models

Kinetic models of natural vibrations are shown in *Fig.5*.
Two natural vibration modes correspond to viscous damped one-degree-of-freedom systems.
A simple harmonic motion of piezoelectric element corresponds to an exciting force along the symmetry axis.

If conditions of these vibrations are equal, amplitudes and phases by the exciting force will be equal. Orbit of the tip will be straight along the exciting force. On this degeneracy condition for phase-difference drive piezoelectric elements can vibrate independently. On the other hand orbit will be elliptical when these conditions are not equal.

Displacement: $x(t)$ of viscous damped one-degree-of-freedom system by a sine-wave exciting force: $f(t) = F_0 \cos \omega t$ is shown as *Formula4*, where $\omega n^2 = k/m$ is a natural angular frequency.

$$x(t) = X \cos(\omega t - \phi)$$

, where $\quad X = \dfrac{X_0}{\sqrt{\left[1 - \left(\dfrac{\omega}{\omega n}\right)^2\right]^2 + \left[2\zeta\dfrac{\omega}{\omega n}\right]^2}}$, $\quad \phi = \tan^{-1} \dfrac{2\zeta\dfrac{\omega}{\omega n}}{1 - \left(\dfrac{\omega}{\omega n}\right)^2}$, $\quad X_0 = \dfrac{F_0}{k}$ $\qquad$ (4)

Exciting force of piezoelectric element can be divided along the two natural vibration modes. Each exciting force is shown as $f_1(t) = f_2(t) = F_0 \cos \omega t / \sqrt{2}$.

When these exciting forces are substitute into *Formula4*, amplitudes and phases of the two vibration systems can be calculated.

4-3. Necessary condition for Single-phase drive

The orbit of the tip will be circular if these amplitudes are equal and phase difference is *90* degrees. Assuming that damping ratios and static displacements are equal, necessary condition for single-phase drive is shown as *Formula5*.

$$\frac{fn_1}{fn_2} = (\alpha \pm \sqrt{\alpha^2 - 1}) \qquad , \text{where} \quad \frac{1 - 2\zeta^2}{1 - 4\zeta^2} = \alpha \qquad (5)$$

When damping ratios are given, two natural frequencies: fn_1, fn_2 for a circular orbit can be calculated. Driving frequency: f_3 of piezoelectric element is shown as *Formula6*.

$$f_3 = \frac{\omega_3}{2\pi} \qquad , \text{where} \quad \omega_3^2 = \frac{2\omega n_1^2 \omega n_2^2}{\omega n_1^2 + \omega n_2^2}(1 - 2\zeta^2) \qquad (6)$$

5. Finite element method
5-1. Natural frequencies

Natural frequencies of the actuator are investigated by *FEM*. In this analysis piezoelectric elements are considered to be fixed at the base, and adhesives are also neglected.

Dimensions and material constants of the members are as *Table1*.

Table1	dimension	density	elastic modulus	damping ratio
Piezoelectric elements	$W1.5 \times H1.5 \times L5mm$	$7.6 \times 10^3 kg/m^3$	$6 \times 10^{10} N/m^2$	0.035
Tip	$R3 \times T2mm$ 1/4circle	$\rho_c \times 10^3 kg/m^3$	$17 \times 10^{10} N/m^2$	0.0005

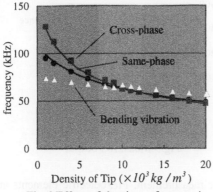

Fig.6 Effect of the tip on frequencies

Effect of the tip on the natural frequencies is shown in *Fig.6*. A density of the tip is imaginary changed. Results by *FEM* are shown as dots, while results by *Formula1* and *2* are shown as lines. Natural vibration modes will be degenerated at $13 \times 10^3 kg / m^3$. Phase-difference drive will be realized on this condition.

There seems to be another vibration mode crossing at $10 \times 10^3 kg / m^3$. This mode is one degree of bending vibration. Desirable orbit cannot be obtained on this condition because these modes are combined elastically.

5-2.Natural vibrations

Characteristics of the natural vibrations are investigated by *FEM*, where density of the tip is $8 \times 10^3 kg / m^3$. Amplitude and phase are shown as dots in *Fig.7*, when cross-phase and same-phase sine-wave signals are given to piezoelectric elements. Amplitude is investigated at the center of the tip, while phase is from driving signal. Resonance frequencies are *72.0kHz* at cross-phase mode and *67.8kHz* at same-phase mode.

Amplitude of the cross-phase vibration shows two peaks at bending and logitudinal natural vibration. The latter is larger than the former. Amplitude of the same-phase mode shows one peak. This is a little larger than that of cross-phase mode. Phase of the cross-phase mode shows a small peak at a bending vibration, becomes larger at longitudinal natural vibration. Phase of the same-phase mode shows monotonous increase. Damping ratios calculated by amplitudes are *0.027* at cross-phase mode and *0.023* at same-phase mode.

Theoretical results calculated by *Formula4* are shown as lines in *Fig.7*, where resonance frequencies and damping ratios given by *FEM*. Both results are almost same except bending vibration. Assuming that damping ratios equal to *0.025*, the ratio of frequencies for the circular orbit calculated by *Formula.5* is $fn_1 / fn_2 = 1 \pm 0.05$. Driving frequency: *f3* calculated by *Formula.6* is *69.8kHz*.

As described before the ratio investigated by *FEM* is $fn_1 / fn_2 = 72.0 / 67.8 = 1.06$. Shown in *Fig.7*, amplitudes are almost equal and phase difference is *90* degrees near the frequency of *69.8kHz*. Single-phase drive can be realized on this condition.

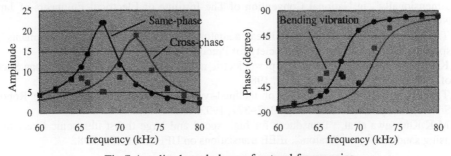

Fig.7 Amplitude and phase of natural frequencies

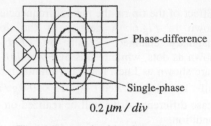

0.2 μm / div

Fig.8 Measured orbit of the Tip

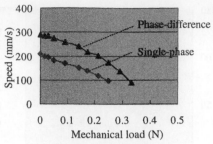

Fig.9 Measured output

6.Experimental studies

The orbits of the tip are observed through laser measurement. Piezoelectric elements are made from a stack of thin ceramics plates. These are polarized along their thickness. The tip and the base are made from metals. These members are attached with adhesives made of epoxy resin. Mass of the tip is controlled by its thickness.

At the phase-difference drive mass of the tip is controlled so that the natural vibration modes will be degenerated. Both piezoelectric elements are driven by two sine-wave signals with a phase difference of 90 degrees. At the single-phase drive mass of the tip is controlled so that the natural vibration modes will be close. One piezoelectric element is driven by a sine-wave signal.

Observed orbits are shown in *Fig8*. Circular orbits are obtained by both driving methods.

Speed and torque are also measured by experimental studies. The actuator is pressed to the rotor, diameter of *30mm*, at the force of *1.5N*. Mass of the tips is controlled again because natural frequencies change by the pressure. Measured output is shown in *Fig.9*, where both driving voltages are *0-7* Volts.

7.Conclusion

In kinetic models natural vibration modes correspond to spring-mass (-and-beam) systems.

Degeneracy condition of natural vibration modes can be controlled by mass of the tip.

On-degeneracy condition circular orbit can be obtained by two driving signals with a phase difference of *90* degrees.

Close-degeneracy condition circular orbit can also be obtained by one driving signal.

References

1.A.Endo et al, "Ultrasonic motor with longitudinal piezoelectric elements placed two-dimensionally", in National Convention of The Institute of Electrical Engineers of Japan, No.735, 1988.

2.T.Akuta et al, "Development of linear actuator with stacked piezoelectric elements", in spring symposium of The Japan Society for Precision Engineering, No.F04, 1991.

3.K.Mori, "Research of high torque piezoelectric motor", in symposium of The Japan Society of Mechanical Engineers, No.1015, 1986.

4.T.Ogiso et al, "Dynamic model of inclined-type piezoelectric motor", in The Robotics Society of Japan, Vol.10, No.3, pp.367~377, 1992.

5.M.K.Kurosawa et al, "Transducer for high speed and large thrust ultrasonic linear motor using sandwich-type vibrators", IEEE translations on UFFC, Vol.45, 1998.

Human Friendly Mechatronics (ICMA 2000)
E. Arai, T. Arai and M. Takano (Editors)

149

Study on driving force of vibrational friction motor

S. Kuroda [a], S. Kuramoto [b], K. Taguchi [a], H. Hosaka [a], K.Kakuta [b], K. Itao [a]

[a] Graduate School of Frontier Science, The University of Tokyo, 7-3-1, Hongo, Bunkyo-ku, Tokyo 113-8656, Japan
[b] Faculty of Science and Engineering, Chuo University, 1-13-27, Kasuga, Bunkyo-ku, Tokyo 112-0003, Japan

The driving characteristics of the vibrational friction motors, hopeful source of power for micro-mechanism, were investigated theoretically and experimentally. First, the effects of the shape and vibration modes of the stator on the driving efficiency were analyzed by calculating the tangential velocity of the stator using FEM without considering the collision between the stator and the rotor. Next, the relationships between the conditions of the collision s urface a nd t he dr iving force w ere measured w ith a n ex perimental a pparatus o f which structural parameters were best tuned according to its dynamic model. With these analyses and experiments, the best conditions for the stator shape, vibration mode, and the collision surface were clarified.

1. Introduction

With mobile communication and computing becoming popular, demands for the miniaturization and higher performance of information equipments are growing. The vibrational friction motor has a high potential as an actuator for micro-sized memories and output devises since it has a simple structure and uses friction force that is a dominant force in micro domain. The prototype motor has been manufactured [1], but the method of design and estimation has not been established yet because the driving mechanism of the motor is complicated due to the sequence motion of vibration, collision, and friction.

Figure 1 shows the structure of the vibrational friction motor. The stator consists of oscillators with PZT film. Figure 2 shows the process of the driving force being generated. The driving mechanism is such that the stator is excited at the resonant frequency by PZT, and the stator t op collides with the rotor obliquely, and the

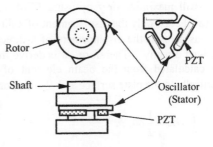

Figure 1. Vibrational friction motor[1]

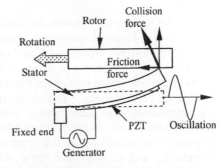

Figure 2. Principle of driving force

frictional force generates the tangential force to the rotor surface. The driving force of this motor depends on both the vibrational condition of the stator and the contact condition between the stator and the rotor. As the first step, the driving force is estimated qualitatively by analyzing only the vibration of stator by FEM. Next, the instrument to measure colliding force is fabricated and the influence of the coefficient of restitution and the coefficient of friction is clarified experimentally.

2. Simulation of driving efficiency by the vibration analysis of the stator
2.1. Simulation method of the driving force

First, only the vibration behavior of the stator is focused. And the driving efficiency of the vibrational friction is evaluated without considering the collision between the rotor and the stator. One oscillator of the stator is considered. The conceptual figure of the oscillator regarded as the cantilever is shown in Figure 3. In this figure, the longitudinal direction of the cantilever defines the x-axis, and the transverse direction defines the z-axis.

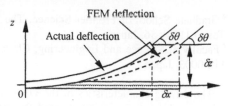

Figure 3. Calculated deflection by linear analysis

If the cantilever oscillates, the x value of the top moves δx. The product of this δx and the oscillating frequency f gives us the rotor rotation velocity $\delta x \cdot f$. The parameter $\delta x \cdot f$ does not strictly agree with the rotation velocity because the top of the cantilever can be separated from and slipped on the rotor. However, $\delta x \cdot f$ and the rotation velocity will be considered to be correlated. We will calculate the vibration shape that will maximize $\delta x \cdot f$ by using FEM. However, because FEM is usually based on the linear theory, only the displacement of z-direction δz is calculated in the cantilever oscillation and as a result the displacement of x-direction is calculated to be zero. Thus, δx cannot be derived directly. Therefore, δx is derive from both $\delta \theta$ and δz since the angle of node $\delta \theta$ can be calculated from the beam element of FEM. In the linear theory, each node moves vertically to the plate surface and θ is proportional to z. Displacement δx is thus given by

$$\delta x = \int_0^{\delta z} \theta dz = \frac{1}{2} \delta \theta \cdot \delta z \qquad (1)$$

Multiplying this δx with the oscillating frequency f gives us $\delta x \cdot f$.

Next the relationship between the stator shape and $\delta x \cdot f$ is studied. Figure 4 shows the shape of the stator adopted in FEM and experimental analyses. This is 30 times the size of the motor of reference (1). One third of the shape (the shaded portion) of the stator oscillator is used in the FEM analysis. To simplify the analysis and the experiment, the stator shape changed only by changing the length a of the cutout part (dotted area). The restriction conditions are that support ends are fixed completely, and all the other ends are fixed in the

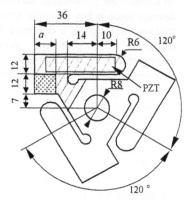

Figure 4. Geometry of stator

translation in x-axis, y-axis and the rotation in z-axis.

2.2. Experimental apparatus

In order to verify the result of the simulation, the measurement apparatus shown in Figure 5 is fabricated. The motor of reference (1) is oscillated by the PZT adhered on the oscillator, but the measurement apparatus is oscillated by the vibrator fixed on the support end of the stator to simplify the apparatus structure. The rotating velocity is measured by measuring the passage of protuberance as a pulse from the laser displacement sensor during the revolution. Eight pieces are attached on the rotor, and a compact disk is used. The material of the stator oscillator is made of acrylic resin plate, and the thickness is 1.5mm.

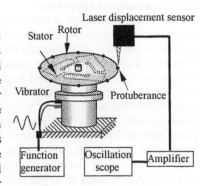

Figure 5. Instrument of driving efficiency

2.3. Simulation and experimental results

For the model that the stator is oscillated at the fixed end, a is varied from 0mm to 23mm. Relationships between a and $\delta x \cdot f$, at the 1st, 2nd and 3rd vibrational modes are shown in Figure 6. In this case, the oscillating amplitude is constant. The parameter $\delta x \cdot f$ is at its maximum value when a=23mm oscillating at the 1st mode. The experimental values for a=0mm, 6mm and 11mm are shown with ◆ □ ▲. The experimental values are in agreement with the calculated values qualitatively, that is, the highest velocity is obtained at the 1st mode for a=0mm, at the 2nd modes for a=6mm, and at the 1st and the 2nd modes for a=11mm. Next, the case of exciting by PZT is investigated. The PZT film generates a distributed bending moment. This is equivalent to a system in which the concentrated moment is applied to both ends of PZT area. Figure 7 shows the relationship between a and $\delta x \cdot f$. In this case, the 3rd mode at a=16mm has the optimized shape. The actual motor is also observed to rotate fastest in the 3rd mode. As a result, this calculated value is in agreement qualitatively with the actual rotating velocity.

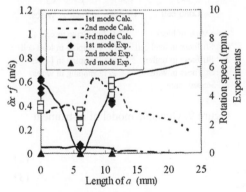

Figure 6. Relationship between a and $\delta x \cdot f$ (*Oscillated at fixed end*)

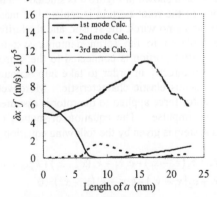

Figure 7 Relationship between a and $\delta x \cdot f$ (*Oscillated by distributed moment*)

3. Measurement of impulsive force
3.1. Experimental apparatus

It is not possible to estimate quantitatively the force generated by collision from calculations only because it is affected by the restitution and the friction in the contact section. Therefore, in order to investigate these phenomena with a macro model, the experimental apparatus shown in Figure 8 is manufactured. The stator portion is treated as a cantilever. The instruments are constructed so that the base of the stator is oscillated, and the top collides with the rotor. As the rotor, a light block is used. This block is supported with piano wires and it collides to a load cell. A thin buffer plate is put between the block and the load cell to avoid double collision. The cantilever is made of acrylic resin plate, and the dimensions are 100mm × 10mm × 1mm. Also, the block is made of aluminum, the diameter of the supporting piano wire is 0.1mm, and 100mm in length. The buffer is made of India rubber. This structure makes it possible to measure the driving force of the rotor without loss because of high stiffness in the vertical direction and free movement in the horizontal direction.

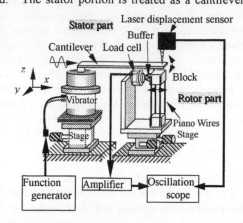

Figure 8. Experimental apparatus for measuring colliding force

3.2. Optimization of instrument

In the measurement of impulse, the measurement accuracy is largely affected by the colliding conditions. Therefore, in order to clarify the optimized measurement condition, the dynamic model of the instrument shown in Figure 9 is studied. The block of the rotor portion is replaced to mass and the piano wires, the sensor, and the buffer are replaced to the spring-damping system. The buffer portion is connected to the sensor portion directly in order to take into account the viscous elastic characteristic. Moreover, the stator force applied to the rotor is assumed as the impulse. The equation of motion of this system is given by the following equation.

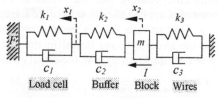

m : Mass of block I : Impulsive force
k_1 : Stiffness in load cell c_1 : Damping in load cell
k_2 : Stiffness in buffer c_2 : Damping in buffer
k_3 : Stiffness in wires c_3 : Damping in wires
F : Sensed force

Figure 9. Dynamic model of instrument

$$m\ddot{x}_2 + c_2(\dot{x}_2 - \dot{x}_1) + c_3\dot{x}_2 + k_2(x_2 - x_1) + k_3 x_2 = 0 \tag{2}$$

$$c_1\dot{x}_1 + c_2(\dot{x}_1 - \dot{x}_2) + k_1 x_1 + k_2(x_1 - x_2) = 0 \tag{3}$$

x_1 is solved under the initial conditions $x_1(0) = 0$, $\dot{x}_1(0) = 0$, $x_2(0) = 0$, $\dot{x}_2(0) = I / m$

The sensor output F is given by Eq.(4), the amount of impulse Ft is given by Eq. (5)

$$F = c_1\dot{x}_1 + k_1x_1 \qquad (4) \qquad Ft = \int_0^\infty F(t)\,dt \qquad (5)$$

In this case, because the load cell have high stiffness, the buffer have high damping, the piano wires have low damping and low stiffness, we calculate under the conditions $k_1 > k_2 > k_3$, $c_2 > c_1$, c_3.

The damping ratio of the buffer portion is expressed as

$$\zeta_2 = \frac{c_2}{2\sqrt{mk_2}} \qquad (6)$$

Since the buffer characteristics affects largely to the impulse force, the relationship between ζ_2 and the sensor output F when the values of c_2, k_2 are changed is calculated, results are shown in Figure 10. In the calculation, each parameters were fixed at non-dimensional value as $m=1$, $k_1=1000$, $k_3=0.01$, $c_1=0.001$, $c_3=0.001$.

The actual measured output from the load cell is only the positive side and 1 cycle (in case of $\zeta_2=0.03$, shown with shaded area) of Figure 10. In case of $\zeta_2 < 1$ ($c_2=0.2$, $k_2=10$), the measured impulse is, at most, twice larger than the actual amount as shown by A in Figure 10. On the other hand, in case of $\zeta_2 > 1$, the amount is theoretically estimated correctly. But, in this instrument, we understood that as the peak level of the load cell output becomes larger, the measurement accuracy improves because the impulse force is almost equivalent to the minimum resolution of the sensor. The peak level is enlarged when c_2 and k_2 are increased. For example, in the case of $c_2=2$, $k_2=1$, $\zeta_2=1$, the sensor output is as shown by B in Figure 10, in the case of $c_2=4$, $k_2=4$, $\zeta_2=1$, the maximum output is improved as shown by C in Figure 10. In the actual experiment, by selecting larger damping ratio and thinner buffer material, c_2, k_2 become larger. But if the buffer is too thin, stiffness and damping ratio of load cell and piano wires etc. would become apparent and the damping ratio ζ_2 becomes smaller than one. So we adopted the buffer about 1mm thick in this experiment.

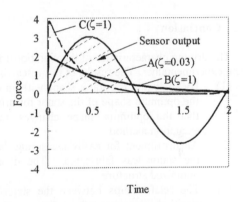

Figure 10. Simulation of sensor output

3.3. Experimental results

By using the instrument of collision force mentioned above, the relationship between the stator/rotor interface and the driving efficiency was investigated. The coefficient of restitution and the coefficient of static friction were examined. In order to change the coefficient of restitution and the coefficient of static friction, 8 different materials of 1mm thick were used and attached on the collision surface on the block of the rotor portion.

First, the relationships between the coefficient of restitution (between the stator and the

154

rotor) e and the impulse were measured. The results are shown in Figure 11. The coefficient of restitution correlates positively with the driving force. This is because the change of momentum in cantilever increases as e increases.

Next, the relationship between the coefficient of static friction (between the stator and the rotor) μ_0 and the impulse is shown in Figure 12. The coefficient of static friction does not correlate with the driving force. This will be because the generated friction force is less than 'static frictional coefficient $\times$ vertical force' since the vertical force is large in colliding phenomena.

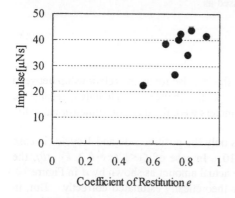

Figure 11. Relationship between coefficient of restitution and impulse

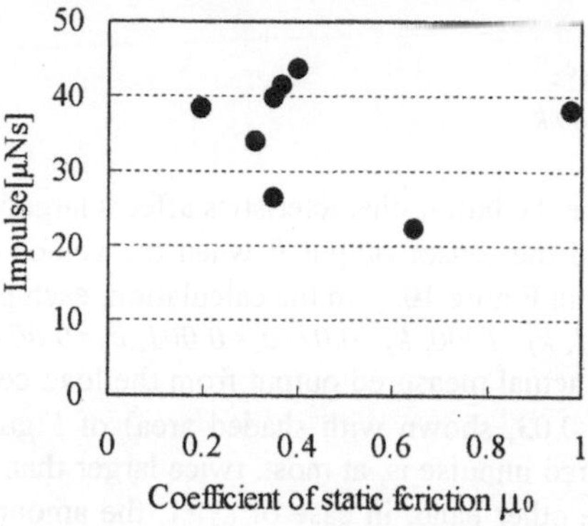

Figure 12. Relationship between coefficient of static friction and impulse

4. Conclusion

The driving characteristics of the vibrational friction motors were investigated theoretically and experimentally. The results of this research are summarized as follows:

(1) By analyzing the stator shape without the collision between the rotor and the stator, the optimum shape of the stator for driving efficiency is indicated. Also it is found that the optimum shape changes according to the vibrational modes and the excitation method.

(2) An instrument for easily measuring the impulse of collision between the rotor and the stator was fabricated. The dynamic model of the instrument lead to the optimized structure.

(3) The relationships between the stator/rotor interface and the driving force were clarified experimentally. By selecting a moderate frictional coefficient and a greater coefficient of restitution between the members, the driving efficiency is improved.

Reference

1. H. Maeda, K. Tani, M. Suzuki, Y. Suzuki, Y. Sakuhara. 'Development of piezoelectric micromotor.' *Micromechatronics (The Horological Institute of Japan)* Vol. 42 No. 4 (1998) : 31-39

Human Friendly Mechatronics (ICMA 2000)
E. Arai, T. Arai and M. Takano *(Editors)*
© 2001 Elsevier Science B.V. All rights reserved.

Design of Ultrasonic Motor for Driving Spherical Surface

Shingo Shimoda[*1], Masayuki Ueyama[*2], Shinya Matsuda[*2], Takashi Matsuo[*2], Ken Sasaki[*1], and Kiyoshi Itao[*1] [a]

[a*1] Dept. of Environmental Studies, University of Tokyo
7-3-1 Hongo Bunkyo-ku Tokyo, 113-8656 Japan
[*2] Minolta Co.,Ltd Takatsuki laboratory

Design of an ultrasonic motor that can drive a spherical surface is presented. The drive unit consists of four piezoelectric stack actuators arranged in a pyramid shape. Circular vibration at the apex generates a thrust when a rotor is pressed against the apex. The direction of the thrust within the tangent plane of the rotor is controlled by the amplitude and phase differences among the four actuators. This enables a spherical drive if a spherical surfaced rotor is used. Theoretical analysis and design method of this actuator is presented. Experimental results have shown that the direction of the thrust was controllable as predicted by the analysis.

1. Introduction

There is a great demand for small and versatile actuators as the size of portable devices become smaller and smaller[1][2][3][4]. Ordinary electromagnetic motors, however, are inefficient when their sizes become smaller than 1 to 2 cm. On the other hand, ultrasonic motors are more efficient in this realm. Ultrasonic motor converts micro mechanical vibration into gross thrust. Usually, the direction of the thrust is fixed.

At Minolta Co.,Ltd Takatsuki Laboratory, an ultrasonic motor based on a new driving method has been developed. This actuator uses two piezoelectric linear stack actuators arranged in a triangular shape; one side is the base and other two sides are the actuators. The circular vibration at the apex of the triangle generates thrust when a rotor is pressed against the vibrating apex. This paper presents a new ultrasonic motor which uses two sets of above mentioned actuator in order to control the direction of the thrust within the tangential plane at the contact point of the actuator and the rotor. If we use a spherical rotor, the rotor can be positioned at arbitrary orientation.

2. The driving method

A wireframe model of the proposed actuator is shown in Fig.1. Four linear electrostatic stack actuators are arranged in a pyramid-like shape. This configuration is a result of arranging two one-degree-of-freedom actuators at 90 degrees angle between them.

One-degree-of-freedom actuator consists of two linear piezoelectric actuators arranged in a triangular shape as shown in Fig.2. The ends of the actuators are connected by a fan-shaped part, and other two ends are fixed on the base. The origin of the stationary coordinate system x-y coincides with the apex. The apex angle is given by θ.

156

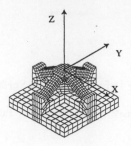

Fig. 1. Model of a actuator

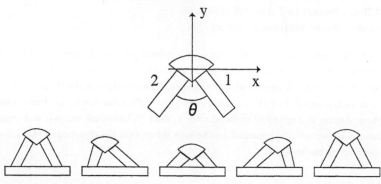

Fig. 2. 2D theory of moving the actuator

When actuator 1 (right side) is driven at its resonant frequency ω, the locus of the tip is given by

$$(x,y) = (-a\sin\frac{\theta}{2}\sin\omega t, a\cos\frac{\theta}{2}\sin\omega t) \tag{1}$$

Where 'a' is the amplitude determined by the voltage applied to the actuator. When the second actuator is also driven at the same resonant frequency but with phase difference of α, the locus of the chip will be the sum of the two vibrations[5].

$$
\begin{aligned}
(x,y) &= (-a\sin\frac{\theta}{2}\sin\omega t + a\sin\frac{\theta}{2}\sin(\omega t + \alpha), \\
&\quad a\cos\frac{\theta}{2}\sin\omega t + a\cos\frac{\theta}{2}\sin(\omega t + \alpha) \\
&= (a\sin\frac{\theta}{2}\sin\frac{\alpha}{2}\cos(\omega t + \frac{\alpha}{2}), \\
&\quad a\cos\frac{\theta}{2}\cos\frac{\alpha}{2}\sin(\omega t + \frac{\alpha}{2}))
\end{aligned}
\tag{2}
$$

This locus becomes an ellipse when $\sin\frac{\theta}{2} \neq 0, \cos\frac{\theta}{2} \neq 0, \sin\frac{\alpha}{2} \neq 0, \cos\frac{\alpha}{2} \neq 0$. Furthermore, the locus becomes a circle if the following conditions are satisfied.

$$
\begin{aligned}
\sin\frac{\theta}{2}\sin\frac{\alpha}{2} - \cos\frac{\theta}{2}\cos\frac{\alpha}{2} &= 0 \\
\cos(\frac{\theta}{2} + \frac{\theta}{2}) &= 0 \\
0 \leq \theta \leq 180, 0 \leq \alpha \leq 180 & \\
\frac{\theta + \alpha}{2} &= 90
\end{aligned}
$$

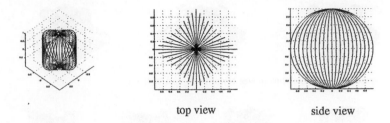

top view   side view

Fig. 3. 3D orbit model

$$\theta + \alpha \quad = \quad 180 \tag{3}$$

Now, two sets of these triangular actuators are joined at the tip with an angle of 90 degrees as shown in Fig.1. Each actuator is driven by the previously described method.

Let "a" and "b" be the two amplitudes of the actuators, then the locus of the chip is given by

$$
\begin{aligned}
(x, y, z) \quad = \quad & (2a \sin \frac{\theta}{2} \sin \frac{\alpha}{2} cos(\omega t + \alpha), \\
& 2b \sin \frac{\theta}{2} \sin \frac{\alpha}{2} cos(\omega t + \alpha), \\
& 2 \cos \frac{\theta}{2} \cos \frac{\alpha}{2} (a + b) sin(\omega t + \alpha))
\end{aligned}
\tag{4}
$$

Further, we assume that

$$\theta = cosnt$$

$$\alpha = cosnt$$

$$a + b = const$$

The locus of the chip is an ellipse or a circle within a vertical plane. The ratio between the amplitudes "a" and "b" determines the direction of this plane. The direction also corresponds to the direction of the thrust generated at the chip. If a sphere is used as a rotor, the sphere can be positioned at arbitrary orientation.

3. Simuration

The actual structure of the actuator has higher order resonant modes besides the dominant modes. We have analyzed the structure by Finite Element Method (FEM) in order to obtain the optimum structure for the spherical actuator. The following points were considered in the analysis.

- There are two dominant vibration modes. 1) Vertical vibration mode: the two piezoelectric actuators stretch and shrink in the same phase, and the tip vibrates vertically(Fig.4). 2) Horizontal mode: the two actuators vibrate in opposite phase, and the tip vibrates in horizontal direction(Fig.5). The resonant frequencies of these two dominant modes should be equal in order to obtain an elliptical or a circular locus. The two frequencies were adjusted by the amount of mass added on the tip.

Fig. 4. The resonace mode the chip moves toward top and bottom

Fig. 5. The resonace mode the chip moves toward right and left

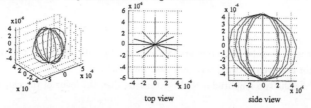

top view side view

Fig. 6. The chip locus in the simuration

- In Fig.4Fig.5, in comparison with the amplitude of the top and bottom mode, the amplitude of the right and left mode is approximately 1/5. Therefore, we decided the angle of two piezoelectric devices $\theta = 120$ degree, to sift the amplitude to the top and bottom mode Eq.4.

- The Eq.4 assumes that the vibrations of the two perpendicularly arranged actuators are independent. The thickness of the base was increased to prohibit the propagation of vibration through the base.

Fig.6 is the locus of the chip, when this actuator is vibrated with the resonance frequency. The chip moves on the circular orbit which radius is $0.5\mu m$ when the phase different α of the piezoelectric devices that face each other is 122 degree.

4. Experiments

4.1. Specification of the prototype

We made the prototype of the actuator based on the simulation as shown in Fig.7. TThe specification of the prototype is as follows:

- Main body size: $15.2 \times 15.2 \times 9.5$ mm

- Piezoelectric actuator size: $5.0 \times 1.5 \times 1.5$ mm

- Apex angle of the chip θ : 120 °

- Mass: 23g

- resonant frequency: 33.5kHz

Fig. 7. The prototype of the actuator

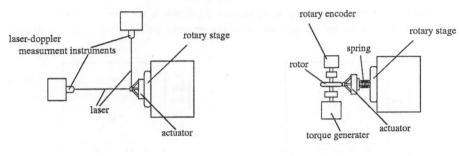

device of experiment 1   device of experiment 2

Fig. 8. Devices of experiments

4.2. Experiment 1

The first experiment was conducted to verify that the direction of the thrust can be controlled by the ratio of the two drive signal amplitudes. Two laser-doppler vibrometers were arranged in a way so that the lines of measurement intersected perpendicularly in the tip as shown in Fig.8. First, the amplitudes for driving the two sets of actuators were set at 40[V] and 0[V]. The phase difference between the two piezoelectric actuators was varied from 0 to 360 degrees with 90 degrees interval. The locus changed from 'vertically thin ellipse', 'round ellipse', 'horizontally thin ellipse', and to 'round ellipse'. This sequence is the same with the sequence obtained for one actuator. Next, the rotary stage was turned 45 degrees to change the orientation of the actuator. The amplitudes were set at 20[V] and 20[V]. The phase difference was varied in the same way. The change of the locus was the same as shown in Fig.9. This means that the direction of the thrust can be controlled by the ratio of the drive signals.

4.3. Experiment 2

Second experiment was conducted to evaluate the thrust. A disk rotor was pressed against the tip of the actuator. The torque generator adjusted the load, and the rotary encoder measured the angular velocity. The left graph in Fig.10 shows the load-velocity characteristics when drive amplitudes of 40[V] and 0[V] were applied, and the right graph shows the result for 20[V] and 20[V]. Both results show similar characteristics.

5. Conclusion

A new ultrasonic drive mechanism consisting of four piezoelectric actuators arranged in a pyramid-like shape has been developed. The apex of the pyramid vibrates in circular locus. This vibration generated

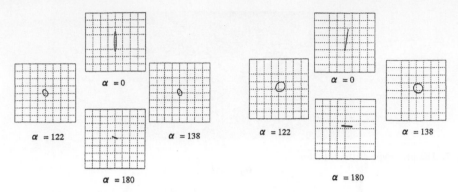

Fig. 9. The locus of the actuator chip at experiment 1(Left side is the case that the voltage of the piezoelectric devices are 20v and 0v, right side is the case that the voltage of the all electric devices are 20v)

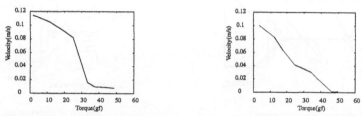

Fig. 10. The the characteristic of the actuator(Left side is the case that the voltage of the piezoelectric devices are 20v and 0v, right side is the case that the voltage of the all electric devices are 20v)

a thrust when a rotor was pressed against the tip of the actuator. Theoretical and FEM analyses have shown that the resonant frequencies of vertical and horizontal vibration modes should be the same in order to obtain a circular locus. This was achieved by adjusting the mass at the tip. Experiments have verified that the direction of the thrust was controllable by the ratio between the two drive signals, as predicted by the theoretical analysis. If a sphere is used for a rotor, it can be positioned at arbitrary orientation by controlling the direction of the thrust. The diameter of the circular vibration at the tip was approximately 1um at resonant frequency of 33.5kHz. The maximum speed at no-load was 0.10 to 0.12 [m/s], and the stall thrust was approximately 45 [gf].

REFERENCES

1. Keisuke SASAE, Kiyoshi IOI, Yasuo OHYSUKI, and Yoshimitsu KUROSAKI. Development of a small actuator with three degrees of rotational freedom. *Journal of the Japan Society for Precision Engineering*, Vol. 61, No. 3, 1995.
2. Kiyoshi ITAO. Infomatic micro system. *Asakura publisher*, 1998.
3. Yasuhiro OKAMOTO and Ryuichi Yoshida. Development of linear actuators using piezoelectric elements. *Electric and Communications in Japan, Part 3*, Vol. 81, No. 11, 1998.
4. Ryuichi YOSHIDA, Yasuhiro OKAMOTA, Toshiro HIGUCHI, and Akira HAMAMATSU. Development of smooth impact drive mechanism (sidm)-proposal of driving mechanism and basic performance-. *Journal of the Japan Society for Precision Engineering*, Vol. 65, No. 1, 1999.
5. Kenji UCHINO. Piezoelectric actuator. *Morikita publisher*, 1986.

Human Friendly Mechatronics (ICMA 2000)
E. Arai, T. Arai and M. Takano (Editors)
© 2001 Elsevier Science B.V. All rights reserved.

161

Development of Master-Slave System for Magnetic Levitation

Naoaki Tsuda, Norihiko Kato, Yoshihiko Nomura and Hirokazu Matsui
Dept. of Mechanical Eng., Faculty of Eng., Mie University, Tsu City, Mie Pref., Japan

Abstract Precise works and manipulating micro objects are afflictive for operators both mentally and physically. To execute these jobs smoothly without feeling wrongness, use of master-slave systems is preferable because position and force are able to be scaled up and down under the systems. In our study we develop a master-slave system whose slave robot is micro and levitated by magnetic forces. The distinction of the levitated robot is that it does not get any other contact forces from outside. Thus we introduce a method using an impedance model for constructing the master-slave system. We confirmed the effectiveness of the positioning control algorithm through experiments.

1. INTRODUCTION

In this study, we develop a kinematically dissimilar master-slave system; as for a slave robot we utilize a magnetically levitated micro robot(Fig.1), and as for a master robot an X-Y-Z Cartesian manipulator(Fig.2). It is desirable that micro components would be manipulated by a micro robot[1] which is not contact with its surrounding, for example using magnetic forces. Since the micro robot is levitated by magnetic forces in stable condition, it does not get any other contact forces from outside. Therefore, we cannot apply conventional force-feed back master-slave systems[2][3] for this system. Thus, we propose a method using an impedance model. Using the model, we assume the dynamics of the slave robot, and we calculate an external contact force that would work on the micro robot when contacting an object. We confirm the effectiveness of the positioning control algorithm through practical experiments as the first step of development of the master-slave system.

2. EQUIPMENT

As an interface to an operator, a force-torque sensor is attached to the z, i.e., vertical

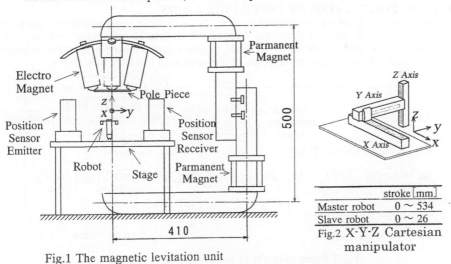

	stroke [mm]
Master robot	0 ～ 534
Slave robot	0 ～ 26

Fig.2 X-Y-Z Cartesian manipulator

Fig.1 The magnetic levitation unit

axis module of the master robot. In our paper, we use position and force data in the vertical direction alone. The magnetic levitation unit, that is, the slave side unit is constructed from a magnetic drive unit, a micro robot, and position sensors. By controlling the current of the electric circuit of each electro magnet, we can stably levitate the robot and move it in three-dimensional space. Levitated micro robot weighs 8.1[g] in total and it can hold up an object with mass of 1[g]. This robot can locate in a rectangular parallelepiped region of 29×29×26[mm]. Because of the characteristic of magnetic force, it is very difficult to levitate the robot stably. We control the position of the slave robot by using PID control system with feed-forward to compensate the nonlinearity with distance.

3. SYSTEM DESIGN OF THE MASTER-SLAVE SYSTEM

Fig.3 shows the master-slave system that we designed. Our master-slave system is classified into impedance control type[4,5], and this type is designed so that both the master and the slave robot have the same impedance characteristics. In the Fig.3, X_m and X_s are the positions of the master and the slave robot respectively. F_m is the force affected by an operator to the master robot and F_{vs} is the calculated external contact force that would work on the slave robot.

As above-mentioned, the slave robot of this system does not get any external contact forces from outside except magnetic forces in stable condition. The reason is that by the computer control magnetic force balances with gravity in vertical axis and all forces similarly balance each other in horizontal axes. Thus we calculate the virtual external contact force(F_{vs}) by assuming a dynamics of the slave robot as:

Slave Impedance Model : $\quad F_{vs}=(M_{vs}s^2+B_{vs}s+K_{vs})\ X_s$ (1)

Here M_{vs}, B_{vs}, and K_{vs} are the virtual mass, virtual viscosity, and virtual spring constant respectively. X_d is calculated by the impedance model from F_d $(=F_m-F_{vs})$ expressed as:

System Impedance Model : $\quad X_d=\dfrac{1}{Ms^2+Bs+K}(F_m-F_{vs})$ (2)

Here M, B, and K are the mass, viscosity, and spring constant respectively that intervene between the master and the slave robot. Configuring M, B, and K to be adequate, X_d converges at a constant value. In addition, K_m and K_s are proportional gains that govern the scale of position between the master and the slave robot.

4. METHOD FOR SETTING OF IMPEDANCE GAINS

In order to set gains of our control system, the following steps should be taken. At first we derive several transfer functions. Second we determine the maneuverability, i.e., how distant the operator wants the slave robot to move when he/she affects the interface with his/her force. Last we assign the poles of the system in the light of stability.

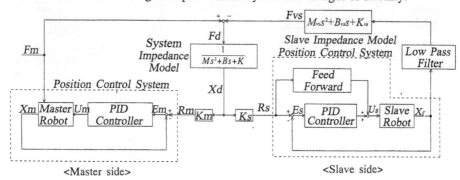

<Master side>   <Slave side>

Fig.3 Block diagram of the master-slave system

First we derive transfer functions from F_m to F_d, X_m, X_s, and F_{vs}. For simplification, we define following four expressions.

System Impedance Model $\quad : G_1(s)=\dfrac{1}{Ms^2+Bs+K}$ (3)

Slave Impedance Model $\quad : G_2(s)=M_{vs}s^2+B_{vs}s+K_{vs}$ (4)

Dynamic model of the master robot $\quad : G_3(s)=\dfrac{A_m}{(T_{m1}s+1)(T_{m2}s+1)}$ (5)

Dynamic model of the slave robot $\quad : G_4(s)=\dfrac{A_s}{(T_{s1}s+1)(T_{s2}s+1)}$ (6)

Then for example we can get the following equations.

$$\frac{X_m}{F_m}=\frac{K_mG_1G_3}{1+G_1G_2G_4K_s}$$ (7)

$$\frac{X_s}{F_m}=\frac{K_sG_1G_4}{1+G_1G_2G_4K_s}$$ (8)

After substituting Eqs.(3)-(6) into Eqs.(7) and (8), we apply the final value theorem for a step input to these equations. For example,

$$\lim_{s\to0} s\cdot\frac{X_s}{F_m}\cdot\frac{1}{s}=\frac{A_s\cdot K_s}{K+A_s\cdot K_{vs}\cdot K_s}$$ (9)

If we ignore the dynamics of the master and the slave robot, the characteristic equation comes to be

$$s^2+2\zeta\omega_n s+\omega_n{}^2=0 \quad where \ 2\zeta\omega_n=\frac{T_{s1}+T_{s2}}{T_{s1}T_{s2}} \ , \ \omega_n{}^2=\frac{K+A_sK_{vs}K_s}{KT_{s1}T_{s2}}$$ (10)

For the desirable maneuverability, if we fix the value of Eq.(9), damping ratio, i.e., ζ in Eq.(10), and gains of position scale(K_m,K_s), we can get K and K_{vs}. In Eqs.(7) and (8), the characteristic equation comes to be

$$D(s)=s^4+\frac{M(T_{s1}+T_{s2})+BT_{s1}T_{s2}}{MT_{s1}T_{s2}}s^3+\frac{M+B(T_{s1}+T_{s2})+KT_{s1}T_{s2}+A_sK_sM_{vs}}{MT_{s1}T_{s2}}s^2+\frac{B+K(T_{s1}+T_{s2})+A_sK_sB_{vs}}{MT_{s1}T_{s2}}s+\frac{K+A_sK_sK_{vs}}{MT_{s1}T_{s2}}=0$$ (11)

Thus we can calculate the other gains(M, B, M_{vs}, B_{vs}) by assigning four poles properly.

5. SETTING OF IMPEDANCE GAINS

We have known from an experiment for identification that the dynamics of the master and the slave robot are described as:

$$G_3(s)=\frac{35000}{s^2+359s+35000}$$ (5)'

$$G_4(s)=\frac{7000}{s^2+350s+7000}$$ (6)'

We configure the specification of the system as:

$X_m/X_s=10$ (12)

$X_s/F_m=2.0\times10^{-3}$[m/N] (13)

Defining $K_m=10$, $K_s=1$, and damping ratio, i.e., ζ in Eq.(10) to be 1, we know K comes to be 114.0 and K_{vs} to be 386.0.

Next we adjust other gains(M, B, M_{vs}, B_{vs}) by the pole assignment. We assign poles intuitively and get actual impedance gains. Using those gains, we observe the behavior of the master and the slave robot when a step load is affected to the operator interface(F_m=200[gf]). Considering the result, we repeat tuning. We started assigning four poles at -15, -120, -225, -330 at even intervals, and at last we decided the best assigning as expressed in Table.1. At that time the impedance gains of the two impedance models are expressed in Table.2 and 3. Fig.4(a) shows the transitions of positions and (b) expresses forces of each robot. We can see that in about 200 [msec] the slave robot settles down at 4[mm] and it keeps its position after that as we designed in Eqs.(12) and (13). The steady point of the master robot is ten times as far as that of the slave robot. Virtual external contact force at the slave robot also settles down at about 150[gf].

6. CHANGEABILITY OF SCALE BETWEEN MASTER AND SLAVE

We showed symbolically in Eqs.(7) and (8) that the position scale between the master and the slave robot can be configured by gains, i.e., K_m and K_s. Here fixing K_s to 1, we change K_m and verify their effects.

Table.1 Pole Assignment

Pole	-15, -20, -130, -330

Table.2 The best impedance gains of $G_1(s)$

M	B	K	ω_n	ζ	Poles
0.2720	39.4	11 4	20.5	3.54	-142, -2.95

$\omega_n = \sqrt{K/M}$: natural frequency, $\zeta = \dfrac{B}{2\sqrt{MK}}$: damping ratio

Table.3 The best impedance gains of $G_2(s)$

M_{vs}	B_{vs}	K_{vs}	ω_{nvs}	ζ_{vs}	Poles
0.0439	18.6	38 6	93.7	2.20	-401, -21.9

$\omega_{nvs} = \sqrt{K_{vs}/M_{vs}}$: natural frequency, $\zeta_{vs} = \dfrac{B_{vs}}{2\sqrt{M_{vs}K_{vs}}}$: damping ratio

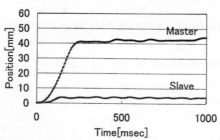

Fig.4 (a) Behavior of Master and Slave robot

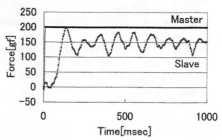

Fig.4 (b) Forces that work on Master and Slave robot

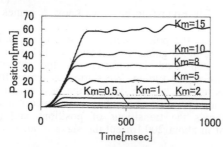

Fig.5(a) Behavior of Master robot

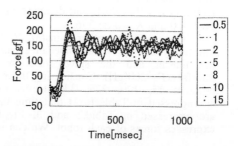

Fig.5(b) Force that works on Slave robot

Fig.5 is the experimental result when K_m is changed to 15, 8, 5, 2, 1, and 0.5. Whichever K_m is selected, the virtual external contact force at the slave robot doesn't vary and the distance from the original point to the point where the master robot settled down is proportional to K_m. Considering this result, we can conclude that we can use our master-slave system with any desirable scale. In addition, our system can be applied to not only expansive operation but also reductive operation.

7. CONSIDERATION OF SYSTEM IMPEDANCE MODEL

So far we concluded that the best impedance gains are as expressed in Tables.2 and 3. Now we focus on the System Impedance Model and investigate its effects to the whole system from the point of view of the stiffness and the viscosity. Concretely we examined the behavior of the master and the slave robot when a step load(F_m=200[gf]) was affected to the operator interface as before. Here we fixed the gains of Slave Impedance Model to the best values. The conditions we applied are as Table.4. Cases.(1)-(4) are the examinations for investigating the effects of the stiffness and Cases.(5)-(7) are of the viscosity.

We show the behavior of the master robot representatively in Fig.6. As the slave robot behaves alike, we omit showing the result of the slave robot. With respect to the stiffness, it seems that the best natural frequency, i.e., ω_n is about 20.5. In case ω_n is less than the best value, the robots behaves more slowly. Oppositely in case ω_n is greater, the robots behaves more quickly and finally oscillates. With respect to the viscosity, the best damping ratio, i.e., ζ is about 3.54. In case ζ is less, the robots behaves more quickly and finally oscillates. Oppositely in case ζ is greater, the robots behaves more slowly. This characteristic is the same of a popular second order vibration system. Thus we can configure the damping or response characteristic of each robot by tuning the one impedance model, that is, System Impedance Model.

8. MANUAL OPERATION

Here we confirm the ability of our system through manual operation, i.e., we investigate the behavior of both the master and the slave robot when an operator handles grasping the interface. Fig.7 shows the experimental result when K_m=10 and K_s=1.

We easily know that the transition of the slave robot is reduced to 1/10 of that of the master robot. This leads to that the operator can handle the slave robot at his/her will.

Table.4 Conditions of experiments

Case	(1)	(2)	(3)	(4)	(5)	(6)	(7)
M	0.0400	0.120	2.00	28.0	(0.272)	(0.272)	(0.272)
B	15.1	26.2	107	400	100	26.0	14.0
ω_n	53.4	30.8	7.55	2.02	(20.5)	(20.5)	(20.5)
ζ	(3.54)	(3.54)	(3.54)	(3.54)	8.98	2.33	1.26

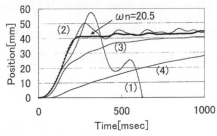

Fig.6(a) Behavior of Master robot about ω_n

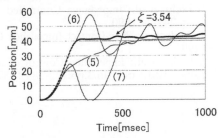

Fig.6(b) Behavior of Master robot about ζ

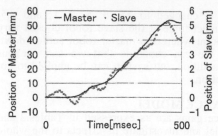

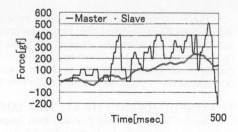

Fig.7 (a) Behavior of Master and Slave robot

Fig.7(b) Forces that work on Master and Slave robot

9. CONCLUSION

In our study we developed a master-slave system whose slave robot does not get any external contact forces. We used the magnetically levitated robot for the slave robot. We confirmed the effectiveness of the positioning control algorithm through practical experiments as the first step of development of the system.

Our results are summarized as:

（1）We proposed a method to calculate the virtual external contact force using an impedance model. By this method we can take a robot which is non-contact with its surroundings for the slave one of the impedance control type master-slave system.

(2) There are two impedance models in our control system. This means that 6 impedance gains should be fixed and there are some other gains that should be set in the system. We showed a guideline for setting them, that is, the gains can be fixed by the maneuverability and assignment of poles. Though there is no way but a trial and error one to decide 6 impedance gains, by this method the number of variables to be chosen can be divided into 2 and 4. And we can tune them as we watch the behavior of robots and presume their response and damping.

(3) We confirmed that the scale between the master space and the slave space could be set optionally. This means that this system is effective for not only an expansive operation, for example micro teleoperation, but also a reductive operation. From another point of view, this system does not need a consideration of the environment of the slave robot for a designer unlike past method.

REFERENCES

1. Tatsuya Nakamura et al.:"A Prototype Mechanism for Three-Dimensional Levitated Movement of a Small Magnet,"IEEE/ASME,Trans.Mechatronics,Vol.2,No.1,MARCH,1997
2. T.Arai et al.:"Bilateral Master-Slave Control for Manipulators with Different Configurations,"JRSJ,Vol.4,No.5,pp.469-479,1986
3. A.Nagai et al.:"On the Remote Mini Manipulator,"SICE,Vol.11,No.1,pp.91-97,1980
4. S.Tachi et al.:"Impedance Controlled Master Slave Manipulation System,"JRSJ,Vol.8, No.3,pp.241-24,1990
5. T.Tsuji et al.:"Non-Contact Impedance Control for Manipulators,"JRSJ,Vol.15,No.4,pp.616-623,1997
6. S.Tachi et al.:"Impedance Control of a Direct Drive Manipulator without Using Force Sensors,"JRSJ,Vol.7,No.3,pp.172-184,1989

Human Friendly Mechatronics (ICMA 2000)
E. Arai, T. Arai and M. Takano *(Editors)*
© 2001 Elsevier Science B.V. All rights reserved.

Precise Positioning of Optical Fibers using Stick-Slip Vibration

Naoto NAGAKI[a], **Tetsuya SUZUKI**[b], **Hiroshi HOSAKA**[a] and **Kiyoshi ITAO**[a]

[a] Department of Environment Studies, The University of Tokyo, 7-3-1 Hongo, Bunkyo-ku, Tokyo 113-8656, Japan

[b] Department of Precision Engineering, School of Engineering, The University of Chuo, 1-13-27 Kasuga, Bunkyo-ku, Tokyo 112-8551, Japan

We propose a method in which micro-optical devices, like optical fibers, which slide on flat surfaces and are largely influenced by frictional force can be positioned by means of open loops. In this positioning method, an optical fiber is vibrated by an actuator. The optical fiber repeats a stick-slip movement and converges on a target position. Then, by attenuating the drive amplitude of the actuator, even when frictional force is indefinite, it is possible to clarify that the optical fiber has been correctly positioned.

1. Introduction

The frictional force of micro-mechanisms sliding on flat surfaces has a greater effect on positioning errors than inertial force does. In order to decrease the positioning errors induced by frictional force in conventional positioning, lubrication for stabilizing friction or feedback control by means of position sensing is commonly performed. However, in the case of micro-optical devices such as optical fibers, such methods often cannot be employed. Then, positioning must be controlled by the open loop. To improve the positioning accuracy of the optical fiber caught in frictional dead zone, authors have, until now, proposed a method by which the optical fiber is vibrated by the actuator. However, when frictional force is indefinite, it is still difficult to perform good positioning. In this study, experiments and simulations are described in detail, and the validity and adaptation range of the above-mentioned method are clarified. Next, we propose the method of positioning optical fibers by changing the vibrating amplitude of the actuator and verify the validity of this method by simulating the optical fibers to the point where friction is indefinite.

2. Positioning Theory

The positioning mechanism of an optical fiber is shown in Fig. 1. An optical fiber is driven in a horizontal direction by an actuator and slides on a glass substrate as the end of the fiber is moved to the target position. The optical fiber bends for the frictional force between the fiber and the substrate, which is why positioning errors arise. Generally, when a minute object slides on a flat surface, a stick slip is generated and the last stop position is changed at random.

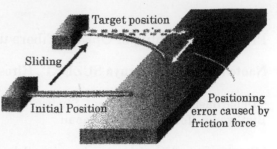

Fig. 1 Sliding positioning of an optical fiber

For this reason, it is impossible to reenact the drive for which the bending by friction was considered. Authors have proposed positioning methods for cases in which the coefficient of static friction is known. As shown in Fig. 2, a cantilever which is caught by friction is moved minutely by sinusoidally vibrating the

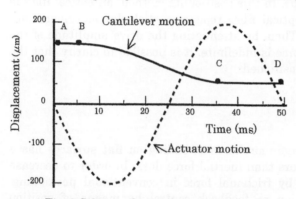

Fig. 2 Schema of precise positioning method

actuator. When an actuator moves in the minus direction, the cantilever first stays at the initial position because of the static frictional force (between A-B, stick state), then begins to slip when the spring-back force overcomes the static frictional force (between B-C). If the speed of the cantilever becomes zero during slip movement, the cantilever will again catch on the surface (between C-D, stick state). Between B-C, the cantilever moves only in the minus direction, and the speed will reach zero after about half of the natural period. So if the cantilever is on the plus side of a target point at this moment, the stop position of the cantilever will approach the target position. Since an actuator continues sinusoidal vibration, the cantilever repeats the cycle of these stick and slip states, and the cantilever approaches the target position gradually.

3. Experimental Apparatus

Experimental equipment is shown in Fig. 3. An optical fiber is attached to the tip of the cantilever actuator with a magnet, and pushed against a glass cylinder. The position where it touches the glass cylinder is the origin. Initial displacement is given to the optical fiber by bending it. When one pulse of sinusoidal voltage is put into a coil, the actuator vibrates continuously, and the tip of the optical fiber repeats a stick-slip movement, and arrives at a position

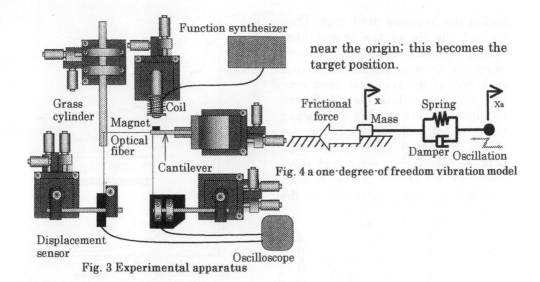

near the origin; this becomes the target position.

Fig. 3 Experimental apparatus

Fig. 4 a one-degree-of freedom vibration model

4. Simulation Method

This positioning method is modeled by a one-degree-of-freedom system with a spring, a mass and a damper (Fig. 4). When this system is forced to vibrate, the movement of the optical fiber is analyzed with the simulation software, *Mathematica*. The actuator is vibrated sinusoidally, and theoretical analysis of the positioning is performed as follows:

○Condition of stick state
$$|X_0 - X_a| < X_s \qquad \cdots\cdots\cdots\cdots\cdots(1)$$
○Driving wave of the actuator
$$X_a = -(F/k)\sin\omega t \qquad \cdots\cdots\cdots\cdots(2)$$
○Equation of motion for slip state
$$m\ddot{x} + c\dot{x} + kx = -F\sin(\omega t + \omega t_1) + F_d \qquad \cdots\cdots(3)$$

X_0 is an arbitrary stop position, X_a is a displacement of the actuator, X_s is a displacement to static friction, F is a drive amplitude, ω is a drive frequency, t_1 is stick time, and F_d is kinetic friction.

5. Positioning in Cases when Frictional Force Is Known

The experimental result (a) and the simulation result (b) at the tip of an optical fiber are shown in Fig. 5, Fig. 6 and Fig. 7. The simulation results and the experiment results are in agreement with each other. That is, the tip of the optical fiber approaches the target position when the drive amplitude of the actuator is equal to half of the frictional dead zone (Fig. 5); the tip of the optical fiber continues past the target position when the drive amplitude is larger than half of the frictional dead zone (Fig. 6); and the tip of the optical fiber doesn't approach the target position when the drive amplitude is smaller than half of the frictional dead zone (Fig. 7), The simulation results for the relationship between the positioning error and the drive amplitude are shown in Fig. 8. The relationships between positioning error and drive frequency are shown in Fig. 9. Both simulations were performed in the

simulation condition

m :	3.3×10^{-7}	(kg)
k :	5.3×10^{-2}	(N/m)
c :	1×10^{-4}	(N·s/m)
Xs :	250	(μm)
f :	64	(Hz)

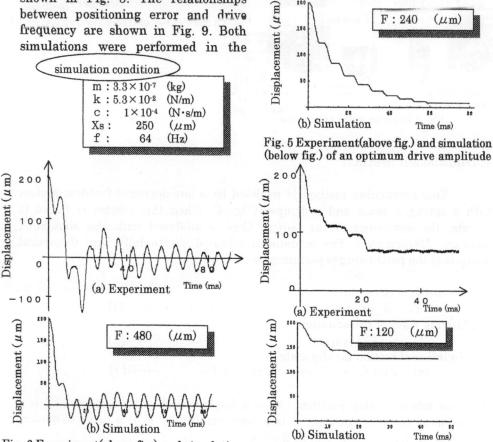

Fig. 5 Experiment(above fig.) and simulation (below fig.) of an optimum drive amplitude

Fig. 6 Experiment(above fig.) and simulation (below fig.) of a large drive amplitude

Fig. 7 Experiment(above fig.) and simulation (below fig.) of a small drive amplitude

conditions in which the width of the frictional dead zone was 500 μm, and the resonant frequency of the optical fiber was 36 Hz. In Fig. 8, the simulation was performed with a drive frequency of 144 Hz. Five points of initial displacement, 0, 50, 100, 150 and 200 μm were adopted and the positioning error for them were plotted in Fig. 8. In this figure, a line connecting two points means that the optical fiber continued to vibrate and wasn't positioned. From this, one can see

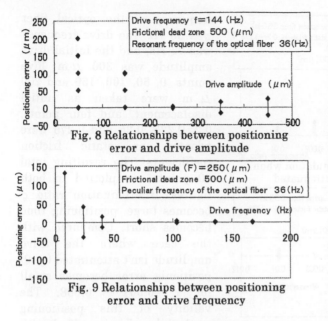

Fig. 8 Relationships between positioning error and drive amplitude

Fig. 9 Relationships between positioning error and drive frequency

that half of the frictional dead zone is at an optimum drive amplitude. In Fig. 9, the simulation at the time of drive frequency change was performed at a drive amplitude of 250 μm. Five points of initial displacement, 0, 50, 100, 150 and 200 μm were adopted and the positioning errors for them were plotted. When the drive frequency was smaller than the resonant frequency of the cantilever, it vibrated greatly, and wasn't positioned, but when the drive frequency was much larger than the resonant frequency the positioning error was small. From this, one can see that good positioning is achieved when the drive frequency is much larger than the resonant frequency.

6. Positioning in Cases when Frictional Force Is Indefinite

In actual positioning systems, the degree of frictional force is unknown and changes from fiber to fiber. In this case, by gradually attenuating the drive amplitude, the optimum amplitude (half of the frictional dead zone) is realized during the attenuation. Then, the second and third formulas used for the simulation; as shown in part 4 of this paper, are changed as follows:

○ Condition of stick state
$$|X_0 - X_a| < X_s \qquad \cdots\cdots\cdots\cdots(1)$$
○ Driving wave of the actuator
$$X_a = -(F/k)e^{-\varsigma \omega t} \sin \omega t \qquad \cdots\cdots\cdots\cdots(2')$$
○ Equation of motion for slip state
$$m\ddot{x} + c\dot{x} + kx = -Fe^{-\varsigma \omega t} \sin(\omega t + \omega t_1) + F_d \qquad \cdots\cdots(3')$$
where, ς is an attenuation ratio of the drive amplitude.

The simulation results for relationships between the positioning time and the attenuation ratio are shown in Fig. 10, and those for the relationships between positioning error and attenuation ratio are shown in Fig. 11. These simulations were performed under the following conditions: the resonant

172

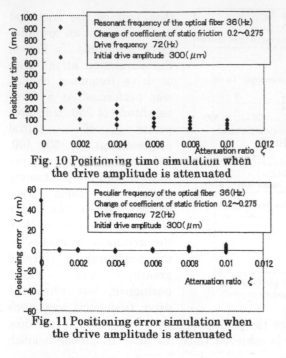

Fig. 10 Positioning time simulation when the drive amplitude is attenuated

Fig. 11 Positioning error simulation when the drive amplitude is attenuated

frequency of the optical fiber was 36 Hz, the drive frequency was 72 Hz, and the initial drive amplitude was 300 μ m. Five points, 0, 50, 100, 150 and 200 μ m were taken as initial displacement, and four points, 0.2, 0.225, 0.25 and 0.275 were adopted as static friction coefficients. The resulting total 20 points are plotted in both figures. If attenuation ratio ς becomes large, positioning time becomes short. Compared with the case where the drive amplitude isn't attenuated (ς =0), positionig error becomes small for ς =0.001 $\sim$ 0.008. The validity of this positioning method has been verified with these simulations.

7. Conclusion

We proposed a method in which micro-optical devices, like optical fibers, which slide on flat surfaces and are largely influenced by frictional force could be positioned by means of open loops. A method for vibrating an optical fiber was proposed, and the optimal conditions for the drive amplitude and frequency in regards to positioning error were clarified through simulation. Results obtained with this resarch are as follows:

(1) Half of the frictional dead zone of the optical fiber is optimum for drive amplitude.

(2) The drive frequency must be larger than the resonant frequency of the optical fiber.

(3) When the frictional force is indefinite, positioning may be performed by gradually attenuating the drive amplitude of an actuator.

Reference

(1) Suzuki, Hosaka and Itao: Theoretical and Experimental Study on Precise Positioning of Sliding Micromechanisms, Journal of the Japan Society for Precision Engineering, Vol.64, No.2, pp.226-230 (1998).

Human Friendly Mechatronics (ICMA 2000)
E. Arai, T. Arai and M. Takano (Editors)
© 2001 Elsevier Science B.V. All rights reserved.

Simulative Experiment on Precise Tracking for High-Density Optical Storage Using a Scanning Near-field Optical Microscopy Tip

T. Hirota[a], Y. Takahashi[b], T. Ohkubo[a], H. Hosaka[a], K. Itao[a], H. Osumi[b],

Y. Mitsuoka[c] and K. Nakajima[c]

[a]Graduate School of Frontier Science, University of Tokyo
7-3-1 Hongo, Bunkyo-ku, Tokyo 113-8656, Japan

[b]Depertment of Precision Mechanics, Chuo University
1-13-27 Kasuga Bunkyo-ku Tokyo, 112-8551, Japan

[c]Seiko Instruments, Inc.
563 Takatsuka-Shinden, Matsudo-shi, Chiba 270-2222, Japan

This paper describes simulative experiment of sub-micron region tracking control for a near-field optical disk storage, utilizing a scanning near-field optical microscopy (SNOM) tip as a highly sensitive head. We introduced a continuous tracking error detection model, and evaluated its equivalent tracking performance experimentally. Sub-sub-micron aperture head's tracking error arising from an electrical and an optical noises could be suppressed to several nanometers assuming that head operating spacing is maintained less than 50 nm and tracking control bandwidth is limited to 5kHz.

1. INTRODUCTION

The recent rapid advance in information technologies keenly demands large-capacity and high-speed access capabilities for digital data storage. It is clear that conventional magnetic or optical storage technology may encounter tough recording density limits caused by super-paramagnetic and/or optical diffraction phenomena. From this perspective, near-field optical recording technology is thought to be one of the most promising ways to overcome these density limitations.

In so many trials to prove high-density recording potential, the first experimental demonstration based on a probe microscopy was performed by Betzig et al.[1] using tapered fiber probe and magnet-optical recording media. Although their experiment was quasi-static one, potential of near-field recording density of a two-digits higher than that of conventional storage was fully demonstrated. Following their study, trials aiming at higher transfer rate under precise head-to-medium spacing control is currently proceeding [2][3]. In regard to this issue, Yoshikawa et al.[3] recently performed near-field signal detection at higher 2-3 m/s relative velocity and up to approximately 10 MHz frequency band adopting an air bearing slider witch mounted sub-micron sized metal planar aperture.

On the other hand, tracking control of ten-nanometer-order accuracy constitutes another challenging subject. For example, assuming 100Gb/in^2 density requires 70nm data mark size and 100nm track pitch; thus, tracking accuracy should be reached to approximately 10nm, which is about 1/6 of required accuracy for today's conventional optical disk storage.

In constructing the tracking control system, it is essential to evaluate the closed loop characteristics of the entire tracking system including sensor head (the position error detector) and tracking actuator. The tracking characteristics using a signal of a near-field head is an unexplored subject that has recently been addressed only using SNOM[4] tip. At this time, studies are focused mainly on the sensitivity of the tracking error signal or influences of other static properties on it. The real time signal, however, contains various kinds of noise arose from the electrical circuitry, laser power source, metal patterned media, or external disturbances. As the result, an evaluation both of the signal-to-noise ratio and total noise power is quite necessary considering the real tracking control band into account.

From this point of view, we performed simulative experiments of a near-field signal based tracking utilizing the SNOM system to evaluate quasi-dynamic characteristics, focusing mainly to the signal-to-noise ratio of the position error signal.

We firstly considered a continuous tracking control model in order to adopt the method of simulative experiment to the real SNOM system. Measured tracking signals were analyzed carefully to evaluate the possibility for realizing several nanometers tracking accuracy with a practical spacing and control bandwidth.

2. MODELING OF THE TRACKING SYSTEM

In regard to tracking error detection methods for near-field optical storage, it is said that the sampled servo method is the most feasible, since other methods for conventional optical storage are based on the principle of optical diffraction or require multi detectors that are not applicable in the near-field optics. In the conventional sampled servo method, as shown in Figure1(a), servo zones are arranged along each track at fixed intervals. Each servo zone consists of displaced pits which are sifted half of the track pitch in the radial direction. When the head

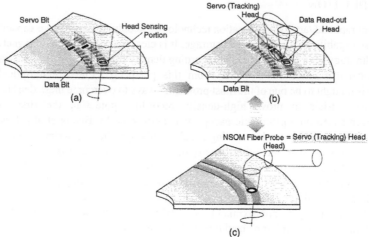

(a)   (b)

(c)

Figure 1. Schematic Diagram of Near-Field Head Sub-Micron Tracking System.

passes over these servo zones, a position error signal (PES) is obtained from the readout intensity of the displaced pit. The bandwidth of PES depends on the interval of the servo zones.

Empirically, it is known that in order to improve tracking accuracy, the control bandwidth should be broadened proportionally. So in cases of very high track density, the number of servo zones per track should be increased, which in turn causes a decrease in data recording efficiency.

In light of these phenomena, we propose a continuous tracking method. As shown in Figure1(b), instead of using displaced pits, a dedicated aperture for tracking is placed on the edge of the data pits. As a result of interference between the near-field and the data pits, scattered light is generated. By detecting this scattered light, we can obtain data readout information as a higher frequency component, and the position error signal as a lower frequency component. This method is based on the uniqueness of near-field optics in that the shape of the sensing spot can be arbitrarily controlled, unlike conventional optics. For tracking error detection, high space resolution of the scanning direction is not needed, so an aperture shape with narrow radial width is adequate. Although the PES obtained using the method described above is also affected by the change of flying height, the signal level of data readout head can be used to estimate the flying height and to compensate for this effect.

As shown in Figure1(c), we proposed a simplified tracking model for our simulative experiment. In this model, instead of considering all data pits, the head follows continuous Saturn's ring-like patterns. In respect to the PES detection and tracking control, this simplification creates no significant difference. The signal level of PES may be doubled, but this effect can be compensated for.

Using a SNOM probe and metal spattered glass substrate with a fine line-and-space pattern, and considering them as a tracking head and a small potion of disk media, respectively, we can perform an simulative experiment of the simplified tracking model mentioned above. The scanning speed of SNOM is very slow compared to the real system, but it is enough to estimate the noise component of PES, as far as the real-time signal is obtained with the bandwidth needed for tracking control.

The details of the specific conditions for our simulative experiment are described below. First, as a scanning mode, we used an interleaved constant distance mode. In this mode, a probe-to-medium spacing is controlled based on topographical information which is pre-measured using contact mode or tapping mode scanning. In spacings of under 20nm, precise spacing control was difficult because of types of contamination such as water adsorption, so the range of spacing control was set to 20nm to 150nm. The bandwidth for signal detection (tracking bandwidth) was 5kHz. This frequency corresponds to 100 times higher harmonics of the rotational frequency of 3000rpm. The actual scanning speed was 4μm/sec at maximum.

3. SIMULATIVE EXPERIMENT

The setup for the simulative experiment used to evaluate the quality of PES is shown in Figure 2. This system is fundamentally similar to the SNOM system which employs a tapered fiber probe. We used a 488nm argon laser as a light source, and the optical lever for spacing control. A lock-in amplifier was not used to obtain a wide bandwidth signal; namely, the current to voltage-converted output signal of the photo multiplier tube (PMT) was directly stored in the digital storage oscilloscope.

Precise measurement of aperture size was difficult, but effective aperture size estimated based on a scanning signal of 100nm line-and-space pattern was about 70nm. The recording media sample was prepared using following procedure. A 20nm Cr thin film was deposited on the glass substrate. Then electron beam exposure followed by ion beam etching was used to fabricate some hundred repetitions of line-and-space pattern.

The SNOM system shown in Figure 2 was used to collect real-time scanning signals based on various probe-to-medium spacings. A 100nm line-and-space pattern was scanned at a speed of 4μm/s. In Figure 3, a representative real-time signal is shown. In this figure, spacing h0 is 30nm(top), 60nm(middle), and 90nm(bottom) respectively. As the spacing increases, the signal component decreases and becomes difficult to recognize. The base signal (f=20Hz) corresponding to a 100nm line-and-space pattern is easily recognized under the condition of h0=30nm.

The collected signals were then processed as follows: First, the signals' power spectrum was calculated using Fourier transformation. Second, based on the spectrum, base signal component and its higher harmonics were eliminated. Then all the components of the spectrum were squared and integrated for the entire control bandwidth. The result was then obtained as a root of the integration. The obtained result corresponds to the total noise energy of the signals.

Figure 4 shows the plots of the signal intensity (not including higher harmonics) and total noise as a function of probe-to-medium spacing. The signal intensity decreases dramatically as probe to media spacing increases. This can be explained as follows: Because the spread of the near-field is on the order of the size of the aperture that generates it, under the condition

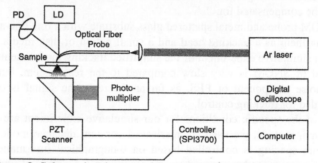

Figure 2. Schematic Diagram of Scanning Near-Field Microscope.

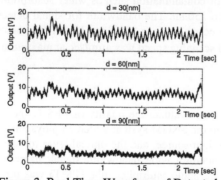

Figure 3. Real Time Waveform of Detected Near-Field Signal (5kHz Cut-off frequency).

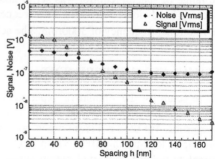

Figure 4. Dependency of Signal and Noise Level on Tip-to-Medium Spacing.

that the probe-to-medium spacing exceeds these sizes, mutual interference between the near-field and media, and thus scattered light intensity as well, decreases quickly.

On the other hand, the decrement of the noise level is very slow. As we will discuss in the following section, the noise is generated in not only the detection circuit, but also in various other sources. On the condition of very low flying height, it is expected that the inhomogeneity of media transparency is the dominant noise source, but as the flying height increases, this noise should be decrease. And electrical noise from the PMT or current to voltage converter should be constant for all ranges of flying height.

4. TRACKING ERROR ESTIMATION

The noise discussed in this section is assumed to contain the following major components, as shown in Figure 5, a) noise from media inhomogeneity, b) noise from laser light source (power drift, undesired oscillation), c) noise from PMT and current to voltage conversion system, and d) noise from the measurement system. The noise generated by the mechanical vibrations of components such as the actuator, flying head slider on witch optical head is loaded, and its supporting mechanism was neglected. Thus, as shown in Figure 6, the tracking error (in this case, precision or uncertainty of PES) is determined by the total noise level.

Now we can estimate tracking error due to the noise component of PES based on the signal to noise ratio (SNR) of the PES. In the previous section we discussed the SNR, with 'signal level' in that discussion corresponding to the difference between the signal level under the on-track condition and on the middle of two tracks. Tracking error ΔEt is written as following equation

$$\Delta Et = (Nt / \Delta S) \cdot (\tau / 2\pi),$$

where τ is track pitch, Nt is total noise (which is defined in previous section), and ΔS is the difference of signal levels in the manner mentioned above.

The result of tracking error estimation based on the method described above is shown in Figure 7. When the flying height is less than 50nm, the tracking error would be less than 10nm. The fiber probe used for this estimation is not designed as a tracking error detector; thus, the spatial resolution is too much high, and the output signal level is relatively small. So

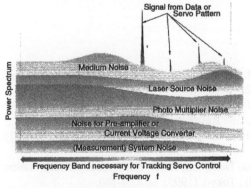

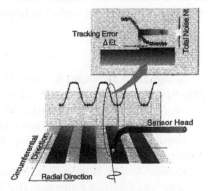

Figure 5. Supposed Spectrum of Noise Component from Various Souces.

Figure 6. Estimation of Tracking Error due to noise of PES.

178

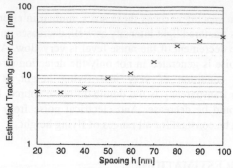

Figure 7. Relationship between Tracking Error and Tip-to-Medium Spacing.

the result should be regarded as an under estimation. In addition, as mentioned above, it is essential to consider the disturbance of the vibration created by the actuator and by the mechanism of the flying head slider.

5. CONCLUSION

A novel continuous tracking control method for near-field optical recording was proposed, and in light of certain assumptions, its performance was evaluated experimentally by the analysis of the SNR of the SNOM signal. The results showed that we could estimate the tracking error due to the noise derived mainly from the opto-electonic conversion system. In this simulative experiment, the optical efficiency of the head was relatively small compared with that of the planer or tapered type aperture; thus, the result should be regarded as an under-estimation.

The results of the estimate for the specific condition were as follows: Under the assumption of the use of a tapered fiber probe with an aperture size of 70nm, control bandwidth of 5kHz, and track pitch of 200nm, required tracking accuracy will be achieved if the flying height of the fiber probe can be set at 50nm or less. On this condition, the tracking error, which arises from the sensor reading noise, will be about 10nm.

The optimization of the sensor location for more efficient data detection, and the more practical evaluation of a tracking control system considering the actuator characteristics are problems that remain to be studied in future.

REFERENCES

1. E. Betzig, J. K. Trautman, R. Wolf, E. M. Gyorgy,, P. L. Finn, M. H. Kryder and C.-H. Chang, Appl. Phys. Lett., 61(2), (1992), p.142.
2. K. Ito, A. Kikukawa and H.Hosaka, NEAR FIELD OPTICS-5,Technical Digest of the 5th International Conference on Near Field Optics and Related, (1998), p.480.
3. H. Yoshikawa, T. Ohkubo, Y. Andoh, K. Fukuzawa, and M. Yamamoto, Opt. Lett. 25(1), (2000),pp.67-69.
4. Takahashi, Oosumi, Hosaka, Itao, Mitsuoka, Nakajima, JSME Conference on Information, Intelligence and Precision Equipment (IIP'99) , Tokyo, (1999), pp.139-140.
5. Japanese patent pending.

Human Friendly Mechatronics (ICMA 2000)
E. Arai, T. Arai and M. Takano (Editors)
© 2001 Elsevier Science B.V. All rights reserved.

Independence / cooperative remote control support system for aerial photographing

Masafumi Miwa, Hiroshi Kojima, and Reizo Kaneko[a]

[a] Department of Opto-Mechatronics, Faculty of System Engineering, Wakayama University
930 Sakaedani, Wakayama, 640-8510, Japan

To ease aerial photographing using an R/C(Radio Control) aircraft, an independence / cooperative remote control support system has been developed. Sensing systems consisted of an acceleration sensor and a gyro sensor to measure the roll angle of an R/C aircraft have been used for this control system. Using this system, stabilization of rolling has been succeeded.

1. Introduction

Aerial photographs are used for analysis of flood flows[1] and photographing of fields of natural disaster, and so on. For example, aerial photographs are used for the flood flow analysis which calculates amount of flowing and velocity vectors of the flood flows, and observations of the volcano activities. Thus, aerial photographs are useful for environmental investigations and disaster investigations. But there are following problems in aerial photographing.

1. In Japan, since an aircraft pilot must have made and declared a detailed flight plan including photographing time, and photographing site beforehand the flight. This made it difficult to carry out urgent observations in disasters, or continuous stationary measurements.

2. In Japan, aviation is strongly regulated that aircraft flight under 300 meters is forbidden, it is impossible to obtain minute observation data of a small area.

3. In almost all cases, aerial photographs are entrust to photographing traders, photographs which researchers truly hopes are not often obtained, and researchers need many days to get photographs. Furthermore the cost of the flying and taking photograph is very expensive.

On the other hand, radio control aircrafts (R/C aircrafts) instead of piloted aircrafts began to be used for aerial photographing systems . The advantages of R/C aircrafts for aerial photographing system are as follows.

1. In comparing with the case of a piloted aircraft, an R/C flight can be carried out with simple processes. It is easy to prepare, and it is possible to carry out photographing and stationary measurements quickly.

2. It is possible to photograph easily from an ultra-low (several 10-s meters) altitude where even a small piloted aircraft can not fly.

3. The flying cost is cheep.

4. Anyone can operate flying and photographing with an R/C aircraft.

5. Researchers can photograph necessary images and evaluate photographs in the field simultaneously.

However, R/C aircraft maneuvering and aerial photographing skills are necessary. At present, researchers can employ aerial photographing traders using R/C aircrafts. But in this case, photographs which a researcher truly hopes are not often obtained, and this is the same case that a researcher employs photographing traders using a piloted aircraft. Thus, we propose an independence / cooperative remote control support system which can be used for aerial photographing by researchers. In this paper, we report the result of control support on roll.

2. Difficulties on controlling R/C aircraft

R/C aircraft control is done with view points of the pilot who is outside of an R/C aircraft. The pilot stays on the ground and maneuvers the R/C aircraft with observing it. This control method is difficult for beginners and needed many skills. Followings are reason for the difficulties.

(a) Well tuned aircraft flies straightly in the stationary sky, but the flying course is veered when the attitude of the R/C aircraft is changed by disturbance such as wind. In order to hold the course, it is necessary to observe the attitude and to take correction rudder. Observation of the attitude and taking appropriate correction rudder are difficult for beginners, and also experts sometimes lost the attitude when an R/C aircraft is at faraway. In such case, a stabilizing system for the attitude is very important.

(b) When an R/C aircraft is at faraway, a pilot see the R/C aircraft as only a point, and sometimes he lost the flying direction of the aircraft. A pilot must recognize the flying direction by another method.

(c) The control stick direction of a R/C controller is defined to correspond to the R/C aircraft's direction. In Fig.1(a), a pilot's right agrees with that of the R/C aircraft, but it doesn't agree in Fig.1(b). This case is dangerous for the beginners, even experts sometimes get into trouble.

Next, we explain the difficulties on the aerial photographing. The R/C aircraft control is done with the view points of a pilot, and this way is the same for the aerial photographing. A pilot maneuvers an R/C aircraft with the view points of himself, and moves it to the desired point for photographing, then take a picture(Fig.2). In this case , the error between the target point and the actual position often occurs.

We propose a method using an installed CCD camera and a transmitter in an R/C aircraft that a pilot can check the image from the R/C aircraft and position the R/C aircraft onto the target point exactly. With this method, maneuvering a R/C aircraft to the desired point becomes easy, but it is very difficult to observe the attitude of the

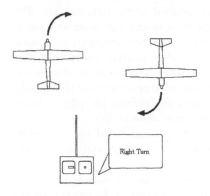

Figure 1. Direction

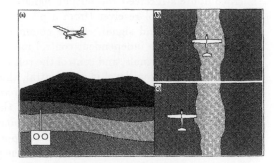

Figure 2. Aerial photographing

R/C aircraft and the image from the R/C aircraft simultaneously. So a stabilizing system for the attitude is necessary for this method. The above difficulties occur from the control method of an R/C aircraft. Thus we propose an independence / cooperative remote control support system. For the problem (a) and stabilization of attitude for aerial photographing, we develop an stabilization system using sensors. For the problem (b), we are going to develop an azimuth detection system using magnetic sensors. And for the problem (c), we propose the combination of a stabilization system and an azimuth detection system, which detects the flying direction of an R/C aircraft and make the coordinate transformation from the ordered direction by a pilot.

3. Experimental Setup

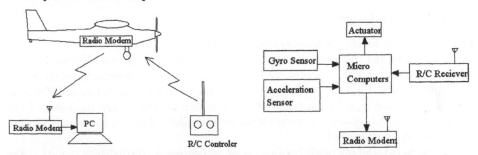

Figure 3. Schematic diagram of experimental setup

Figure 4. Block diagram of control system

Fig.3 shows the experimental setup. We have installed the acceleration sensor(ANALOG DEVICES: ADXL202), the gyro sensor(Futaba Corporation: GY-501), and the microcomputers (Hitachi: 3048F, and MicroChip: PIC16F84) in the R/C aircraft (OK Model: High Wing trainer). The radio modem(Futaba Corporation: FDA-01) has been also installed to send output signals from sensors and a roll control signal from the microcomputers. On the ground, a PC has been connected to a radio modem and stored data from the R/C aircraft.

Fig.4 shows the construction of the control system installed in the R/C aircraft. The microcomputers received the output signals from the sensors and the command signal from a radio control receiver (from a pilot). The microcomputers changed the control method by command signal. If the command signal level is neutral, microcomputers control the roll (the independent control mode). The microcomputers calculate the roll angle with sensor signals, and control the roll with proportional control mode (P control). If the command signal level is not neutral, the command signal from the pilot is taken priority, i.e. the pilot maneuvers the R/C aircraft(the cooperative mode).

Fig.5 shows the inclinometer consisted of the acceleration sensor and the microcomputers. The acceleration sensor detects component $g \cdot cosw$ of gravitiational acceleration. The roll angle w is calculated from this component.

Fig.6 shows the inclinometer consisted of the gyro sensor and the microcomputers. The gyro sensor outputs the angular velocity. The roll angle w is calculated by integrating the sensor output.

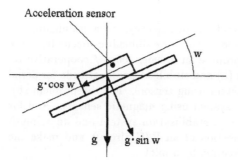

Figure 5. Acceleration sensor   Figure 6. Gyro sensor setup

4. Experimental Result

4.1. Roll control using an acceleration sensor

We carried out the roll control experiments using the inclinometer consisted of the acceleration sensor and the microcomputers. The results of experiments succeeded to hold the level flight. Fig.7-(1) shows the output signal of the inclinometer, and Fig.7-(2) shows the command signal from the pilot. The region (a),(b),(c) are the parts of the independence control, others are maneuvered by the pilot. Between the region (a) and (b) shows a 180 ° turn period.

In the region (a), the roll angle became about 30 degrees by disturbance (wind) at the arrow part, the P control made correction rudder, and the R/C aircraft returned to the level condition in almost 4 seconds.

In the region (b), the airframe tilts after one 180° turn period, and correction was made and the airframe returned to the level.

In the region (c), the P control could not hold level flight.

4.2. Roll control using a gyro sensor

We developed another inclinometer consisted of a gyro sensor and the microcomputers, and carried out the roll control experiments using this inclinometer. Fig.8 shows the result

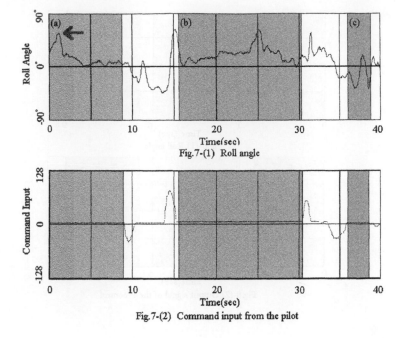

Fig.7-(1) Roll angle

Fig.7-(2) Command input from the pilot

Figure 7. Roll control using an acceleration sensor

of the experiment. Fig.8-(1) shows the output of the inclinometer, Fig.8-(2) shows the output signal of the P control. In this experiment, we used the rudder as the disturbance to tilt the airframe.

In the region (a), the pilot held the airframe as level and started the P control.

In the region (b), the pilot tilted the airframe to left by rudder, the roll angle became about 40 degrees. The pilot release rudder stick at the arrow part, and it returned to the level condition in almost 1 seconds by the P control.

In the region (c), the pilot tilted the airframe to right by rudder, the P control moved both ailerons, and it returned to near the level condition, but the error angle (about 10 degree) remained.

5. Conclusion

We developed two types of inclinometer using an acceleration sensor and a gyro sensor, and installed them to an R/C aircraft, and carried out the roll control experiments. The results are shown in the following.

1. In the experiment using the acceleration sensor, the roll angle became about 30 degrees by the disturbance (wind), the P control made correction rudder, and it

184

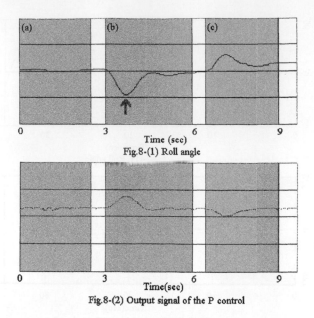

Fig.8-(1) Roll angle

Fig.8-(2) Output signal of the P control

Figure 8. Roll control using a gyro sensor

   returned to the level condition in almost 4 seconds.
2. In the experiment using the gyro sensor, the pilot tilted the airframe by rudder, the roll angle became about 40 degrees, and it returned to the level condition in almost 1 seconds by the P control.
3. In both experiments, the P control sometimes failed to hold level, but the priority was given to the command from the pilot, and the pilot corrected the error roll angle and kept the safety flight.

Using this system, the independence control(P control) succeeded, the corrections have been made automatically, and the load of a pilot is reduced. When a pilot wishes to maneuver the R/C aircraft or the independent control is out of control, the pilot can maneuver the R/C aircraft with normal control procedure, and the independent control is easily to be reentered. This system will be useful as the efficient and cheap remote control support system for aerial photographing, an environmental investigation, a disaster investigation, and so on.

REFERENCES

1. T.Utami and T.Ueno,Journal of Hydraulic, Coastal and Environmental Engineering, No. 503/2-29 (1994) 1 .

Human Friendly Mechatronics (ICMA 2000)
E. Arai, T. Arai and M. Takano *(Editors)*

185

Control strategies for Hydraulic Joint Servo Actuators for mobile robotic applications

Jörg Grabbel, Monika Ivantysynova

Institute for Aircraft Systems Engineering, Technical University of Hamburg-Harburg
Nesspriel 5, D-21129 Hamburg, Germany

This paper presents a new joint integrated rotary servo actuator design for heavy load mobile manipulators and robots. The current state of actuator technology in mobile machines is discussed briefly. A new solution for a pump controlled joint integrated actuator is proposed. Two different control strategies are discussed in its advantages and disadvantages.

1. INTRODUCTION

The continued progress in automation technology during the past twenty years was clearly dominated by innovations in the field of stationary robots. Since the economic advantage of process automation is proven, process automation is now pushing towards mobile applications, e.g. at building sites or freight centers. While for several reasons the stationary industrial robots are driven electro-mechanically, this concept does not appear to be suitable for mobile applications, especially for the 10 or even 100 times higher pay loads. Here hydraulically driven actuators are ideal due to their minimum ten times higher power density. Since a Diesel engine is used as primary energy source for all mobile machines the disadvantage of energy conversion does not weigh too high.

2. JOINT INTEGRATED SERVO ACTUATOR

Nearly all existing actuators in today's mobile machinery are based on differential cylinders operating in a constant pressure or load sensing net and controlled by proportional valves. Disadvantages of this design are:

- energy losses due to valve control, causing relatively low efficiency rates, and a not negligible amount of waste heat, relating to further problems with heat removal,
- large space required for differential cylinders, especially at the end-effector, where joints with more than one degree of freedom may be needed,
- limited rotary angles, what reduces operation flexibility.

These problems can be widely reduced by introducing two concepts:

1. *pump control* instead of valve control, using a closed hydraulic circuit. This will widely reduce energy losses and waste heat.
2. *rotary actuators*, replacing the differential cylinders at the end-effector. This will result in compact design and swivel angles of more than 180°, the typical barrier of cylinder drives.

Pump control means that the servo valve is replaced by a small servo pump. The use of swash plate axial piston pumps allows to achieve short response times and excellent dynamic behaviour, comparable to servo valve controlled actuators. Due to the low inertia of the swash plate used for volume adjusting in the displacement machine, this design is ideal for being the control element in the actuation system. Therefore a single pump in a constant pressure or load sensing net, respectively, is replaced by a number of small pumps. It can be shown that this will have *less weight* than a single pump at identical power output by applying linear scaling laws for a series of servo pumps (Ivantysynova 1998). Assuming that with increasing size the linear scaling factor is λ. Then mass and volume increase with λ^3, output power increases only with λ^2, while maximum speed decreases with λ^{-1}. Regarding further, that for pump control the totally needed power is considerably less than in valve controlled systems, the advantages of this concept are obvious.

A solution for wider rotational angles can be achieved by integrating a rotary actuator directly into the joint of a heavy load manipulator. Applying closed loop automatic control the complete concept may be called »joint integrated servo actuator«. The basic principle of the pump controlled joint integrated servo actuator is shown in figure 1. It consists of a servo pump (1), driven by small electric motor (2), rotating at very high velocity. The pump is connected to the vane type swivel motor (3) by a closed hydraulic circuit. An integrated charge pump (5) supplies the electro-hydraulic servo pump adjustment system (7) and charges the low pressure line to a typical level of 20 bars to increase the load stiffness. An integrated micro controller (6) is supplied with positional and/or velocity commands from a central control unit, operated by the driver and provides all controls for pump adjustment as for velocity and positional control of the hydraulic axis.

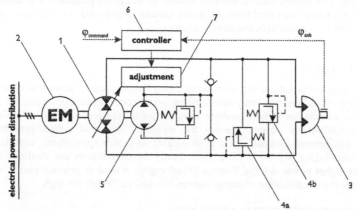

Figure 1: principle of the pump controlled joint integrated servo actuator

Since the joint integrated servo actuator takes its full advantages at the end effector joints, a different concept to conventional hydraulics, the 'power by wire' technology is proposed. This technology avoids long lines and its disadvantages by transferring the energy electrically to the actuators (Ivanty-synova et al 1995). Instead of using a central power unit or centrally installed pumps, each actuator uses its own pump, installed locally at actuator. This means the line lengths can be widely reduced by moving the pump right next to the motor and building a compact unit with highly increased hydraulic eigenfrequency and stiffness. This compact unit brings also some other advantages with it as in case of malfunction or maintenance this unit can easily be replaced and repaired. The complete testing of these units can be easily done prior to installation and allows a control parameter adjustment very comfortable.

3. DYNAMIC CHARACTERISTICS

For analysis of the dynamic behaviour and for closed loop controller design a state space model of the pump controlled joint actuator needs to be created. It can be derived by using two basic equations, the continuity equation and the balance of torque at the joint motor shaft. A complete mathematical model of the actuator was explained in Grabbel and Ivantysynova (1998). Looking at the fact, that the servo pump eigenfrequencies are much higher than the eigenfrequencies of the main circuit, the dynamic behaviour is basically defined by the eigenfrequencies of the main circuit. Further assuming the load torque as a disturbance, reducing friction to pure (linear) viscous friction and finally reducing the model to velocity control, a completely linear state space model of second order can be derived. The third order model for positional control contributes a third eigenvalue of zero, representing the integrating character of the hydraulic motor.

The system is dynamically characterized by a dominating, conjugated complex pole pair, which represents the hydraulic eigenfrequencies of the main hydraulic circuit. This pole pair defines substantially the dynamic behaviour of the system for applied velocity and positional control. The influencing parameters are on one hand side the pipe and motor fluid volume and load torque and inertia for deriving the eigenfrequency, while on the other hand internal leakage and friction losses for the damping ratio. As a result from analysis the dominating pole pair of the main hydraulic circuit is extremely low damped which is a typical property of all pump controlled systems. This low damping ratio is responsible for significant oscillations in the main circuit, if no other measures (by means of closed loop control or added hydraulic components) are implemented. The closed loop system will then become instable already at low gain, when single proportional control is used.

Especially relevant for the controller design is the fact, that load torque and inertia will change during operation. What is quite normal for all mobile machines carrying different loads, makes the controller design difficult, what will discussed later. Since the eigenfrequencies are dependent on the load torque and inertia, a variation in the load inertia, will also lead to a variation in the hydraulic eigenfrequencies. Combined with the very low damping ratio this will basically rule all efforts for an adequate controller design.

4. CONTROL STRATEGIES

For automated process management and precise, collision free positioning the following demands have to be fulfilled by designing a suitable control strategy:

- overcritical damping ratio for collision free positioning,
- high bandwidth for short work cycle duration.

It has to be pointed out, that an overcritical damping ratio is the most important demand, all other goals are of secondary importance. Considering the fact that for a significant number of applications velocity control is required additionally, the controller design can be divided into three major steps:

- velocity control,
- positional control,
- increase of system damping.

A two cascade control concept is proposed (figure 2). The inner cascade is the velocity control loop, while the outer cascade is the positional control loop. Positional and velocity commands for axis and trajectory control will than be submitted by signal wiring from a centralized control unit, where power-by-wire-module appears as a smart unit, what only needs signals of desired position and maximum velocity. Investigations have shown the achievement of overcritical damping to be the most difficult part. For velocity control a PI-controller has shown to be effective, where the integrative part is necessary to compensate control deviation in the velocity control loop, because an integrative part is missing in the plant. This control structure is shown within figure 2, where for the integrative part a limited semi-integrator (Berg 1999) is used, where the system is prevented from carrying out limit cycles without any typical integrator problems like wind-ups or defining switching conditions.

It can be shown, that carefully designed velocity controller, including measures to increase the damping ratio to desired values, would allow to use simple proportional control for the positional loop without losing (further) bandwidth. As mentioned before, the basic disadvantage of pump controlled hydraulic systems is the extremely low damping ratio of the dominant pole pair of about $d = 0.01$ .. 0.2. Since overcritical damping ($d > 1$) is required to prevent the system from overshooting the damping has to be increased by suitable measures.

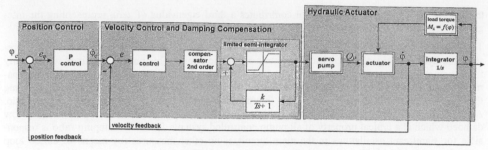

Figure 2: two cascade control strategy

Since the design of an energy efficient actuator is a main goal, typical hydraulic means to increase damping, e.g. using a bypass throttle, have to be avoided, as this would also increase the losses. On the other hand, a number of control measures to increase damping are known. Two of those are suggested:

- *Compensation* of the low damped pole pair by a complementary pair of zeros (cancellation), re-placing the old pair by a new one with overcritical damping.
- *Pole placement* by stated feedback or reduced state feedback allows free choice of eigenfrequency and damping ratio, limited only by saturation and eigenfrequency of the control element.

Both methods have been explained in earlier papers (Ivantysynova etat 1998, Grabbel 2000). Since here test rig results are only shown for the compensator, this design is briefly explained.

It consists of a of complex conjugated pair of zeros and two real poles. The purpose of the pair of ze-ros is to cancel the low damped pole pair of the plant, while the two poles will replace the cancelled pole pair by forming a new dominant pole pair. For this case the compensator transfer function gives

$$G_{comp}(s) = \frac{\frac{1}{\omega_{cz}^2}s^2 + \frac{d_{cz}}{\omega_{cz}}s + 1}{\frac{1}{\omega_{cp}^2}s^2 + \frac{d_{cp}}{\omega_{cp}}s + 1} \tag{1}$$

where ω_{cz}, d_{cz} - the compensator zero pair eigenfrequency and damping ratio,

ω_{cp}, d_{pz} - the compensator pole pair eigenfrequency and damping ratio.

However, a complete cancellation of the low damped pole is not possible by a fixed parameter con-troller design due to the variation of the plant's poles, what is basically provoked by varying pay loads (causing varying torque T_{load} and inertia Θ_{load}). This means, even with closed loop control the low damped pole will remain, practically, since the compensating pair of zeros is fixed, while the plant's pole pair is moving with varying load. A dominant behaviour of the compensation pole pair can only be achieved by reducing the compensator's eigenfrequency to values below those of the lowest hy-draulic eigenfrequency of the plant. This means explicitly

$$\omega_{cp} \overset{!}{<} \omega_{hydr., min.}$$

5. EXPERIMENTAL RESULTS

This test rig for all experimental results consists of an adjustable arm, where the length and position of the pay load can be varied during operation to create different pay load mass and inertia while running a manoeuvre (Figure 3). It allows the verification of the controller performance and positional preci-sion in a real situation as well as recording leakage and friction behaviour of the used hydraulic mo-tors. All sensor signals are recorded in a PC-based measuring system, where also the control algo-rithms are implemented. The angle is measured by TTL angle sensor with a resolution of 0.01°. This signal is differentiated to the velocity signal within in PC. The torque at the motor shaft and pressure in line A and B are measured as well.

Figure 3: test rig for experimental verifications

Test rig results have shown the compensator design to work acceptable within its desired parameters. A step response is shown in figure 4. Here the velocity signal shows a recognizable ripple. The reason can be found in the incomplete cancellation of the underdamped poles. Another reason is the method to generate the velocity signal. Since the velocity signal is calculated from the measured position, the accuracy and time delay of the velocity signal depend on the method, the sampling time and the resolution of the angle sensor. For the shown measurements the time delay of the speed signal is approx. 35 ms, what will consequently lead to a time delay in the velocity control and to further oscillations.

Following statements can be made to summarise the compensator design

- For insufficient compensation the transition behaviour shows underlaid oscillations of the insufficiently compensated eigenfrequency, what has no effect on the total transition time.
- The dynamic is limited by the lowest frequency of hydraulic pole pair (derived from the highest pay load). The dominant eigenfrequency of the compensator must be chosen below this frequency.
- The main goal, overcritical damping can be achieved at acceptable bandwidth.

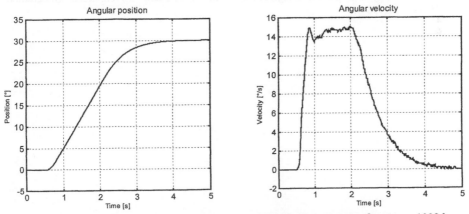

Figure 4: Test rig results - Design for m_{design} = 1500 kg, Step response for m_{load} ≈ 1000 kg

6. CONCLUSION

This paper introduced a new concept for directly driven servo joints, the integrated servo joint actuator, mounted directly into the joint of a mobile manipulator or robot. Combined with the advantages of a closed hydraulic circuit this appears to be a very promising system solution, especially at the end-effector, where these joint drives can be combined to compact unit with two or three joint axes. Compared to conventional actuators less space and energy is needed while wider rotational angles and a higher flexibility in operation are provided.

Simulations and experimental verification have shown, that the introduced compensator control will be able to achieve a satisfying increase in damping while the overall performance is acceptable. However, basic disadvantage of this control concept is the reduction of the closed loop eigenfrequency compared to the open loop eigenfrequency. Other concepts, e.g. pole placement or pressure feedback, will achieve better results. Future investigations will focus on pole placement and load identification for an improvement of control bandwidth.

REFERENCES

1. H. Berg, Robuste Regelung verstellbarer Hydromotoren am Konstantdrucknetz, Dissertation, Gerhard-Mercator-Universität Duisburg (1999).
2. J. Grabbel and M. Ivantysynova, Integrated Servo Joint Actuators for Robotic Applications. 6th Scand. International Conference on Fluid Power, Tampere, Finland (1999).
3. M. Ivantysynova and J. Grabbel, Hydraulic Joint Servoactuators for Heavy Duty Manipulators and Robots. 2nd Tampere International Conference on Machine Automation ICMA '98. Tampere, Finland (1998).
4. M. Ivantysynova, Die Schrägscheibenmaschine – eine Verdrängereinheit mit großem Entwicklungspotential, 1. Internationales fluidtechnisches Kolloquium, Aachen (1998).
5. M. Ivantysynova, O. Kunze and H. Berg, Energy saving hydraulic systems in aircraft – a way to save fuel. 4th Scand. International Conference on Fluid Power, Tempere (1995).
6. J. Mäkinen, Ellman and R. Piché, Dynamic simulation of flexible hydraulic-driven multibody systems driven using finite strain beam theory. Proc. 5th Scandinavian International Conference of Fluid Power. Linköping/Sweden (1997).
7. J.-C. Maré and P. Moulaire, Expert Rules for the Design of Position control of Eletrohydraulic Actuators. 6th Scandinavian International Conference on Fluid Power, Tampere, Finland (1999).
8. J. Mattila and T. Virvalo, Computed Force Control of Hydraulic Manipulators. 5th Scandinavian International Conference on Fluid Power. Linköping, Sweden (1997).
9. J. Roth, Regelungskonzepte für lageregelte elektrohydraulische Servoantriebe. Dissertation. RWTH Aachen (1984).

Human Friendly Mechatronics (ICMA 2000)
E. Arai, T. Arai and M. Takano *(Editors)*

Integration of Micro-Mechatronics in Automotive Applications using Environmentally accepted Materials

F. Ansorge, K.-F. Becker, B. Michel, R. Leutenbauer, V. Großer, R. Aschenbrenner, H. Reichl

Fraunhofer Institute Reliability and Microintegration; Developement Center for Micro-Mechatronic and System Integration

Argelsrieder Feld 6, D-82234 Wessling
Tel. +49-8153-9097-525 Fax.: +49-8153-9097-511 e-mail: ansorge@izm.fhg.de

Micro-Mechatronics & System Integration is a multidisciplinary field, which offers low cost system solutions based on the principle of homogenizing system components and consequent elimination of at least one material component or packaging level from the system. These system approaches show, compared to the existing solutions, a higher functionality, more intelligence and better reliability performance. The number of interconnects necessary to link a motor, or sensor, or an actuator to the digital bus decreases, as the smartness of the devices increases.

The paper presents human friendly system solutions and manufacturing technology for mechatronic systems, developed at Fraunhofer IZM using well adapted resources and human friendly materials. To reduce package volume, advanced packaging technologies as Flip Chip and CSP are used, for increased reliability and additional mechanical functionality encapsulation processes as transfer-molding, a combination of transfer- and injection molding or modular packaging toolkit based on LTCC are selected.

The first system is a multi-chip-module for motor control used for automotive applications as window-lifts or sun-roofs. For the module at least three interconnection layers were eliminated using novel concepts as intelligent leadframes and integrated plugs. The package resists harsh environment as present in automotive applications. The use of lead free components will also be a topic.

Another mechatronic package called TB-BGA or StackPac resp., is a 3-D solution, containing a pressure sensor and the complete electronics necessary for control and data transfers. This package involves CSP-technologies on flex, with an increase of functionality per volume unit by direct CSP-to CSP mounting, eliminating organic substrate layers and large signal distances of unamplified signals from sensor output and thus reducing resources needed. Packages like this can be integrated in medical applications.

The mechatronic packages will be discussed in detail. Especially cost, reliability performance and according "Design for Reliability" show the potential of micro-mechatronic solutions in automotive and industrial applications.

1 Introduction

The development of electromechanic assemblies has so far been carried out mainly by different manufacturers who go their own separate ways. Assembly and fine tuning of the individual components and their integration into an overall unit normally takes place at the very end of the manufacturing process.

Due to increased system intelligence many electronic systems comprise an additional control unit which receives information from different sensors and selects the individual electromechanical actors. These central control units need a lot of space and are fairly inert due to the long transmission paths of the electric signals. On top of that, the cabling of all the different components involved in the course of the assembly can be a complex and expensive business; it is definitely a major headache for mobile applications.

Sub-systems on the other hand, which mechatronically integrate both actor and sensor facilities as well as all information processing (software) in the component itself, deliver an optimized use of existing space, higher speeds of data transmission between actor/sensor functions and μ-processor technology. All this results in substantially higher performance capabilities. This would reduce transmission distances and data volumes. It would also steer the individual systems towards greater independence and autonomy.

The overall system is characterized by the communication of the individual sub-systems both with each other and the central control unit (main controller), Figure 1. This process can be conducted via accident-insensitive bus systems. One or two lines connecting all sub-modules with the main controller or a (thus defined) master module in the form of a ring or a string are already enough to ensure the free flow of all information.

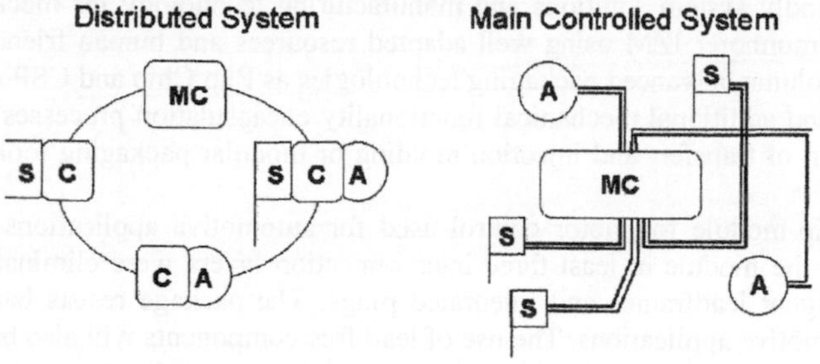

The basic sensor data were already processed and converted in the sub-module. Only the data required by the control logic system is transmitted to the main control unit. As a consequence, you have electromechanical systems with "subsidiary intelligence" – a mechatronic system.

This multi-directional feedback mechanism results in a dominant technology, based on subtle analysis of the functions needed and optimum separation of functionality of the parts involved. Mechatronics is defined here in this sense as the application of intelligent sub-modules. A more comprehensive definition would define it along such lines as the following: Mechatronics is the synergistic integration of engineering, electronics and complex control logic into development and working procedures, resulting in mechanic functionality and integrated algorithms [1, 2].

Figure 1: Juxtaposition of a sub-system with bus technology (left) and a conventional control unit (right). (MC Main Controller; C Controller; A Actor; S Sensor).

2 Basic conditions for the development of mechatronics

The development of mechatronics is a vital key for the future of electronics. Ever more complex applications require the processing of ever increasing data volumes: a corresponding increase in flexibility and functionality is the only possible response developing micro-mechatronics.

Synergistic co-operation between the individual departments concerned is also an indispensable condition. Electronic and mechanical simulation and designs must be compatible for a proper realization of the thermomechanical concept. The package fits to an optimum place within a macroscopic system of mechanical parts, which are especially designed to work with electronics as intelligent sub-system. It is of equal importance to generate a software map of all requirements and functions, if at all possible already at the planning and development stage; this is the ideal way to shorten development times while optimizing the system eventually produced. To yield this added functionality within one package the effort of specialists from various fields of technology as physics, chemistry, electrical engineering, microelectronics, software design, biology, medicine etc. are needed.

2.1 Target temperature range for automotive applications

Another important requirement is the question of an increase in the reliability of electronic and mechatronic components. Electronic components in the area of the car engine, for example, must be able to perform consistently at temperatures of up to 200 °C and at extreme temperature cycles, see Figure 2.

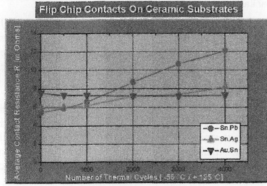

Figure 2: Flip Chip Reliability Data for Variations in Solder Metallurgy

The assemblies used in these areas must be cost-effective and yet at the same time able to perform for the entire technical life span of the automobile. They need to be assembled in a light and compact pattern. All this can be delivered by sub-modules as parts of subsidiary intelligence with interconnection levels which have been reduced to a bare minimum.

Both, materials used and processes applied must be carefully attuned to each other in order to meet all requirements of and in particular to achieve the highest possible standards of reliability. This means that, beyond the quality of the individual component, it is the way the components are combined and put together which determines the functionality and quality of the integrated set-up.

2.2 Novel concepts for polymer materials

Focusing on micro-mechatronic applications there are a few additional demands to encapsulants commonly used for packaging. The encapsulants need a wide range of temperatures, a high resistance against harsh environment and the integration of moving and sensing elements without losing package functionality.

Typical materials used for microelectronics encapsulation are epoxy resins, where the chemical basis is a multifunctional epoxy oligomer based on novolac. These materials do have T_g's beyond 200°C and so they have the potential for short term - high temperature application. The evaluation of encapsulants for optimized long-term stability is one of the topics the micro-mechatronics center is focusing on.

Other areas of research important for the manufacturing of miniaturized mechatronic systems are wafer level encapsulation and the handling of thin and flexible base materials. The use of thermoplastic hotmelt materials for micro-mechatronic packaging allows a decrease of processing temperature, the use of cost effective low temperature materials.

Further potential for micro-mechatronics lies in the use of advanced thermoplastic materials. In the past they have been used rather for electronics housing than for the direct encapsulation of

194

microelectronics. The use of thermoplastic materials does not only allow the integration of additional mechanical functionality (plug housing, clip on mounting, ...) simultaneously with microelectronics encapsulation, but these materials do not cross-link and thus do have a high potential for recycling.

Research of Fraunhofer IZM is performed in the fields of direct encapsulation of microelectronics using thermoplastic polymers, thermoplastic circuit boards a.k.a. MID-devices for both, Flip Chip and SMD components and on the material compatibility of classic encapsulants and novel thermoplastic materials. The suitability of thermoplastics for the process of direct encapsulation, meanwhile, is subject to a number of conditions. In order to preserve a multifunctional package, both technologies would have to be joined in the forming of two- or multi-component-packages. (see Figure 3)

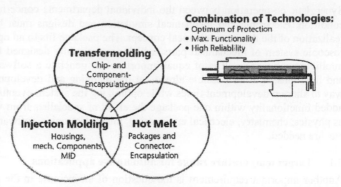

Figure 3: Combination of Packaging Materials for optimized package functionality

For the purposes of material and process selection, many considerations are relevant, e.g., the processing in the course of standard operations, a homogeneous interconnection technology and procedures under conditions designed to minimize stress factors. It is fairly difficult, meanwhile, to make universally valid propositions about the suitability of certain technologies and materials for mechatronic products. Such statements can only have value if they refer to individual products and the clearly identified purposes they are supposed to serve [3].

3 Mechatronic applications

Three of the main areas for the application of mechatronic sub-systems are heavy engineering, car manufacturing and medical technology. The products of these three areas have one thing in common: they consist of difficult overall systems with complex control units. Depending on the complexity of the application in question, the mechatronic design needs a defined design and interconnection technology. The following examples illustrate the potential range of intelligent sub-systems in existing and future applications. They also show the most important current trends.

3.1 Modular construction systems for mechatronics

In the course of a project sponsored by the BMBF, the Fraunhofer Institute for Reliability and Microintegration (IZM) developed a modular construction system for sub-systems. Spatially stackable single components include sensor elements, signal processing, actors and bus technology. The components are mounted taking into account the application they are supposed to serve. They produced standard components and can be mounted by soldering, leading to a minimized number of different interconnection technologies.

The standardized dimensions in different size categories enable the user to combine the system components, which are part of the modular construction program, in a most flexible way by defining his own functionality parameters [4]. Connection to the actorics can be effected by a direct integration of the components.

Based on molding-on-flex-technology the StackPac offer its advantages. This technological concept allows the insertion of individual ICs by contacting through either wire bonding or flip chip technology and integration of SMD-components. Multichip modules can be created using different contacting technologies.

Molding the package widens the range of possible package designs from simple patterns to additional functional geometries which are compatible with the modular construction system as depicted in Figure 4. The same technology can be used to design chip size

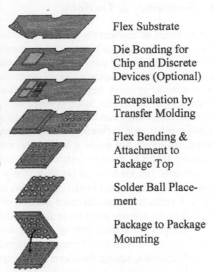

Flex Substrate

Die Bonding for
Chip and Discrete
Devices (Optional)

Encapsulation by
Transfer Molding

Flex Bending &
Attachment to
Package Top

Solder Ball Placement

Package to Package
Mounting

Figure 4: StackPac production process

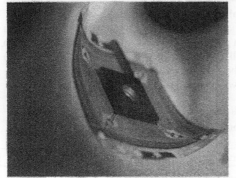

Figure 5: CavityPack: Molding on Flex approach to Sensor Packaging

packages (CSP's). The necessary combination with the sensorics can be created by using a CavityPack (see Figure 5).

The systems have already been used successfully in engineering applications. StackPac's achieve excellent reliability scores, in particular for high-temperature applications, using the mentioned novel encapsulants. This allows a reduction of development and qualification costs by using standard modules, which makes the use of advanced technology in confined spaces economically viable.

3.2 Mechatronic solutions for the automobile

3.2.1 High Power Applications

In the course of a project jointly run by BMW AG and the Fraunhofer-IZM, an H-bridge circuit with control logic, bus connection and sensor facility has been developed. The use of power ICs required a concept for the heat dissipation of the package. The reduction in the number of interconnecting levels was another purpose of this mission. The power semiconductors have more advantages: in comparison with relays, they use substantially less space and are extremely well suited for the software selection process. This can turn out to be a particularly decisive advantage to conventional solutions, once the 42-V-on-board-network has been installed.

Thermal management functions are fulfilled perfectly by the heat sink comprising the inserted copper lead frame and the engine which has been mounted right at the bottom side of the package. The motor block has the effect of an active cooling element. This integrated thermal concept allowed the substitution of the conventional trigger system by spatially optimized electronics. The high degree of integration resulted in a reduction of spatial requirements by about 50%.

4 Summary & Outlook

Future fields of application for mechatronics will include robotics and industrial engineering, i.e. the construction of machinery and production plants. Intelligent robots who are programmed to learn on their own, to communicate with each other by means of wireless communication and to develop solutions to complex tasks autonomously can be used in either the exploration of space/ other planets or more earth-bound functions such as repair/maintenance works in inaccessible machinery or the service labor in hospitals.

Medical technology can be regarded as one of the markets with the biggest potential for the use of micro-mechatronic technology. The increased use of such technology in the area of minimally invasive surgery has made a substantial contribution. This development has been supported by progress achieved in microsystems technology: the miniaturization of cameras and electromotoric instruments has made it all possible. Only micro-mechatronics allows us to reach the ultimate stage in the integration of existing systems to reliable and cost-effective sub-modules [9].

Table 1 reflects the requirement profiles of the individual application fields. The trend towards mechatronics persists in all these fields and is even gaining momentum. According to the research study Delphi 98, a large part of what is today still a vision of the future will be converted into reality by micro- machine technology already in the period between 2001 and 2007: the integration of sensorics, actoric functions and controllers will make it possible [6].

Table 1. Mechatronics requirements and trends
(+++ crucial, ++ very important, + important, □ optional, ↗ increasing).

Requirement / use	Engineering	Automotive	Medical – Tech	Private
Share of Sensorics	++	++	++	++
Miniaturisation	+	+	+++	+
Modularity	++	++	□	++
Cost effective	++	+++	□	+++
Market potential	↗	↗	↗	↗

5   Literature

[1] Control and Configuration Aspects of Mechatronics, CCAM Proc.; Verlg. ISLE, Ilmenau, 1997;

[2] F. Ansorge, K.-F. Becker; Micro-Mechatronics – Applications, Presentation at Microelectronics Workshop at Institute of Industrial Science, University of Tokyo, Oct. 1998, Tokyo, Japan

[3] F. Ansorge, K.-F. Becker, G. Azdasht, R. Ehrlich, R. Aschenbrenner, H.Reichl: Recent Progress in Encapsulation Technology and Reliability Investigation of Mechatronic, CSP and BGA Packages using Flip Chip Interconnection Technology, Proc. APCON 97, Sunnyvale, Ca., USA

[4] Leutenbauer R., Großer V., Michel B., Reichl H.; The Developement of a Top Bottom BGA (TB-BGA) IMCM 98, IMAPS Denver, Colorado, April 98; Proceedings IMAPS, 1998, pp. 247-252

[5] H. Oppermann, H. Köhne, H. Reichl; Mikrosystemintegration für Heimanwendungen; VIMP; München, Dezember1998

[6] Delphi `98, Studie zur globalen Entwicklung von Wissenschaft und Technik, Fraunhofer Gesellschaft, München, 1998

[7] Fraunhofer Magazin 4.1998, Fraunhofer Gesellschaft, München

[8] F. Ansorge; Packaging Roadmap – Advanced Packaging, Future Demands, BGA's; Advanced Packaging Tutorial, SMT-ASIC 98, Nuremberg, Germany

[9] Töpper, M. Schaldach, S. Fehlberg, C. Karduck, C. Meinherz, K. Heinricht, V. Bader, L. Hoster, P. Coskina, A. Kloeser, O. Ehrmann, H. Reichl: Chip Size Package –

Vision and Sensing

Human Friendly Mechatronics (ICMA 2000)
E. Arai, T. Arai and M. Takano *(Editors)*
© 2001 Elsevier Science B.V. All rights reserved.

Robust estimation of position and orientation of machine vision cameras

Jouko Viitanen

VTT Automation, Machine Automation
P.O.Box 1302, FIN-33101 Tampere 10, Finland

This paper describes a method for robust estimation of the position and orientation (POSE) of a machine vision camera. Special attention has been paid on the easy scalability and portability of the system. Discussion is also included about the advantages and disadvantages of a measurement system that allows large angular deviation from the typical arrangement where the camera is positioned in the direction of the normal of the surface to be measured. A practical implementation for the purpose is described and user experience reported.

1. INTRODUCTION

A common problem for measurements using machine vision equipment is the determination of the position and orientation of the camera with respect to the world coordinate system. One possible approach for the task is described in another paper submitted to this conference [1]. It allows also the camera parameters to be determined in the same process. However, that method, like many others, too, requires one to use a special 3D calibration target that is difficult to scale to different environments, and which allows only limited deviation from the normal direction of the measured surface. Several prior methods also suffer from the fact that such multivariable optimization methods typically have a risk to be stuck at local minima, unless such very time consuming methods as e.g. simulated annealing are used. Some previous works can be found in [2,3]. Typically, calibration of the cameras (i.e. determination of the focal length and the aberration parameters) can be performed off-line in a laboratory, where the time consumption is not critical. After the calibration, the number of unknown variables has reduced down to six, i.e. the POSE parameters. For the determination of these, a simple and easily scalable planar calibration target can be used, as will be shown in this paper.

Another common problem, especially in small parts assembly, is the fact that the space is very limited inside the assembly cell. Therefore, it would be a great advantage, if the measurements could be performed from the side of the cell, perhaps so that the optical axis of the camera has a largish angle with respect to the normal of the table of the assembly cell. That can be done if the perspective distortions are corrected, and, in fact, in the following we can see that this tilted arrangement may even have some advantages compared to a viewing direction parallel to the normal.

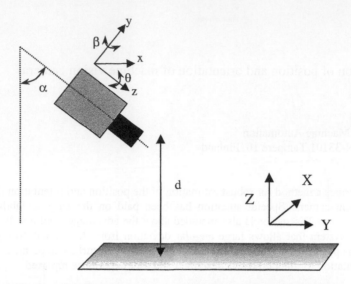

Figure 1. The coordinate system for a machine vision measurement system where the camera coordinate system has major rotation α only around the world X axis. The camera coordinate system is shown by lowercase x, y, and z , while the world coordinate system is shown by uppercase X, Y, and Z. The plane Z = 0 is assumed to coincide with the measurement plane, and d is the distance of the optical center point of the camera from this plane.

2. THE MEASUREMENT PROBLEM

The task in the measurements using computer vision is to perform a coordinate transformation from the (known) camera surface coordinates to the world coordinates, in a setup illustrated in Figure 1. The easiest case positions the camera so that the camera and world z coordinate axes are parallel, i.e. for objects with no variation in z, the object and image planes are parallel to each other. This arrangement is sometimes not possible, however, in most cases, we are able to restrict the rotational deflections, so that a large angle only exists with respect to one of the world coordinate axes. The arrangement of Figure 1. shows a typical case where there is major rotation only with respect to the X axis, and the camera is approximately aligned so that the rotations with respect to the Y and Z axes are small. Such a case applies often to e.g. machine vision systems in a robotic assembly cell, where a camera cannot be positioned within the movement range of the robot, and the measurements are performed in a reference plane where the bottom floor of the cell corresponds to Z = 0.

In this description we limit the case to 2D measurements within one plane of the world coordinate system, as shown in Figure 1; however, if the heights of the objects on the surface are known, the measurements can also be applied to 3D objects that have surfaces parallel to the bottom plane. We also assume that the geometric distortions of the optics of the camera are compensated in an off-line calibration process, e.g. using the methods in [1,3], so that the ideal pinhole model for the camera can be used in the following formulation. One has to remember, however, that the pinhole model only applies to distances that are far compared to the focal length, otherwise the corrections from the full lens equation has to be added.

3. POSE ESTIMATION

In order to be able to do the coordinate transformation accurately, we need to know the three positional coordinates $\{x,y,z\}$ of the camera in the world coordinate system, as well as the three rotation angles $\{\alpha,\beta,\theta\}$ with respect to the three world coordinate directions (see Figure 1.). When these parameters are known, the well known perspective matrix coordinate transformation equations can be used [4]:

$$
\begin{bmatrix} kx \\ ky \\ kz \\ k \end{bmatrix} = \mathbf{M} \bullet \begin{bmatrix} X \\ Y \\ Z \\ 1 \end{bmatrix}
\tag{1}
$$

where the capital X, Y, and Z denote the world coordinates, and the small x, y, and z the corresponding coordinates of the image on the camera detector; k is the scaling coefficient, and the matrix $\mathbf{M}$ contains the POSE -dependent elements. In practise, we would like to know the world, instead of the camera coordinates, so the problem returns to the determination of the elements of the inverse matrix $\mathbf{M'}$ of the equation:

$$
\begin{bmatrix} X \\ Y \\ Z \\ 1 \end{bmatrix} = \mathbf{M'} \bullet \begin{bmatrix} kx \\ ky \\ kz \\ k \end{bmatrix}
\tag{2}
$$

The elements of $\mathbf{M'}$ for the complete case with rotations α, β, θ, can be found in e.g. [3]. However, in this paper we restrict to a case where β and θ are assumed to be small, and only α is large. This is a very practical limitation, as the following consideration proposes. Because the task in our case is to perform metric measurements within the plane shown in Figure 1. in world coordinates, and thus a suitable calibration target would contain a known length between two markers, we would like to use such an optimization process, where the sensitivity for the optimized variables would be large, and for those variables that can be fixed close to the final value, the sensitivity would be small. Fixing of some variables would be preferred in order to reduce the optimization process and keep the results robust. Now, in the tilted arrangement of Figure 1, we can readily see that at a position $\beta = 0$, the sensitivity of the scaling factor k for variable β is at a minimum. The same applies to the position $\theta = 0$. This can also be found by solving the first derivatives of the full perspective transformation with respect to β and θ.

In practise, this means that close to an orientation where $\beta = 0$ and $\theta = 0$, small changes in β (Pan) or θ (Swing) angles do not result in large errors in the planar measurements. In fact, the corrections can be performed by simple linear approximations after the optimization loop. However, the sensitivity of the X direction offset for β is large, but this can be compensated when the calibration marker distance from the world origo is known. Correspondingly, the sensitivity of the Y direction offset for θ is large, but this can also be compensated by the known calibration marker position.

The desired condition $\beta \cong 0$, $\theta \cong 0$ can be arranged easily with a simple calibration target. One example is a target that has two markers at a predetermined distance from each other. If this is positioned at the world coordinate system on the measurement plane so that the two markers are aligned with the Y axis, and their X direction distances from the camera origo are approximately equal, then the condition can easily be achieved in the camera attachment by adjusting the camera swing so that the markers appear approximately on the same horizontal line in the camera image, and their x-direction distances from the optical center appear approximately equal.

Assuming that $\beta = 0$ and $\theta = 0$, the elements of the inverse transformation matrix $\mathbf{M'}$ reduce to the following set: [4]

$$m_{11} = d + f \cos \alpha$$

$$m_{22} = f + d \cos \alpha$$

$$m_{24} = -df \sin \alpha \tag{3}$$

$$m_{42} = \sin \alpha$$

$$m_{44} = f \cos \alpha$$

$$m_{12} = m_{13} = m_{14} = m_{21} = m_{23} = m_{31} = m_{32} = m_{33} = m_{34} = m_{41} = m_{43} = 0$$

where f is the focal length of the camera lens, and d the distance of the center of the sensor plane from the measurement plane, in a configuration where the image plane is flipped around the optical center point of the lens, into the same side as the measurement plane [4]. Now we see that the only unknown variables needed for the determination of the position and the orientation of the camera are the distance d and the rotation angle α with respect to the x axis. We can also find the expressions for X and Y from the inverse matrix equation:

$$X = (d + f \cos \alpha) kx \tag{4}$$

$$Y = \frac{(f + d \cos \alpha) y}{y \sin \alpha + f \cos \alpha} - \frac{df \sin \alpha}{y \sin \alpha + f \cos \alpha} \tag{5}$$

When the number of unknown parameters is only two, sufficient accuracy in the POSE determination can be achieved even by using exhaustive search, within a few seconds of processing time. The minimum number of calibration points needed is four, but an increased number can be used for smaller uncertainty.

The cost function in the optimization can be e.g. the sum of the absolute values of the differences between the estimated and true lengths between the two pairs of calibration points that are positioned on the surface Z = 0. In practise, these points can be located at the ends of two steel rulers that are laid on the surface. The Y-direction difference between the rulers need not to be given, but for best accuracy, the spacing should be close to the maximum field of view of the camera in the Y direction. This simple cost function can be expressed as:

$$C(\alpha, d) = |((\tilde{X}_2 - \tilde{X}_1)^2 + (\tilde{Y}_2 - \tilde{Y}_1)^2)^{1/2} - ((X_2 - X_1)^2 + (Y_2 - Y_1)^2)^{1/2}| +$$
$$|((\tilde{X}_4 - \tilde{X}_3)^2 + (\tilde{Y}_4 - \tilde{Y}_3)^2)^{1/2} - ((X_4 - X_3)^2 + (Y_4 - Y_3)^2)^{1/2}| \qquad (6)$$

where the four world X and Y coordinates of known calibration points are denoted with corresponding subscripts, and the corresponding measured estimates are shown with tildes. Any common optimization methods can be used, and a good global minimum will be received from the algorithms without heavy processing.

3.1 Practical considerations

It is important to notice that if the tilt angle of the camera is large, we see from equation (4) that the derivative of the cost function increases when α increases from zero (the typical straight staring position corresponds to $\alpha=0$). Therefore we can also achieve correspondingly better angular accuracy in the POSE determination, assuming similar inaccuracy in the estimated X_n and Y_n. However, we seldom measure POSE alone; the more common task of x-y measurements within the plane Z = 0 does not get any improvement from the tilt. Also, we have to remember that the spacial resolutions for targets at different y-positions within this plane become very inequal when tilting is applied.

The minimum cost of the optimization run is a good practical indicator of the success of the calibration process. It cannot be used as an estimate of the final accuracy, but accidental mistakes in, e.g. the placement of the calibration targets, can typically be found due to increased final cost. Other indicators are the differences of the manually measured camera height d and angle α from the optimized height and angle.

After α and d have been obtained from the optimization routine, the small linear corrections for β and θ are now easily obtained in the correct scale from the measured differences in the calibration ruler rotations around the corresponding coordinate directions (if the rulers were well aligned in the X direction). For multiple camera systems along the X axis, each camera can be individually calibrated.

3.2 Sub-pixel calibration

Measurements in a computer vision system cannot be performed in better than integer pixel accuracy, unless we have some prior knowledge of the target shapes and intensities. However, calibration of a measurement system has to be more accurate than the actual measurements. Therefore sub-pixel measurement in the calibration is performed: the calibration targets are formed from two areas of different constant reflectivity, and the sub-pixel position is achieved from interpolation that utilizes measured intensities of the detector pixels. At least 1/10 pixel accuracy can be achieved, if the signal to noise ratio of the camera is good.

4. EXPERIMENTAL RESULTS

The measurement system was applied to an outdoor application, where the targets were edges of small sand pits on a levelled ground field. A two-camera system was positioned near to the side of the field, so that the angle α of the cameras with respect to the horizontal direction were about 40 degrees. The measurement width for one camera was about 3 meters in the y direction, and 2.8 meters in x direction at the far side, and 1.8 meters at the closer side. The fields of view of the two cameras were overlapping so that in the closer side there was no gap. The horizontal pixel count of the camera was 758 and vertical 480. The measurement task was to find the x-direction positions of the edges of the sand pits, so the theoretical pixel resolution varied from 2.4mm/pel (close) to 3.7mm (far). Therefore, theoretically we would result in a mean error from 1.2mm to 1.85mm, depending on the y position. From a test run of 20 events, we got an average error of 2.4mm and maximum error of 5mm. The reference measurements were performed using a steel tape measure. In this situation, there was only a few seconds of time for each measurement, so the greatest error came from the fact that for the two measeurement, the operators occasionally selected a different target position. Indoor laboratory tests gave results close to the theoretical specifications. The test results were found satisfactory for the specific application, where the demand was for fast operation and rapid recalibration of the system, after it was moved. A competitive approach with a tachymeter and a manually placed target prism on the measured position was more accurate, but the whole measurement with such a system lasted about 30 seconds per measurement, while the camera-based system gave the result in about 2 seconds, and needed no target placement.

The time required for moving and recalibrating the camera-based measurement system was about 10 minutes, including camera re-placement, positioning of the two steel calibration targets, and running the optimization routine.

5. CONCLUSIONS

As a conclusion, a computer vision system was described that can be used for dimensional measurements in such demanding conditions where rapid recalibration, tilted viewing direction, and easily portable calibration targets are needed. A simple optimization routine was described that can be applied for systems that have distorted perspective views.

REFERENCES

1. M. Kutila, J. Korpinen, and J. Viitanen, Camera calibration in machine automation, International Conference on Machine Automation, 27.-29. 9. 2000, Osaka, Japan.
2. G.D. Wei and S.D. Ma, Implicit and explicit camera calibration: Theory and experiments, IEEE Trans. on PAMI, vol 5 (1994), no.16, 469 - 480.
3. J. Heikkilä, Accurate camera calibration and feature based 3-D reconstruction from monocular image sequences, Acta Universitatis Ouluensis C Technica 108, Oulu, Finland (1997).
4. B. Correia, J. Dinis, Camera model and calibration process for high-accuracy digital image metrology of inspection planes, proc. SPIE, vol. 3521, Boston, Massachusetts, U.S.A. (1998).

Human Friendly Mechatronics (ICMA 2000)
E. Arai, T. Arai and M. Takano (Editors)
© 2001 Elsevier Science B.V. All rights reserved.

Deflection Compensation of Flexible Manipulators Using Machine Vision and Neural Networks

Ilkka Koskinen[a], Jarno Mielikäinen[b], Heikki Handroos[a], Heikki Kälviäinen[b], Pekka Toivonen[b]

[a]Department of Mechanical Engineering
[b]Department of Information Technology
Lappeenranta University of Technology, PO Box 20, FIN-53851 Lappeenranta, Finland

Abstract

This paper introduces a concept of positioning deviation compensation of a large scale flexible hydraulic manipulator by integrating a machine vision system and applying neural networks. The method has the advantage of being able to apply simple geometry based rigid inverse kinematic models in high precision flexible applications. Basically the method consists of a simple control scheme, an inverse kinematic model and a neural network based deviation compensation module. The neural network is trained with data obtained with a machine vision system. The method has been tested in real-time on a large scale flexible manipulator in two dimensions. The results have been found applicable in real life problems.

1 Introduction

In order to place the end-effector in a certain point in the workspace, the end-effector position has to be defined in joint angles or actuator strokes of the robot arm. The values are calculated by the robot controller and set as reference values. The joint angles and actuator strokes are calculated by the robot controller by mapping the Cartesian space to joint or actuator space so that the end-effector position is defined by a set of joint angles or actuator strokes. The mapping is called inverse kinematics.

The introduction of flexible manipulator structures causes problems in inverse kinematics calculations. Applying flexibility matrices in the mapping is complicated, as it requires knowledge of the affecting forces. By making a dynamic analysis of the system, the problem can be solved [4], [5], [6]. However, dynamic analysis requires high calculation capacity and cannot be easily applied in real-time applications like a robot controller.

If the manipulator flexibility can be compensated , lighter manipulator structures can be applied. Lightness of structures is a common requirement among vehicle mounted manipulators such as lumber hoists and other maneuverable cranes.

New mathematical methods such as neural networks introduce appealing ways [3], [7], [8] of estimating the deviation at a given point in the manipulator workspace. Unfortunately, the neural networks require a sets of teach data, before they can be used to estimate the deviation In the case of a large manipulator the teach data matrix has to be large, before the manipulator workspace can be covered.

2 Manipulator

This work is based on large flexible log (figure 1) crane PATU 655, manufactured by Kesla. The crane is an ideal example of large flexible manipulator, as it is light compared to the heavy loads it is intended to handle. The structure is designed to be easily mounted on a log truck, it is compact during transportation and has a large workspace. Thus, the manipulator is light, very flexible and little attention is paid on joint precision. The manipulator is hydraulically driven and can lift loads up to 500 kg. It has a maximum reach of 4850 mm, but during transportation it can be folded thanks to its quadrangle joint. In the experiment only lift and jib cylinders were used to manipulate the crane. The manipulator controller is PC based, running on dSpace 1103 DSP card.

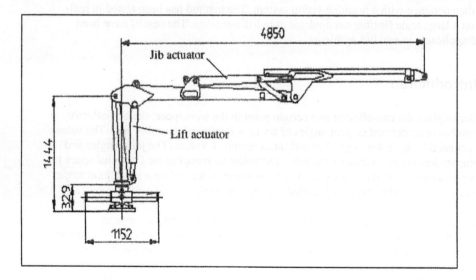

Figure 1. The manipulator

3   Rigid Inverse Kinematic Model

In this case a large log lifting crane was to be modeled. The joint matrix method was complicated, because of the link quadrangle of the boom. It was, thus, decided to make a simple geometric model of the boom [1]. The inaccuracies of the model would be compensated along with other deviations [2] . As the model was very simple it could easily be implemented into the control loop, without having to decrease the control frequency.

The model was built so, that coordinates were fed into the model and actuator strokes were given as solutions. This way the inverse kinematic module could be added to the closed control loop as a independent unit. The user could set the end-effector in Cartesian coordinates, which would then be transformed to actuator strokes. Thus, the inverse kinematic model acts as an interface between the users Cartesian world and the manipulators workspace which can only be defined by actuator strokes.

4   Machine Vision

By utilizing a machine vision system, the cumbersome routine of training data obtaining can be automated. It was chosen to install the camera on the boom tip. From the boom tip the machine vision system could be devised in such a way, that it could have a clear view of the measuring plate field at any given actuator configuration.

The measuring plate field was a mat painted steel plate with blue circles. The blue circles were 4 centimeters apart, with a maximum deviation of 0.1 millimeters. The measuring plate was placed parallel to the boom, so that the blue circles could be clearly viewed with the camera.

Due to the fact that the camera was installed on the boom tip, the camera tilted along with the jib boom. Although, the camera angle could be calculated with an inverse kinematics routine, it was chosen to install an red laser pointer on the boom tip near the camera. As long as the laser pointer and two blue circles could be the viewed by the camera, the boom angle and the position of the boom tip could be defined in Cartesian coordinates.

The camera was a Sony DXC-9100P videocamera, which was connected to a standard Pentium II PC with a Matrox Meteor II grabber card. The actual grabbing algorithm was written in ANSI-C utilizing the Matrox Imaging Library and Aamer Haques mathematical routines [9].

5   System Integration

To obtain the coordinates a connection to the manipulator control scheme was necessary. Thus, a simple communication protocol was devised. The control loop was in command of the manipulator positioning. A read flag was set when the boom tip was at the desired position. The machine vision system ran parallel to the control loop and monitored the read flag. When the read flag was set the set point values, and actuator strokes were acquired from the control system by utilizing the DSpace MLIB library functions (figure 2).

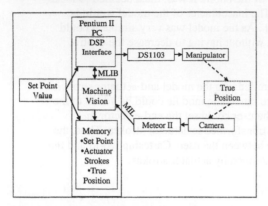

Figure 2. Localization scheme

With the obtained information the machine vision algorithm defined the true position of the manipulator tip (figure 3).

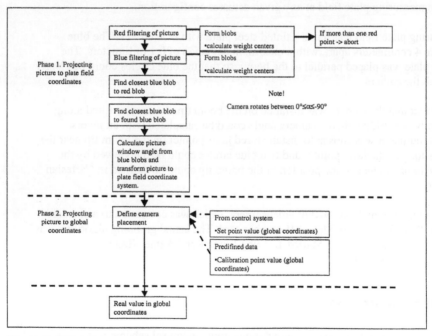

Figure 3. Obtaining coordinates with machine vision.

The real values, actuator strokes and set points were then written into data file for later use.

6 Deviation Compensation

The obtained set of data is used to teach a neural network to approximate the deviation in actuator strokes at a given point in the work space. The applied neural network is a feed forward MLP-network with 2 and 5 hidden layers. Each actuator had its own separate compensation network.

For the lift actuator, the training resulted in a well correlating network already in a third degree polynomial. However, the jib actuator didn't do as well even with much higher polynomials. The trained networks were then implemented into the control loop as individual modules.

7 Results

As a result the positioning deflections are reduced from the maximum of ±40 mm to max ±10 mm without decreasing the speed of the system (figure 4 and 5). The compensation is invisible to the end user, but can be turned on and off at will. Such a system is easy to construct and applicable also in other types of cranes and robot arms. The measuring system and neural network teach processes are automatic and easy to transport and use for untrained personnel. Thus, a major benefit of the system is that it can be used in the field by the end user whenever necessary.

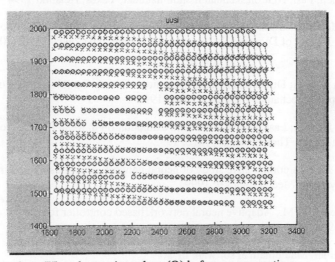

Figure 4. True values (X) and set point values (O) before compensation.

210

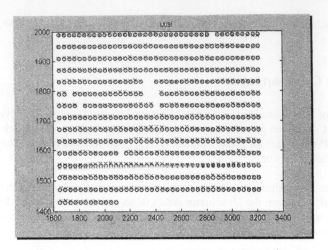

Figure 5. True values (X) and set point values (O) after compensation.

References

[1] Craig, John J.; Introduction to Robotics; Mechanics and Control; 1986; ISBN 0-201-10326-5

[2] Rouvinen, Asko; Use of Neural Networks in Robot Positioning of Large Redundant Manipulators; 1999; ISBN 951-764-368-3

[3] Koikkalainen, Pasi; Neurolaskennan mahdollisuudet; Tekes; Helsinki 1994

[4] Arteaga, Marco A.; On the Properties of a Dynamic Model of Flexible Robot Manipulators; Journal of Dynamic Systems, Measurement and Control, Vol 120, March 1998

[5] Rubinstein D. & al; Direct and inverse dynamics of a very flexible beam; Computer methods in applied mechanics and engineering; 1995

[6] Li, Chang-Jin & al; Fast Inverse Dynamics Computation in Real-Time Robot Control; Mech. Mach. Theory Vol 27, No 6, 1992

[7] Efe, M. O. Et al; A comparative study of neural network structures in identification of nonlinear systems; Mechatronics 9; 1999

[8] Abdelhameed, M. M.; Adaptive neural network based controller for robots; Machatronics 9 (1999);

[9] Haque, Aamer, Mathematical software in C; www.science.gmu.edu/āhaque/math/m_index.html; 17.1.2000

Human Friendly Mechatronics (ICMA 2000)
E. Arai, T. Arai and M. Takano *(Editors)*
© 2001 Elsevier Science B.V. All rights reserved.

Camera calibration in machine automation

Matti Kutila, Juha Korpinen, and Jouko Viitanen

VTT Automation, Machine Automation
P.O. BOX 1302, FIN-33101 Tampere, Finland

This paper deals with utilization of a camera calibration method in machine automation. Currently available calibration methods are described shortly, and a new type of calibration object which is suitable for use in, e.g. small parts assembly cells, is presented. Also the method of searching the calibration points is given new consideration. Experimental results are presented on the tests using the calibration method with the novel target.

1 INTRODUCTION

Simplified camera models, so called pinhole models, normally assume that light rays go through the lens without bending. In real life light rays bend in the lens which causes geometric distortion (aberration) to the image. Aberrations are especially amplified in short focal length lenses. In the following, aberrations caused by the non-linearity of the lens are called radial distortion. Another type of distortion which is typically called tangential, is the result of the offset between the optical axis of the lens and the normal of the image plane, together with possible tilt of the image plane. Radial distortion is formed symmetrically around the location of the optical axis and therefore accurate correction of the radial distortion requires that tangential distortion is corrected first.

The purpose of the camera calibration is to define the accurate camera model, which also takes into account the distortions in the image. When the accurate camera model is known, it is possible to remove the effects of the distortions from the image and perform accurate geometric measurements based on it.

Several different kinds of calibration methods are proposed in literature. We have tested and investigated especially the method introduced by Heikkilä [2] which uses four calibration steps. We have further developed the method by adapting it to use a new kind of calibration object, which makes it easier to utilize the method in machine automation industry. Only one image of this object is needed for executing the calibration process. Other new developments are the modules of software for searching the calibration points and executing the calibration process. These were written in the ANSI C programming language. The software makes it easy to embed the calibration method in machine vision applications in e.g. automation cells.

The advantage of the new calibration method is that it is fully automatic, while previously several manual steps were needed, and several pictures of test images had to be taken. If the calibration object is located inside an automatic assembly cell, the camera parameters can even be calculated while the cell is running online.

2   CALIBRATION METHODS

2.1   Camera model

The purpose of the camera calibration is to build an accurate mathematical camera model. Model building starts from pinhole model:

$$u_p = K_x \frac{f}{Z} X + u_0 \qquad (1)$$

$$v_p = K_y \frac{f}{Z} Y + v_0 \qquad (2)$$

where (X, Y, Z) are world frame coordinates, (K_x, K_y) are coefficients for converting millimeters to pixels, (u_0, v_0) is the location of the optical axis, (u_p, v_p) coordinates according to the pinhole model, and f the focal length.

The pinhole camera model is valid only when lenses with long focal length are used. If more accurate measurements are wanted, the radial distortion has to be taken into account. The second and fourth order terms in the polynomial approximation that is used for modelling the radial distortion have been proposed by Heikkilä [2]:

$$\Delta u_r = K_x \frac{f}{Z} X \left(k_1 r^2 + k_2 r^4 \right) \qquad (3)$$

$$\Delta v_r = K_y \frac{f}{Z} Y \left(k_1 r^2 + k_2 r^4 \right) \qquad (4)$$

where $r = \frac{f}{Z} \sqrt{(K_x X)^2 + (K_y Y)^2}$, (k_1, k_2) are radial distortion coefficients and $(\Delta u_r, \Delta v_r)$ are the effects of the radial distortion (differences from the linear model).

There also other kinds of radial distortion models proposed in literature. For example Correia et al. [1] use the third and the fifth order terms in their model.

Heikkilä [2] has modelled tangential distortion with the following model:

$$\Delta u_t = 2t_1 \frac{f^2}{Z^2} K_x K_y XY + t_2 \left(r^2 + 2 \frac{f^2}{Z^2} K_x^2 X^2 \right) \qquad (5)$$

$$\Delta v_t = t_1 \left(r^2 + 2 \frac{f^2}{Z^2} K_y^2 Y^2 \right) + 2t_2 \frac{f^2}{Z^2} K_x K_y XY \qquad (6)$$

where (t_1, t_2) are tangential distortion coefficients and $(\Delta u_t, \Delta v_t)$ are effects of the tangential distortion.

The following accurate camera model combines the distortion models to the pinhole camera model:

$$u = u_p + \Delta u_r + \Delta u_t \qquad (7)$$

$$v = v_p + \Delta v_r + \Delta v_t \qquad (8)$$

2.2 Calibration routine

Many different kinds of calibration methods are presented in literature. The main difference between the methods is that many of them calculate only radial distortion coefficients and omit the tangential distortion. Tsai's RAC method is probably the most commonly used one in camera metrology [3]. With Tsai's technique it is possible to define radial distortion very accurately, but before using it, tangential distortion has to be removed. Another well known calibration technique is Weng's two step calibration process [3]. Actually, the method developed by Heikkilä [2] has similarities to Weng's method. The main idea is to first define a "good initial guess" for the nonlinear optimization routine by using a linear camera model.

During the camera calibration process, the focal length (f), the position of the optical axis on the image plane (u_0, v_0), the radial distortion coefficients (k_1, k_2) and the tangential distortion coefficients (t_1, t_2) are defined. When those parameters are known, it is possible to define the image which corresponds to the one produced by the pinhole model camera. The external parameters of the camera, its position and orientation relative to the world frame coordinates are also calculated.

Heikkilä's [2] calibration process is based on four successive steps. The actual parameter calculation is performed in three steps and the image correction is performed in the last step. At the first step, coarse camera parameters are calculated using the linear camera model. More accurate camera parameters are then calculated during the nonlinear optimization routine. The third step removes the inaccuracies caused by perspective projection.

3   CALIBRATION OBJECT

3.1   Object

Figure 1. The calibration object which is suitable for positioning into an automatic assembly cell.

In the design of the calibration object, special attention has to be paid on the compactness (see figure 1). The idea was that it should be possible to place the object inside into the most commonly used automatic assembly cells. The area taken by the pyramidal calibration object is about 221 mm x 221 mm and it's height is 138 mm. A three-dimensional calibration object makes it possible to calibrate the camera by using only one image. The holes in the object are used as calibration points. The material of the object is aluminum and it has been coated matt white for reducing specular reflections. The holes are drilled by an inaccuracy of 0,01 mm.

3.2 Calibration point searching

Not only the accuracy of locations of the calibration points on the object is important, but also the accuracy when searching them from the image. The positions of the points have to defined by sub-pixel accuracy. The easiest way for finding the centers of the calibration points is to calculate their central moments, i.e. weighted averages:

$$\bar{u} = \frac{\sum_v \sum_u uf(u,v)}{\sum_v \sum_u f(u,v)} \tag{9}$$

$$\bar{v} = \frac{\sum_v \sum_u vf(u,v)}{\sum_v \sum_u f(u,v)} \tag{10}$$

where u and v refer to the coordinates in an image and $f(u, v)$ to the corresponding gray level. Note that the weight at the outside of the calibration point has to be 0. For good accuracy, we have to have plenty of pixels per calibration point. It has been found that they should be at least 10 pixels wide.

4   EXPERIMENTAL RESULTS

The tests of the calibration method have been divided to three parts. First we tested the ability of the method to define radial distortion parameters. Then we tested its suitability for recognizing the offset of the optical axis, and finally we checked the correctness of the external parameter calculation routine. A video camera with an 8 mm lens was used in the tests. We utilized the calibration software in Matlab code made by Heikkilä [2], and the calibration object shown in Figure 1.

A raster pattern was imaged in the radial distortion test. The distances between the raster points were equal, which means that the same number of pixels should be found between the points. Dividing the number of pixels by the distances of the points in millimeters gives us the spatial resolution. If there is radial distortion, it should cause variations to the resolution at different parts of the image area. Figure 2 shows the variation of the spatial resolution in the image area, measured in the horizontal direction. The upper line shows the resolution in the uncorrected image and the lower line in the corrected image that utilized the calibration method. In the corrected image, there are only small deviations in spatial resolution, which is mainly due to the measuring inaccuracies during the test. The effects of the aberrations have decreased clearly.

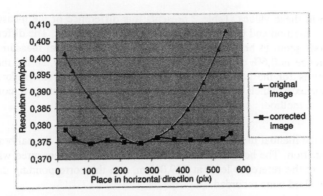

Figure 2. The measured spatial resolutions in the horizontal direction of the image.

The ability of the method to find the position of the optical axis from the image was examined by comparing the search results to an opto-mechanical method developed at VTT. The method is based on an iterative search of target points from the images of the camera while it is rotated around an axis that is close to the direction of the optical axis. The target points have to be far enough from the camera. The results of the measurements are shown in Table 1. The result of the calibration method has been calculated as an average of four measurements.

Table 1
X-Y positions of the optical axis measured with different methods.

	Nominal	Optomechanical method	Calibration method
Measure-ments	(320, 240)	(310.7, 223.9)	(313.5, 219.5) (312.3, 224.6) (309.6, 222.6) (311.0, 224.6) (309.7, 224.1)
Result	(320, 240)	(310.7±3.5,223.9±3.5)	(311.2±0.8,223.1±1.0)

The ability of the method to define the external parameters (position and orientation) of the camera was tested by measuring the distance between the calibration object and the camera. The inaccuracy of the handmade measurements can be ±2 mm. The calibration method gives the distance to the primary principal point of the lens which is difficult to determine without having the prescription data of the lens (the secondary principal point is, of course, well known). However, if we move the camera to different heights, and find the differences of the measured heights to a known reference position, then there is no dependence on the principal point separation in these values. Table 2 presents this difference of the measurements given by the calibration method and those given by manual measurements. The differences between these two methods seem to be within the error limits of the manual measurements, and the maximum difference is 0,6% of the value of the reference position, so the method seems to give reliable results for the camera height. Corresponding tests were performed for object rotation along one axis, which also gave positive results, so these tests gave confidence on the performance of the method.

Table 2
Comparisons of the manual measurements with the values given by the calibration method in position determination. The height variations of the camera were measured when the camera was lifted up from the reference level at 269,0 mm, and the corresponding differences from the calibration method were recorded.

Manual measurement (mm)	Calibration method (mm)	Difference (mm)
30	31,7	1,7
45	46,2	1,2
65	66	1
110	110,9	0,9

5 CONCLUSIONS

We presented improvements to a camera calibration method, including a compact calibration object, and possibility to perform the calibration with only one captured image. The tests of the method show that significant improvement of the accuracy was achieved for cameras that use short focal length lenses. In several tests the method was found to give reliable results when the novel calibration object was used. We also performed a few successful tests for determining the position and orientation of the camera with the same method.

REFERENCES

1. B. Correia, J. Dinis, Camera model and calibration process for high accuracy digital image metrology of inspection planes, proc, SPIE, vol 3521, Boston, Massachusetts, U.S.A. (1998).
2. J. Heikkilä, Accurate camera calibration and feature based 3-D reconstruction from monocular image sequences, Acta Universitas Ouluensis C Technica 108, Oulu, Finland (1997).
3. H. Zhuang, Z. Roth, Camera-Aided Robot Calibration, CRC Press, Florida, Florida Atlantic University. (1996).

Human Friendly Mechatronics (ICMA 2000)
E. Arai, T. Arai and M. Takano (Editors)
© 2001 Elsevier Science B.V. All rights reserved.

Real Time Machine Vision System in Micromanipulator Control

J. Korpinen[a], P. Kallio[b], J. Viitanen[a]

[a]VTT Automation, Machine Automation
P.O. BOX 1302, FIN-33101 Tampere, Finland

[b]Tampere University of Technology, Automation and Control Institute
P.O.Box 692, FIN-33101 Tampere, Finland

ABSTRACT

In this paper a system for controlling a micromanipulator having several degrees of freedom is presented. One CCD camera is used together with a microscope and a high speed DSP system to provide the correct position of the actuator in a microscopic space. The vision algorithm is based on a modified *CMA* method for producing the exact XY-position. The relative Z-coordinate is calculated using a modified *depth-from-defocus* method. A software implementation is presented and performance for the vision system is given together with the test results.

1. INTRODUCTION

Micromanipulation technique is a widely studied area. Several areas of biotechnology are using or planning to use micromanipulation as one of the techniques, when new drugs are studied or when the effect of e.g. toxic substancies is studied in human cells or organisms. Also the assembly of microelectromechanical components needs accuracy and dexterity which has not yet been available. The physical sizes of these kinds of microcomponents are in the area from less than a micrometer to a few hundreds of micrometers. This raises the problem of not only detecting the particles or components, but also detecting the active part (*pipette*) of the micromanipulator itself. Also controlling the manipulator using conventional sensors is difficult. In this paper we present one technique on how to detect the pipette end of a micromanipulator.

2. METHODS FOR 3D POSITION DETERMINATION

We used a method called Chamfer Matching Algorithm (CMA) [2,3] to detect the XY-position of the pipette. It produces is a robust way to use models in detecting predefined forms in a digital image. CMA is not always the fastest way to detect the appearance of predefined objects in an image. For this reason we modified the original algorithm.

There are several different techniques on how to calculate the depth in a three dimensional world coordinates (x,y,z) using a machine vision system in a macroscopic world. Normally these techniques use e.g. lasers (TOF type of systems or structured light), or stereo camera systems. We used an algorithm modified from the depth-from-defocus method [4]. The

218

calculation of the 3D coordinates is done in two phases. First the (x,y) position is detected using the CMA and second the relative z coordinate is calculated using the depth-from-defocus method. The result is an exact (x,y) position and a relative z coordinate.

3. VISION METHODS USED

All of the methods used are adaptations from different original methods. The adaptation was done because the original methods were not efficient enough to provide real time speed.

3.1. Image preprocessing

The first task in processing is to grab a suitable background image (see *Figure 1*). This image is then stored in memory. After this a new image (see *Figure 2*) is grabbed with the manipulator (in this case modelled with a pen) visible in the FOV (Field Of View). These subsequent images are then subtracted from each other and the result is shown in *Figure 3*. The subtraction can be expressed as follows:

$$S(i, j) = |I_1(i, j) - I_2(i, j)| \tag{1}$$

where S is the result image with dimensions i*j.

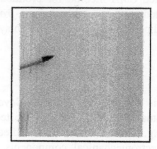

Figure 1. Background image

Figure 2. Subsequent image, with the actuator (e.g. pen)

Figure 3. Result image.

The result is an image where the actuator is seen more clearly. Normally this results in an output image, which can be segmented by using just one global threshold value.

3.2. Segmentation

Following the subtraction a binarization is executed by using a global threshold value. This value is user defined and it can be selected through the user interface. The segmentation (e.q. binarization) can be expressed as follows:

$$B(i, j) = 1|0; S(i, j) > T_{val}, i = i, N, j = 1, M \tag{2}$$

where B is the result image and T_{val} is the selected threshold value.

In an ideal situation this will lead to an image, where the actuator is the only object (see *Figure 4*).

Figure 4. Threshold Image

3.3. Edge Detection

Edge detection is used for two different purposes. The first one is to create a suitable model polygon for the CMA (explained in chapter: *Chamfer Matching Algorithm – CMA*) and the other one is for the distance transformation, which is also needed for the CMA. We use the Sobel edge detector for creating a continuous edge.

3.4. Creating the model

The Sobel gradient is not always the ideal method in model polygon generation, considering the number of points it gives. Sobel creates quite wide edges thus the number of edge points increases. To avoid this problem we modify the number of points by using an automatic point selector function. The function creates a raster envelope which then automatically selects the points to be included in the model. The raster function is shown in *Figure 5*.

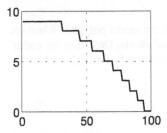

Figure 5. Column spacing between selected columns, when creating the model polygon.

The raster function determines increase in column spacing, when we move away from the left side of the image.

3.5. Chamfer Matching Algorithm – CMA

The Chamfer Matching Algorithm (CMA), first proposed by Barrow et. al. [4], is a template matching method for finding predefined objects in a two dimensional image. For the CMA, a distance image is generated. A fast method for calculating this distance transformation is the so called Chamfer 3-4 [2].

The matching measure **E** for elements e(i,j) in a given position (i,j) can be expressed as follows:

$$e(i, j) = \sum_{m=0}^{M-1}\sum_{n=0}^{N-1} d(i+m, j+n)p(m,n); 2 \leq M \leq I, 2 \leq N \leq J \tag{3}$$

where **D** (with elements d(i,j)) is the distance transformed edge image, with dimensions I*J and **P** (with elements p(m,n) is a binary image for the model polygon, with dimensions M*N.

The best fit for the model polygon is found in the position (i,j), which gives the minimum matching measure E_{min} over the image, i.e.

$$E_{min} = \min_{i,j} e(i,j); 0 \leq i \leq I, 0 \leq j \leq J \tag{4}$$

3.6. CMA in real time environment

As CMA is computationally intensive method for calculating the position of a predefined object, a hierarchical method is often used. This Hierarchical Chamfer Matching Algorithm (HCMA) [2] is based on scaling down the original image and the model polygon. The scaled model is fitted to the scaled images; therefore only a small area needs to be checked when the final position is calculated in the original unscaled image.

Instead of scaling we used decimation in finding the best fit for the model polygon. When we know the size of the model polygon, we can calculate predefined points. Detecting the width (W) and length of the model (L) does this calculation. The decimation factor (R_s) used in decimation can then be expressed as follows:

$$R_w = I/W \ , \ R_L = J/L \tag{5},(6)$$

we used a common value (R_s) for both directions, selected by (7)

$$R_s = R_L | R_W; R_L < R_W \tag{7}$$

Also a Region Of Interest (ROI) was used. The ROI size was defined as follows:

$$S_{ROI} = 2*R_L | 2*R_W; R_L > R_W \tag{9}$$

where S_{ROI} is the width and length of the ROI.

3.7. Calculating the relative Z – coordinate

We use a modified method from the *depth-from-defocus* method. In this method we calculate the pixel energy in the center (E_c) and in the edge (E_e) of the pipette. The pixel energy (E_p) for an image position $f(i,j)$ can be expressed as follows:

$$E_p = (f(i,j))^2 \tag{10}$$

To avoid noise, we use an area taken from and center of the pipette and from the edge line.

$$E_c = \frac{1}{N}\sum_{i=n}^{n+N_1}\sum_{j=m}^{m+M_1} f(i,j)^2 \; , \; E_e = \frac{1}{N}\sum_{i=n}^{n+N_1}\sum_{j=m}^{m+M_1} f(i,j)^2 \; , \; N = N_1 * M_1 \tag{11),(12),(13}$$

where N is the number points in selected area, n is starting row, m is starting column.

Calculating a norm from the background (Nb) does also compensating variations in illumination. For the relative the Z – coordinate, we must calculate the difference (dE) between these two energy values:

$$dE = E_e - E_c \tag{14}$$

The background norm (N_b) is calculated from a position, where the pipette is not seen. This value is then compared to a predefined energy level. The norm can then be expressed ad follows:

$$E_b = \frac{1}{N}\sum_{i=n}^{n+N_1}\sum_{j=m}^{m+M_1} f(i,j)^2 \; , N_b = E_b / E_{pre} \tag{15),(16}$$

finally the relative Z – coordinate Z_{rel} is calculated by (17)

$$Z_{rel} = dE / N_b \tag{17}$$

Following this relative change and comparing it to the value we got in the first image, we can monitor the relative z coordinate. The relative z coordinate then makes it possible to use a closed loop control in three-dimensional world to compensate the drift of the manipulator.

4. TESTS AND RESULTS

The test system consists of a microscope with an internal illumination source and a CCD camera attached to the microscope. An objective with magnification factor of ten is used to provide one micrometer spatial resolution. Image grabbing and processing is done on a proprietary DSP-system, builtin VTT.

The tests showed that the inaccuracy of the system was ± 1 pixel in XY-direction, giving a ± 1 μm absolute accuracy in XY-plane. The relative Z-coordinate was also used. The tests also proved that using feedback signal obtained from the vision system reduces the drift, this can be seen in figure 7. The result of the accuracy tests can be seen in *Figure 6*.

222

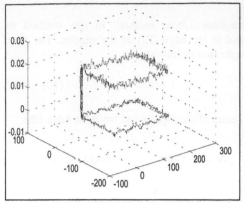

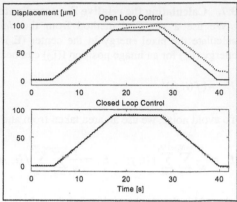

Figure 6. A measured micromanipulator trajectory in two different planes in 3D. The X-Y plane units are micrometers; the Z plane units are relative arbitrary units.

Figure 7. Open Loop Control vs. Closed Loop control. Closed Loop Control removes the drift caused by the piezos.

The calculation time was 67 ms. Giving 15 Hz Frequency for the feedback signal. This is without code optimization.

5. CONCLUSION

It has been found that using a vision system and CMA, it is possible to build a real time system for controlling a micromanipulator. Using modified method's based on *CMA* and *depth-from-defocus* together with DSP-board gives a feedback signal frequency high enough, for a closed loop control. This removes the drift caused by the piezos. Using several ROI's gives us the possibility to monitor and control several manipulators in the same FOV. Controlling several manipulators gives us the possibility to manipulate objects so, that gripping; holding and releasing will be easier.

REFERENCES

[1]   J.H. Takala, J.O. Viitanen, A Model-Based Vision System of Work Machine Manipulator, Proc. Of the ICMA '94. Vol. 1 (1994). pp 243-256.

[2]   G. Borgefors, On Hierarchical Edge Matching in Digital Images Using Distance Transformations. Dissertation TRITA-NA-8602, The Royal Institute of Technology, Stockholm, Sweden 1986.

[3]   H.G. Barrow, J.M. Tenenbaum, R.C. Bolles, H.C. Wolf, Parametric Correspondence and Chamfer Matching: Two New Techniques for Image Matching, Proc. Of the 5th Annual Conference on Artificial Intelligence, IJCAI, August 1977. pp 659-663.

[4]   A. Pentland, T. Darrel, M. Turk, and W. Huang, A Simple, Real Time Range Camera, CH2752-4/89/0000/0256$01.00 IEEE, 1989

Human Friendly Mechatronics (ICMA 2000)
E. Arai, T. Arai and M. Takano *(Editors)*
223

Object tracking by fusing distance and image information

Koichi HARA and Kazunori UMEDA

Course of Precision Engineering, School of Science and Engineering, Chuo University
1-13-27 Kasuga, Bunkyo-ku, Tokyo 112-8551, Japan

In this paper, combination of multiple ultrasonic sensors and vision sensor are discussed as the sensing system for mobile robots, and methods of object tracking are proposed by fusing distance and imagery information. They are formulated using the framework of Kalman filter. As a technique of object tracking, a method to detect change of moving direction of a moving object is proposed. Methods of object tracking by using images are established utilizing subtraction and normalized correlation of images, and are verified by experiments.

Key words: Mobile Robot, Vision, Ultrasonic Sensor, Object Tracking, Kalman Filter

1. INTRODUCTION

Vision is necessary for mobile robots to avoid obstacles and reach the destination in natural environment. Vision system of human beings has two different functions : intensive and wide-angle observations. The vision system of a mobile robot should have the same ability[1]. Umeda et al. [2] proposed the combination of ultrasonic sensors as wide-angle observation and vision sensor as intensive observation, and formulated the fusion of two different data using the framework of Kalman filtering for measuring moving obstacle's motion. In this paper, we improve the proposed sensing system and methods in [2] and propose the methods for the object tracking. Methods of object tracking by using images are verified by experiments.

2. OVERVIEW OF OBJECT TRACKING SYSTEM

2.1. Overview of sensing system

Figure 1 shows the overview of the sensing system. The sensing system consists of multiple ultrasonic sensors and a CCD camera. The ultrasonic sensors are placed around the system. The CCD camera is placed above the center position of the system. We suppose the camera can pan and tilt according to the motion of a moving object.

The system architecture is improved from the sensor system in [2] so as to perform object tracking with higher flexibility and precision. The system has H8 microcomputers (HITACHI) and one personal computer (hereafter, PC) as data processing units. Ultrasonic sensors are controlled by H8, since H8 can write some programs on ROM, and can run various processes. H8 transmits measured distance data to the PC when the PC requires. On the other side, an image processor is controlled by the PC.

Multiple ultrasonic sensors are turned on alternately to avoid the crosstalk. Consequently, it is about 50msec (2 cycles) to measure all sensor data with the measuring range of 3m. The most effective point of using H8 is that H8 can hold about five hundred data from ultrasonic sensors for a while on RAM. Therefore it can hold about 3.5sec's data, and the improved sensing system can process two kinds of sensor data at the same time.

The data processing of the sensing system is illustrated in Figure 2. The PC controls the image processor, and obtains data including the position of the moving object on images. On the other side, H8 controls multiple ultrasonic sensors, and transmits the distance data to the PC at constant intervals. Image data and distance data are fused by Kalman filtering. Flow of the processes is as follows (see Figure 2).

(1) PC orders the image processor to obtain the image data.
(2) Image processor transmits the image data to PC.
(3) PC orders CCD camera to change the direction using image data.
(4) PC obtains the camera parameters.
(1)' H8 orders multiple ultrasonic sensors to obtain the distance data.
(2)' Multiple ultrasonic sensors transmit the distance data to H8.
(3)' H8 transmits the distance data to PC at constant intervals.

2.2. Overview of object tracking

Figure 3 illustrates the object tracking with the proposed sensor system. Multiple ultrasonic sensors find moving objects, and measure rough position and motion of each object. Motion in the measuring direction of each ultrasonic sensor is measured using the change of measured distance data, and motion in the vertical direction is measured using the change of measuring range.

The CCD camera utilizes the data from the multiple ultrasonic sensors, and turns toward the nearest object from the sensing system. The camera is controlled so that the moving object is kept on the center of the image of the camera. While the moving object is in the range of the CCD camera, the CCD camera tracks the moving object and can add more precise measurement of motion.

Two methods of sensor fusion for object tracking are constructed. One is fusion of the two different sensor data, and the other is fusion of multiple ultrasonic sensor data. They are formulated using the framework of Kalman filtering.

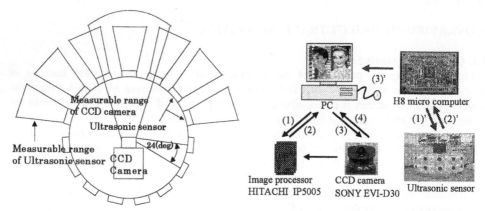

Figure 1. Overview of sensing system   Figure 2. Data processing of sensing system

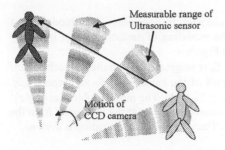

Figure 3. Measurement of motion by proposed sensor system

3. FORMULATION OF OBJECT TRACKING

In this chapter, we formulate the object tracking shown in section 2.2 by using the framework of Kalman filtering[2] under following three conditions.
- one moving object
- moving object is a material point
- velocity of the moving object is nearly constant in horizontal direction, and nearly zero in vertical direction

3.1. State equation

On the above assumptions, state variables can be set $\mathbf{x}=[x\ v_x\ y\ v_y\ z]^\mathrm{T}$, where $[x\ y\ z]^\mathrm{T}$ is the position vector of the moving object, and $v_x,\ v_y$ are velocity components for x and y. z is the vertical component of the position vector, and can be supposed to be constant. State equation for the motion of the moving object is defined as

$$\mathbf{x}_{i+1} = \Phi\mathbf{x}_i + \Gamma\omega_i, \tag{1}$$

$$\Phi = \begin{bmatrix} 1 & \Delta t & & & \\ 0 & 1 & & & \\ & & 1 & \Delta t & \\ & & 0 & 1 & \\ & & & & 1 \end{bmatrix}, \quad \Gamma = \begin{bmatrix} 0 & 0 & 0 \\ 1 & 0 & 0 \\ 0 & 0 & 0 \\ 0 & 1 & 0 \\ 0 & 0 & 1 \end{bmatrix},$$

$$\omega_i = \begin{bmatrix} \omega_{vxi} \\ \omega_{vyi} \\ \omega_{zi} \end{bmatrix}, E[\omega_i] = \begin{bmatrix} 0 \\ 0 \\ 0 \end{bmatrix}, V[\omega_i] = \begin{bmatrix} \sigma_{vxi}^2 & 0 & 0 \\ 0 & \sigma_{vyi}^2 & 0 \\ 0 & 0 & \sigma_{zi}^2 \end{bmatrix}.$$

x_i is the value of state variables $\mathbf{x}$ at time i, and Δt is the sampling time. ω_i is the disturbance for the velocity of the moving object. The average of ω_i is zero, and its covariance matrix is defined as above with known variances.

3.2. Modeling of CCD camera

To adopt Kalman filtering, modeling of measurement is necessary. The modeling of ultrasonic sensors is shown in [2]. Here we show the modeling of measurement of image by CCD camera. Measuring model of image is shown in Figure 4.

Each axis of the coordinate system of the camera Σ_c is originally set parallel to each axis of robot coordinate system Σ_R. The angles of pan and tilt of the camera are θ_P, θ_T respectively. The position vector of the lens center of the camera in Σ_R is $\mathbf{d_c}$. The moving object at $[x\ y\ z]^T$ is projected on $[u\ v]^T$ in the image plane which is assumed to be set at the distance f from the lens center. f is focal length of the lens of the camera. Then the following equation is derived.

$$
d \begin{pmatrix} \cos\theta_P & \sin\theta_P & 0 \\ -\sin\theta_P & \cos\theta_P & 0 \\ 0 & 0 & 1 \end{pmatrix} \begin{pmatrix} 1 & 0 & 0 \\ 0 & \cos\theta_T & -\sin\theta_T \\ 0 & \sin\theta_T & \cos\theta_T \end{pmatrix} \begin{pmatrix} u \\ v \\ f \end{pmatrix} + \mathbf{d}_C = \begin{pmatrix} x \\ y \\ z \end{pmatrix} \tag{2}
$$

By eliminating unknown variable d,

$$
\begin{bmatrix} u \\ v \end{bmatrix} = \frac{f}{-\sin\theta_T\sin\theta_P x'-\sin\theta_T\cos\theta_P y'+\cos\theta_T z'}\begin{bmatrix} \cos\theta_P x'-\sin\theta_P y' \\ \cos\theta_T\sin\theta_P x'+\cos\theta_T\cos\theta_P y'+\sin\theta_T z' \end{bmatrix} \tag{3}
$$

where $x'=x-x_c, y'=y-y_c, z'=z-z_c$. By differentiating eq.(3) by $[x,y,z]^T$, Jacobi matrix H, which is utilized in Kalman filtering, is obtained as

$$
H = \frac{f}{\left(-\sin\theta_T\sin\theta_P x'-\sin\theta_T\cos\theta_P y'+\cos\theta_T z'\right)^2}
$$

$$
\times \begin{bmatrix} -\sin\theta_T y'+\cos\theta_T\cos\theta_P z' & \sin\theta_T x'-\cos\theta_T\sin\theta_P z' & -\cos\theta_T\cos\theta_P x'+\cos\theta_T\sin\theta_P y' \\ \sin\theta_P z' & \cos\theta_P z' & -\sin\theta_P x'-\cos\theta_P y' \end{bmatrix} \tag{4}
$$

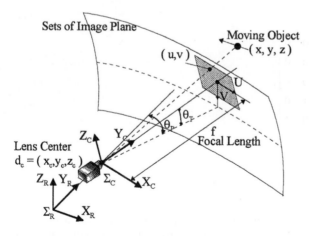

Figure 4. Measuring model of image

3.3. Detection of change of moving direction

The moving object e.g., a man, may change the moving direction. Here we show the method to detect the change of moving direction. This method uses estimated values by Kalman filtering, and measured values by multiple ultrasonic sensors and CCD camera. The validity of the fusion of the estimated values and the measured values is evaluated by chi-square test. If the test is not satisfied, a new state equation is formulated using the latest two measured data, and estimation of position is continued by Kalman filtering. This is illustrated in Figure 5. The sum of square of residuals S_i for the test is calculated by

$$S_i = (\overline{\mathbf{x}}_i - \hat{\mathbf{x}}_i)^T \mathbf{M}_i^{-1} (\overline{\mathbf{x}}_i - \hat{\mathbf{x}}_i) + (z_i - \mathbf{H}_i \hat{\mathbf{x}}_i)^T \mathbf{R}_i^{-1} (z_i - \mathbf{H}_i \hat{\mathbf{x}}_i) \tag{5}$$

where $\overline{\mathbf{x}}_i$ is the estimated value before adding measured value, and $\hat{\mathbf{x}}_i$ is the estimated value after adding the measured values. $\mathbf{M}_i$ is covariance matrix of $\overline{\mathbf{x}}_i$, and z_i is the measured value by ultrasonic sensors or the measured value $[u_i \ v_i]^T$ by the CCD camera. H_i is the Jacobi matrix between z_i and $\mathbf{x}_i$. R_i is the covariance matrix of z_i.

4. OBJECT TRACKING BY USING IMAGES

Subtraction and normalized correlation are utilized to extract a moving object from images and tracks it. The effective point of using subtraction is that the influence of background is little, and it is possible to decide template area of the moving object from out of range of the CCD camera automatically. Figure 6 shows the flow of the processes.

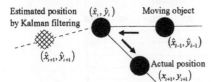

Figure 5. Detection of change of moving direction

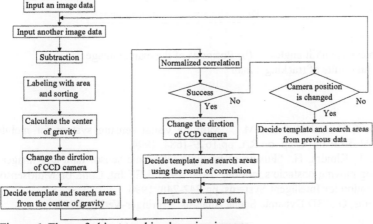

Figure 6. Flow of object tracking by using images

228

5. EXPERIMENTS OF OBJECT TRACKING BY USING IMAGES

We show experimental results to evaluate the object tracking method with images proposed in chapter 4. Sensing system in the experiments is constructed with an image processor IP5005 (HITACHI), a PC (Pentium III 500MHz, Linux), and a CCD camera EVI-D30 (SONY). Experiments were performed in an office room (our laboratory room). A man walked in front of the sensor system from left to right for about 3m, and then turned back. The distance between the sensor system and the man was about 2.5m. Figure 7 shows the experimental results. The system found the man at "start" point in Figure 7(a) and then turned to the man and started tracking. Figure 7(b) shows the positions of the tracked man in the images. It is shown that the camera continued to hold the man at the image center. It is also shown that measurable range is extended by pan of camera. The experimental results show that the proposed method is appropriate for object tracking.

6. CONCLUSION

In this paper, we proposed the method for object tracking using the sensing system with multiple ultrasonic sensors and a CCD camera with pan and tilt motion. The methods were formulated using the framework of Kalman filtering. As a technique for object tracking, we proposed a method to detect change of moving direction of a moving object. And then we realized the methods of object tracking by using images utilizing subtraction and normalized correlation of images.

More experimental verification is the most important future work.

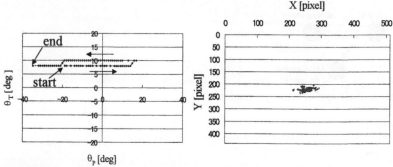

(a) Trajectory of camera's pan/tilt angles   (b) Positions of the man in images
Figure 7. Experimental results of tracking a man

REFERENCES

[1] Yagi, Y., Okumura, H. and Yachida, M. : "Multiple visual sensing system for mobile robot," Proc. 1994 IEEE Int. Conf. on RA, pp.1679-1684, 1994.
[2] Umeda, K., Ota, J., Kimura, H.: "Fusion of multiple ultrasonic sensor data and imagery data for measuring moving obstacle's motion," Proc. 1996 IEEE Int. Conf. on Multisensor Fusion and Integration for Intelligent Systems, pp.742-748, 1996.
[3] Zhang, Z., Faugeras, O.: "3D Dynamic Scene Analysis," Springer-Verlag, 1992.

Human Friendly Mechatronics (ICMA 2000)
E. Arai, T. Arai and M. Takano *(Editors)*
© 2001 Elsevier Science B.V. All rights reserved.

Face Detection and Face Direction Estimation Using Color and Shape Features

Yuichi ARAKI   Nobutaka SHIMADA   Yoshiaki SHIRAI

Dept. of Computer-Controlled Mechanical Systems Osaka University
2-1, Yamada-oka, Suita, Osaka, 565-0871, Japan

Abstract: This paper describes the detection of faces and the estimation of the face direction from an image where their sizes, positions and poses are unknown. As the candidate regions of the face, we first extract the regions whose upper part is hair color and lower part is skin color. To make sure that there is a face actually in the face candidate region, we next extract facial features, pupils, a nostril and a mouth, using color, shape and the geometrical conditions of the facial features. After the face detection, the face direction is estimated on the basis of the position of the pupils and face contour. We show the effectiveness of this method by experimental results.

1. Introduction

Face detection and face direction estimation are important for face recognition. In personal identification with surveillance cameras, for example, it is necessary to detect the face whose size, position, and pose are unknown. After the face detection, the face direction estimation is useful for the correct face recognition because we can select the face image of the most desirable direction from the face images taken by the multiple cameras.

Many methods have been proposed in the field of the face detection. One of them is based on the matching of facial template images. However, the size and pose of the face are limited because it takes terrible computation cost to consider all sizes and poses of the template image. On the other hand, the methods based on a skin color can detect any sizes and poses of the face. Because it is difficult to detect the face from a skin color background, the methods use in addition a head shape information[1] or a hair color information[2]. Moreover it is necessary to make sure that there is a face actually in the region detected by the methods in order to reject the false detection. To make sure whether there is a face actually or not, the approach to extracting facial features such as pupils, a nostril and a mouth is considered. For the facial features extraction, the method based on the geometric face model is proposed[3]. However, the method assumes the nearly frontal face.

In this paper, we propose the new method of the face detection. We first extract the face candidate regions by using the skin and hair color information and then seek the facial features to make sure that there is a face actually in the face candidate region. To extract the facial features of the face whose pose is unknown, we extract the candidates of

Figure 1. Two color regions

Figure 2. Combined regions

Figure 3. Face candidate regions

Figure 4. Template image for the separability

the facial features by using the color and shape information and then match them to the face models of three directions: front, left and right. If there are the facial features in the face candidate region, we regard it as the face region. Otherwise we eliminate the region. By the matched face model type, we can classify the face direction into three: front, left and right. In case of the frontal face, we estimate the face direction more precisely on the basis of the position of the pupils and face contour. We show the effectiveness of this method by the experimental results.

2. Extraction of the face candidate regions

For the skin color pixels, we extract the pixels satisfying all of the following conditions:

$$0.333 < r < 0.664, \qquad 0.246 < g < 0.398, \qquad r > g, \qquad g \geq 0.5 - 0.5r \qquad (1)$$

where r=R/(R+G+B) and g=G/(R+G+B). These conditions are experimentally determined by extracting the skin color pixels manually for 15 persons in our laboratory and 23 persons in the television images. The skin color regions are made by connecting the skin color pixels.

For the hair color pixels, we use the luminance signal Y=0.30R+0.59G+0.11B. If a pixel is not in the skin color region and its Y is smaller than a threshold (90 in the following experiments), it is regarded as the hair color pixel. The hair color regions are made by connecting the hair color pixels.

In Fig.1, the skin color regions and the hair color regions are respectively shown with gray and black color. We extract the combined regions whose upper part is the hair color region and lower part is the skin color region. In Fig.2, the combined regions are shown. If the aspect ratio and the size of the combined region satisfy the conditions as the face region, we extract the rectangle region enclosing the skin color region of the combined region as a face candidate region, shown as the white rectangle in Fig.3.

3. Extraction of the facial features

3.1. Separability of the dark facial feature

In the face candidate region, we first extract the candidates of the facial features. The common feature of them is dark. To obtain them, we extract the dark regions by connecting the dark pixels. Circularity is another common feature of the pupil and the nostril. To extract the dark circular region, we use the degree of the separability. At the dark pixels, two regions are defined by making three concentric circles (Fig.4). We compute the ratio of the between-class variance of the brightness and the total variance of the two regions[4]. The ratio is called as the separability. The higher the ratio of the

  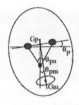 

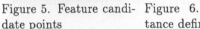

Figure 5. Feature candidate points Figure 6. Distance definition Figure 7. Angle definition Figure 8. Extracted facial features

dark pixel is, the more circular the dark region is. In each dark region, the pixel having the maximal separability is extracted as a feature candidate point (Fig.5).

3.2. Matching the feature candidate points to the face models

The feature candidate points are matched to four face models in the order of the frontal face model with a nostril, the frontal face model without a nostril, the left or right directions model (profile model) and the profile model with an occluded pupil. If the feature candidate points are matched to a certain face model, we stop matching and regard them as the facial features.

3.2.1. Model A (Frontal face model with a nostril)

We define the variables used in the model A and B in Fig.6 and 7 where G_p and G_m are respectively the center of the gravity of the pupils and that of the mouth. We first extract the sets of the feature candidate points satisfying the Eqs.(2) and (3).

$$-\pi/8 \leq \theta_p \leq \pi/8, \qquad W_s/5 \leq D_p \leq W_s/2 \tag{2}$$

$$D_{nx} < D_p/2.7, \qquad D_p/2.7 < D_{ny} < D_p, \qquad \theta_{pn} < \pi/2 \tag{3}$$

If there are the sets, we seek the dark region satisfying the Eq.(4) for each set.

$$1.2 < D_{my}/D_{ny} < 2.0, \qquad \theta_{pm} < \pi/12 \tag{4}$$

If there are the sets having the dark region, we select the set having the maximal sum of the separability of the feature candidate points as the facial features (Fig.8).

3.2.2. Model B (Frontal face model without a nostril)

In this model, we first extract a red color region as the mouth region and then we seek the pupils based on the mouth position. For all pixels in the dark regions, we compute the hue by using the Eq.(5).

$$\theta = arccos \left(\frac{0.5 \times (2R - G - B)}{\sqrt{(R-G)(R-G) + (R-B)(G-B)}} \right) \tag{5}$$

The lower θ represents the closer to the red color. In Fig.9, the darker pixel represents the lower θ. We select the dark region having the minimal average of θ and the proper aspect ratio as the mouth region. Based on the mouth region, we obtain the candidates of the pupils (Fig.10). If there are the sets of the feature candidate points satisfying the Eq.(6), we select the set having the maximal sum of the separability as the pupils.

$$-\pi/6 \leq \theta_p \leq \pi/6, \qquad W_s/5 \leq D_p \leq 2W_s/3, \qquad \theta_{pm} < \pi/9 \tag{6}$$

The conditions (6) are more generous than those of the model A.

232

| Figure 9. θ value | Figure 10. Candidates of the pupils | Figure 11. Right direction model | Figure 12. Search region of the pupil |

3.2.3. Model C (Profile model)

This model consists of the left and right direction models. We first suppose the face direction based on the center of the gravity of the skin color region (G_s) and that of the hair color region (G_h). If G_s is on the left of G_h, we suppose that the face direction is right. Otherwise we suppose left. The positions of the pupil and the nostril are estimated based on the mouth region as the same way of model B. Fig.11 illustrates the right direction model. In case of left, it is horizontally reversed. In this model, we extract the sets of the feature candidate points satisfying the Eq.(7) instead of the Eq.(6).

$$\pi/9 < \theta_{pn} < 17\pi/36, \qquad \theta_{npm} < 2\pi/9, \qquad 0.6 < D_{nm}/D_{pn} < 1.5 \qquad (7)$$

If there are the sets, we select the set having the maximal sum of the separability as the facial features.

3.2.4. Model D (Profile model with an occluded pupil)

We also use the mouth region extracted above in this model. We extract the feature candidate points satisfying the Eq.(8) as the candidates of the nostril.

$$4\pi/9 < \theta_{nm} < 2\pi/3, \qquad D_{nm} < 0.3H_s \qquad (8)$$

where H_s is the height of the face candidate region. We select the point having the maximal separability as the nostril. Based on the positions of the mouth and the nostril, we set the search region of an occluded pupil by the Eq.(7). Fig.12 shows the search region with a white fan shape. If the ratio of the area of the hair color region and that of the search region is more than a threshold (0.3 as the threshold experimentally), we suppose that a pupil is occluded by the hair. Otherwise we consider that there is not a face in the face candidate region.

4. Estimation of the face direction

By the matched face model type, we can classify the face direction into three; front, left and right. If the model C or D is matched, the face direction is already known. In case of front, we estimate face direction more precisely by using the face direction estimation model (Fig.13). For the preprocess, we rotate the face region to equalize the vertical positions of the pupils. If the distance between the pupils is longer than a threshold ($W_s/2.5$: W_s is the width of the face region), we determine the L_x and R_x by using the skin color region. The L_x and R_x are respectively the horizontal distance between the G_p and the left side of the face region and the horizontal distance between G_p and the right side. Otherwise we need to extract the facial contour because we suppose that there are the skin color backgrounds in the face region. The example of this case is shown in Fig.15 where the left side of the face region doesn't correspond to that of the actual facial contour. In this case, we approximate the facial contour by two a-quarter ellipses. We

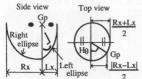

Figure 13. Face direction estimation model

Figure 14. Intensity of the contrast

Figure 15. Fitting of the ellipses

Figure 16. Extracted facial contour

obtain the contrast image by the Sobel operator for all pixels in the face region. Fig.14 shows the contrast image. We suppose that the center of each ellipse is around G_p and the major axis is vertical. Based on the positions of the facial features, we can set the initial a-quarter ellipses. The initial length of the minor axis is the distance between the pupils and that of the major axis is $1.2 \times (Gm_y - Gp_y)$ where Gm_y is the vertical position of G_m and Gp_y is that of G_p. We make the sets of two a-quarter ellipses by changing the length of the minor and the major axis of the initial ellipse as illustrated in Fig.15. We fit them to contrast image. As the facial contour, we select the set of two a-quarter ellipses whose sum of the edge contrast along the contour is maximal. Fig.16 shows the extracted facial contour. The L_x and R_x are respectively the length of the minor axis of the left ellipse and the right ellipse.

After we obtain Lx and Rx, we estimate the face direction (θ_H) with the following equation.

$$\theta_H = \arcsin \left(\frac{|R_x - L_x|}{R_x + L_x} \right) \tag{9}$$

5. Experimental result

We made experiments with the images whose size is 360×240. In all images, there was only one face. It took 0.2 seconds to detect one face and estimate the face direction.

We made two experiments to verify the effectiveness of the model B. We used only the model A in the first experiment and we used both the model A and B in the second experiment. We took 1901 face images of 15 persons in our laboratory. In those images, there was no profile face. The detection rate of the face was 83.4% in the first experiment and it was 87.2% in the second experiment. Most of the rests 12.8% were detected as the face candidate regions but the incorrect points were extracted as the facial features. We classified the face direction into six directions manually and obtained the detection rate in each direction (Table 1). In Table 1, the model B is shown to be effective especially in down direction. We suppose that the reason is because the nostril is often invisible in down direction, no set of the feature candidate points is matched to the model A and

Table 1

Effectiveness of model B(left:Only model A, right:Model A and Model B)

	Half left and right		Front	
Up	75.7%	86.7%	80.0%	85.4%
Front	91.1%	93.9%	96.5%	96.8%
Down	74.4%	95.0%	74.1%	93.4%

234

For example, 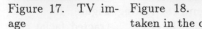 respectively represent a front, a right, and a left face direction.

Figure 17. TV image | Figure 18. Image taken in the corridor | Figure 19. Fault Detection | Figure 20. Two faces detection

because the pupil positions are limited in the search region, the fewer incorrect points are extracted as the pupils.

In the third experiment, we took the face images of the various kinds of persons and illuminations by using the television images. We took 911 face images of 11 males and 12 females. 18 persons were indoors and 5 persons were outdoors. The example of the experimental results is shown in Fig.17 where the rectangle and three crosses represent respectively the face candidate region, the pupils and the nostril. The line beside the face represents the estimated face direction from top view. The detection rate of the face was 83.9%. The reason of the reduction of it was due to a mustache, the shade or a wide open mouth. Without those cases, it was 85.2%.

In the forth experiment, we verified the robustness against the face directions by using 4404 face images of 13 persons in the corridor. In those images, the faces were oriented to various directions. Fig.18 shows the example of the results. The detection rate of the face was 73.4%. The reason of reduction of the detection rate was due to fault extraction of the mouth region (Fig.19). We suppose that the reason is because it is difficult to distinguish from other dark regions as regards the aspect ratio in case of profile face.

In the above experiments, we used the images where there was only one face. A few faces can be detected if they are not near each other (Fig.20).

6. Conclusion

This paper has demonstrated the new method of the detection of the face whose size, position and pose are unknown and the estimation of the face direction. For the face detection, we first extract the face candidate regions using the skin and hair color information and then we extract the facial features by using the face models of three directions. By using these face models, our method is independent of the face pose. Currently we assume the dark hair color. For future works, we will detect the faces without the assumption and recognize the detected faces.

REFERENCES

1. Ryoichi Nagata and Tsuyoshi Kawaguchi: "Detection of Human Faces in a Color Image with Complex Background", Technical Report of IEICE.PRU99-111(1999-11).
2. Haiyuan WU,Qian CHEN,and Masahiko Yachida: "Face Detection From a Color Image", Technical Report of IEICE.PRU94-108(1995-01).
3. Shi-Hong Jeng, Hong Yuan Mark Liao,Chin Chuan Han, Ming Yang Chern, Yao Tsorng Liu: "Facial Feature Detection Using Geometrical Face Model: An Efficient Approach", Pattern Recognition,Vol.31,No.3,pp.273-282,1998.
4. K.Fukui and O.Yamaguchi: "Facial Feature Point Extraction Method Based on Combination of Shape Extraction and Pattern Matching", Trans.IEICE(D-II),Vol.J80-D-II,No.8,pp.2170-2177,1997.

Human Friendly Mechatronics (ICMA 2000)
E. Arai, T. Arai and M. Takano *(Editors)*

Contour Based Hierarchical Part Decomposition Method for Human Body Motion Analysis from Video Sequence

Kengo Koara, Atsushi Nishikawa and Fumio Miyazaki [a]

[a]Graduate School of Engineering Science, Osaka University, 1-3, Machikaneyama-cho, Toyonaka, Osaka, 560-8531, Japan

In this report, we propose a method that can divide the body region into characteristic elements without any model of link parameters or shape based on the distribution of high-curvature points of the object's contour in the video sequence (the hierarchical part decomposition method). The shape model of human body can be generated from the decomposed result, and it can be used for handling more general body motion analysis. Experiment results of analyzing the kip motion under real environment show that the proposed method can divide the body into characteristic elements and can find postures of the body from the video sequence.

1. Introduction

Image-based human motion analysis is widely used for a lot of applications such as gait analysis, form analysis of sports science, media industry and so on. Most of the practical methods are based on visual tracking of markers put on the body surface. Such methods, however, have some problems which are not desirable in case of sports form analysis, e.g., markers may disturb body motion and such markers must be prepared beforehand. The method using colored markers has some difficulties to choose marker colors which are not contained in the background scene. Many methods have been proposed that involve marker-less measurement of human body motion. For example, methods have been proposed and commercialized that involve manual instruction of characteristic points on the human body for each video frame by a human operator. These methods, however, impose a heavy load on the human operator. Some methods are based on model matching. Those methods use 3D CAD models[1] or body structure models [2–4]. These approaches are efficient because any operations by human operator are not needed, however, appropreate models should be prepared.

By the way, the main point of measuring human body motion is how to find characteristic parts of the object. Kakadiaris et al.[5] proposed a contour decomposition method for articulated objects. Their method, however, seems to assume that the shape of each element does not change dramatically during the motion, which means that it is not suitable for objects with deforming elements such as human body with dynamic gymnastic motion.

From this point of view, we are exploring the method that can divide human body region into characteristic elements without models such as link parameters, and does not require any operations by human operator.

In this report, at first, we describe a method of extracting the human body from video sequence using snakes based active contour model. Then, we propose a method that decomposes a human body contour into characteristic parts hierarchically using temporal information of high curvature points on the body contour extracted from a natural scene. Finally, we apply the proposed method to analyze a human moving on a horizontal bar.

Here, we assume that the camera is fixed in the environment, the target object (human body) is only one and the target is considerably larger than any other moving objects in the camera view. In our method, all the processing is done off-line because, in case of sports form analysis or creating the model of body elements, real-time measurement is not required.

2. Extracting moving human body from natural scene

In this section, we describe how to find the human body from the scene. Here, we regard the human body as a closed region, and use an active contour model (based on snakes[6]) to find a contour of the body region which is extracted by subtraction from the background image.

2.1. Generating the background image

Now suppose that a camera is fixed on the tripod and the intensity of the scene does not change dramatically. In this case, it seems that the background pixel value (intensity) of each pixel appears most frequently in the pixel intensity histogram of the video sequence because the frequency of the value which comes from moving objects is usually less than that of the background. From this point of view, we use the median value of the intensity histogram of the sequence as the background value.

2.2. Extracting the initial body region

We can easily get the moving regions by subtracting the background from a target frame. However such regions contain not only the target human body but also noise and the other moving things in the background. We extract the human body region (below referred to as the *initial body region (IBR)*) as follows: for each frame, (i)the background subtraction image is thresholded, (ii)the labeling process is applied to the resulting image so that the small regions which may come from noise are removed, (iii)adjacent regions are connected, (iv)as a result, the largest region is picked up as the human body.

2.3. Finding the contour of the body region

The extracted body region has lack of information due to occlusion or slight difference in pixel intensity between the target and the background. To cope with this problem, we use snakes[6] based active contour model to find one closed contour of the body region. We impose two external forces – "Inward force" and "Sticking force" upon snakes. "Inward force" leads each node of the snakes toward the inside of the closed region which helps getting the contour of hollow shaped object. "Sticking force" stops each node of the snakes from over-shrinking. By introducing these forces, the contour model can catch the edge of hollow shaped object.

The contour information obtained from snakes is described as a polygon (closed curve)

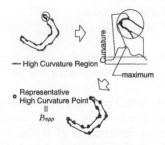

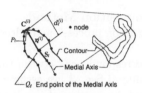

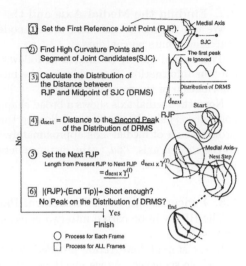

Figure 1.   Representative high-curvature points (P_{repp}).

Figure 2. Detecting the tip points of the contour from the medial axis.

Figure 3.   Procedure of the hierarchical part decomposition of the human body contour.

with the finite number of nodes. By doing this process for all frames of the video sequence, a sequence of body region contours can be obtained.

3. Finding Characteristic Points on Human body by the Hierarchical Part Decomposition Method

The method we propose here uses the distribution information of high curvature points on the contour of the body from the image sequences, which is not sensitive to the noise of extracted contours. It is based on the observation that the curvature of the human body contour is high at the bending joints, pinched parts, and tips of the body. We regard such parts as the characteristic points of moving human body, and use them to divide the object into significant elements hierarchically. We refer to this as the *Hierarchical Part Decomposition Method (HPD)*. Here, we assume two conditions: the length of each link is invariant, and the object has no branches.

3.1. Extracting the Representative High Curvature Points (P_{repp})

The angle between two successive lines created by connecting adjacent nodes on the contour is referred to as the "curvature" (at the common node). If the curvature at a node is greater than the product of mean exterior angle of polygon with N-vertices ($2\pi/N$, where N is the number of nodes of the contour) and an appropriate coefficient, the corresponding point is regarded as a candidate of the high curvature point. As such points appear continuously along the contour, we choose the maximum curvature point of these successive nodes as the *"representative high curvature point"* (P_{repp}, see Fig. 1).

3.2. Finding the Medial Axis and the Tip of the Contour

We use the medial axis and the tip node of the contour as reference information for matching points on contours.

The medial axis is calculated as follows: at first, reconstructing bitmap image from contour information, then, thinning the bitmap image, finally, creating the line segment information of the resulting skeleton.

Now, the medial axis shows a broad shape of the target. From this point, the tip of the contour (P_i) should satisfy the following three conditions (see Fig. 2). ①P_i exists on the prolongation of medial axis approximately. ②Contour direction at P_i is almost vertical to the medial axis. ③$d_t^{(i)}$ (the distance between P_i and the end point of medial axis Q_t) is relatively small.

3.3. Procedure of the Hierarchical Part Decomposition Method

The HPD can be parted into two processes: "integration process" and "decomposition process".

In the step of "integration process", we find the correspondence between the tips over the whole sequence. As we mentioned at the beginning of Sec. 3, we assume the target has no branch, so that the medial axis is only one and the number of tips is always two. Thus we decide the correspondence by evaluating the distance between the tips at two successive frames and selecting the paired tips that minimize the distance.

The next step, "decomposition process" (see Fig. 3) is hierarchical. We refer to the reference points at the n-th hierarchical process as the "n-th" RJPs(Reference Joint Points). We, at first, choose one of the two time-sequences of tips on the contour as the "first" RJPs. On the other hand, we call the other sequence the "final" RJPs. Please notice that, at this stage, we have one "1st" RJP and one final RJP per each frame. After that, we calculate the total length of medial axis for each frame and then obtain its mean value. Finally, for each frame f, we also calculate the ratio of the medial axis length at the frame f to the mean length, denoted by $\gamma_l^{(f)}$.

Under these preliminaries, we can summarize the hierarchical decomposition process as follows:

① $n \leftarrow 1$

② For each frame, we search for the high curvature point (referred to below as HCP) along the contour from the n-th RJP to the final RJP. If HCP is found, then we look for the nearest node that exists on the other side of the contour and refer to the segment with the ends at these paired nodes as "SJC (Segment of Joint Candidate)" (see Fig. 3). For all SJCs, we then calculate the distance between the n-th RJP and the midpoint of the SJC.

③ We divide the distance described above by $\gamma_l^{(f)}$ for normalizing, and then, calculate the distribution density of the "normalized distance" over all the frames. Incidentally, we use Gaussian function with appropriate standard deviation σ to calculate the distance distribution density.

④ We regard the distance from the n-th RJP to the "second" peak of the distance distribution density as the distance to the next RJP, denoted by d_{next}. The reason for selection of the second peak of the distribution is the following. The peaks of the distance distribution density are derived from characteristic points such as joint parts because SJCs concentrate near such points. However, if a joint is bending, the peak which appears farther than the first bending point is not reliable. Besides, the first peak should also be ignored because it might be derived from the influence of SJCs near the present RJP.

⑤ For each frame f, we find the point on the medial axis whose distance from the n-th RJP is $d_{next} \times \gamma_l^{(f)}$. These points are selected as the "next" RJPs and used for the next hierarchical process. That is, $n \leftarrow n + 1$ and back to step 2.

⑥ The processing is repeated until the searching length for SJCs is shorter than a threshold or a new RJP is no longer found.

3.4. Application to human body motion analysis

The HPD cannot directly handle loop shaped or branched objects. To cope with the problem, at first, we only select the frames with the "long and thin" and "branch-less"-shaped body. Such frames can be selected by using the information of the repeat number of the thinning process for finding medial axis of the body region and hand tip positions. Then we apply the HPD to the selected frames to decompose the body region. We can generate the shape model of the body elements from the contours and decomposed results. Details of the process in relation to generation of the element models can be found in [7]. Such element models can be used to measure the body motion over all the frames.

4. Experiment

We applied the proposed method to an image sequence with a human moving on the horizontal bar (the kip motion: see Fig. 4). The total number of frames is 50 {0-49}. The kip is a fundamental gymnastic bar movement, and its motion is done on the 2-dimensional plane. In this experiment, we placed a video camera on the vertical line of the movement plane so that the movement is regarded as 2-D motion.

Snakes-based contour finding method succeeded in extracting a closed body region from the silhouette, although the initial body region had a lack coming from occlusion caused by the stay of the horizontal bar.

The HPD was applied to the frames {00-06, 08, 10-18, 20-27, 29}. These frames were selected to avoid the loop shape (e.g., frame No. 35, 40 in Fig. 4) and the arm which overlaps with torso (e.g., frame No. 45, 49). Fig. 5 shows the human body contours divided by the HPD. We can see that the target human body was parted into some link elements — arm, torso, upper and lower parts of the leg.

The contour model of link elements was generated from these divided contours, and the postures of the rest of frames were estimated by fitting the model into the contour of the whole body. Fig. 6 shows the body shapes reconstructed by using the estimated postures and the contour model.

5. Conclusion

We proposed a method for decomposing a moving human body into characteristic elements hierarchically based on the distribution of high curvature points on the body contours without shape models, and we applied the method to measurement of the kip motion.

The experimental results showed that the HPD could divide a human body into link

240

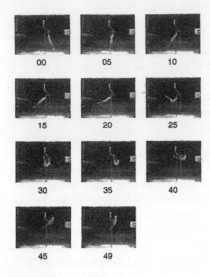

00   05   10

15   20   25

30   35   40

45   49

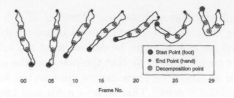

● Start Point (foot)
○ End Point (hand)
◎ Decomposition point

00   05   10   15   20   25   29
Frame No.

Figure 5. Divided human body contours

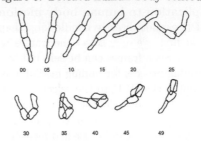

00   05   10   15   20   25

30   35   40   45   49

Figure 4. Video sequence of the kip motion

Figure 6. Reconstructed human body motion

elements from the selected frames, and the postures of each element were found by fitting the generated shape model. Our method could obtain the tendency of the whole motion sequence. Therefore, the method is suitable for the applications which focus on the timing of the motion such as sports coaching [8].

Although, in this paper, we assumed a non-branched object, the proposed method can be extended to analyze more general human body motion because branched object can be decomposed into non-branched elements[9].

REFERENCES

1. Lerasle et al.: "Human Body Tracking by Monocular Vision", In *Proc. ECCV '96* pp. B518-527, 1996.
2. Mochimaru et al.: "Two Dimensional Human Motion Measurement using the Texture Mapped Model", In *Proc. MIRU'96*, pp. II-127-132, 1996 (in Japanese).
3. Maylor K.Leung and Yee-Hong Yang: "First sight: A human body outline labeling system". *PAMI*, Vol. 17, No. 4, pp. 359-377, April 1995.
4. Masanori Yamada, Kazuyuki Ebihara, and Jun Ohya: "A new robust real-time method for extracting human silhouettes from color images". In Proc. of *FG'98*, pp. 528-533, 1998.
5. Kakadiaris et al.: "Active Part-Decomposition, Shape and Motion Estimation of Articulated Objects: A Physics-based Approach", In *CVPR '94*, pp.980-984, 1994.
6. Kass et al.: "Snakes: Active Contour Models", *Int'l J. Computer Vision*, pp.321-331,1988.
7. Koara et al.: "Hierarchical Part Decomposition Method of Articulated Body Contour, and its Application to Human Body Motion Measurement", IROS 2000, Nov. 2000 (in press).
8. Nakawaki: A Multi-model Approach for Improving Skill Transfer, Ph.D Thesis, Graduate School of Engineering Science Osaka Univ., 1999.
9. Koara et al.: "Image-based Marker-less Measurement of Human Body Motion – Extending to Handle Branched Objects–", In *Proc. of 43rd Annual Conf. the Inst. of Sys., Cntl. and Info. Eng.*, pp. 281-282, 1999 (in Japanese).

Human Friendly Mechatronics (ICMA 2000)
E. Arai, T. Arai and M. Takano (Editors)
© 2001 Elsevier Science B.V. All rights reserved.

Man Chasing Robot by an Environment Recognition Using Stereo Vision

Yoshinobu SAWANO[*], Jun MIURA[**], Yoshiaki SHIRAI[**]

* Fujitsu Co.ltd, Kawasaki, Kanagawa 211-8588, Japan
** Dept. of Computer-Controlled Mechanical Systems,
   Osaka University, Suita, Osaka 565-0871, Japan
Email : {sawano,jun,shirai}@cv.mech.eng.osaka-u.ac.jp

Abstract

This paper describes a control method of chasing person by a vision guided mobile robot in an unknown indoor environment. The robot obtains the environment information by stereo vision and makes a map. From both the map and the observed person position, the robot decides the optimal path to chase the person while avoiding obstacles. Since the observation by stereo vision may include uncertainty, we model the uncertainty to calculate the reliability of the map to be obtained by integrating multiple observation results.

1. INTRODUCTION

One of the main issues of autonomous mobile robot studies is sensor data processing and planning when the robot has uncertainty in position estimation or in sensing. If the robot has a map of the environment, the robot can plan a path and observes landmarks recorded on the map for localization on the path [1]. By matching the observed landmarks to the map, the robot can move safely. There is a study in which a man guides a robot to the destination in an unknown environment [2]; the robot first makes the map of the environment and, then it moves autonomously. In our approach, the robot detects a person using vision. It also observes an environment by stereo vision to make a map. The robot then decides a path to chase the person while avoiding obstacles. The robot does not necessarily move on the path the target has moved, but chooses the optimal path.

Position estimate only by dead reckoning includes uncertainty and this uncertainty is accumulated as the robot moves. To reduce the uncertainty, the robot observes the environment by stereo vision. This observation has also uncertainty (or an error). For example, the robot sometimes observes nonexistent obstacles or sometimes fails to observe an obstacle due to the error in stereo matching. To recognize the environment more reliably, we develop the following method.

By modeling the error of observation based on the stereo vision characteristics, the robot evaluates the reliability of the observation result. If an obstacle is detected, its probability of existence is determined depending on the distance from the robot. The nearer to the robot the obstacle is, the larger the reliability is. We then model the positional uncertainty due to quantization error in stereo vision by using a two-dimensional normal distribution. We update the existence probability of each obstacle using the Bayes' theorem. If the robot detects an obstacle several times, then the existence probability of the obstacle becomes high. On the other hand, if the robot cannot detect an obstacle several times, the probability becomes low. The map contains the existence probability in the environment. The robot makes the map by repeating the observation. We also apply the Extended Kalman Filter [3] to reduce the uncertainty in motion and obstacle position.

Stereo Vision

Camera for tracking

Figure 1. Mobile Robot

The simulation of this stochastic model shows the effectiveness of the uncertainty model. The experiment with the real robot shows the usefulness of the proposed method. Figure 1 shows a robot with three wheels. The robot can control a steering wheel, a camera head, and speed. A stereo pair of cameras is fixed on the camera head. Using the steering input and the odometer reading, the robot estimates its position by dead reckoning.

2. DETECTING OBSTACLES BY STEREO VISION

2.1 Stereo Vision System

We use a stereo vision to obtain range data of the environment. In stereo vision, it is necessary to find the correspondence between feature points in the left and the right images (see Figure 2). After smoothing the input image to suppress noises, we make edge images (see Figure 3) ; edges with high contrast are used as the feature points. A pair of points is considered to match if the sum of absolute difference (SAD) of the intensity value of 5 x 5 windows W around the points is small enough and minimum among SAD values computed for the possible disparity range. SAD is given by the following

$$\sum_{(i,j)\in W} I_R(i,j) - I_L(i+d,j) \quad (1)$$

where $I_R(i,j)$ and $I_L(i,j)$ represent the intensity values of the point (i,j) in the right and the left images, respectively, and d represents the disparity. Figure 4 shows a result of the stereo method. In this figure, darker points indicate larger disparities.

Figure 2. Input Image

Figure 3. Edge image

Figure 4. Disparity image

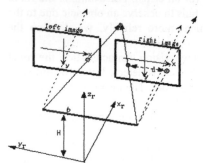

left image

right image

Figure 5. Geometry of Stereo Vision

2.2 Uncertainty of Observation

Once a set of matched feature points are obtained, their three-dimensional positions are computed by triangulation. If we set each parameter as Figure 5, these are given as

$$\begin{bmatrix} x_r \\ y_r \\ z_r \end{bmatrix} = \frac{b}{d}\begin{bmatrix} f \\ -x \\ -y \end{bmatrix} + \begin{bmatrix} 0 \\ -\frac{1}{2}b \\ H \end{bmatrix} \quad (2)$$

where $x_r=(x_r,y_r,z_r)^t$ is a three-dimensional position; $x=(x,y)$ is the point of right image; H is the height of the cameras from the; b is the distance between the cameras (called *baseline*); f and d are the focal length and the disparity, respectively.

We consider the three-dimensional position error due to the quantization error in the image. Using the Taylor series expansion and neglecting higher-order terms, we linearize equation (2). We then model the uncertainty of x_r as a two-dimensional normal distribution. The relation between the covariance matrix of disparity space $(x_d=(x,y,d)^t)$ and that of real space is given as

$$x_r = Ax_d + c, \quad (3) \qquad \Lambda_r = A\Lambda_d A^T, \quad (4)$$

where

$$A = \frac{b}{\hat{d}^2}\begin{bmatrix} 0 & 0 & -f \\ -\hat{d} & 0 & \hat{x} \\ 0 & -\hat{d} & \hat{y} \end{bmatrix} \qquad c = \frac{b}{\hat{d}}\begin{bmatrix} f \\ -\hat{x} \\ -\hat{y} \end{bmatrix} + \begin{bmatrix} 0 \\ -\frac{b}{2} \\ H \end{bmatrix}$$

3. DETECTING TARGET POSITION

We assume that the target person wears a colored cap and his height is known. We use the auto-tracking function of the camera (EVI-D30:SONY) to track the target. This camera can track the target quickly using color information. From the position of the target in the image and the pan and tilt angles, the target position $x_p=(x_p,y_p)$ is given as (see Figure 6)

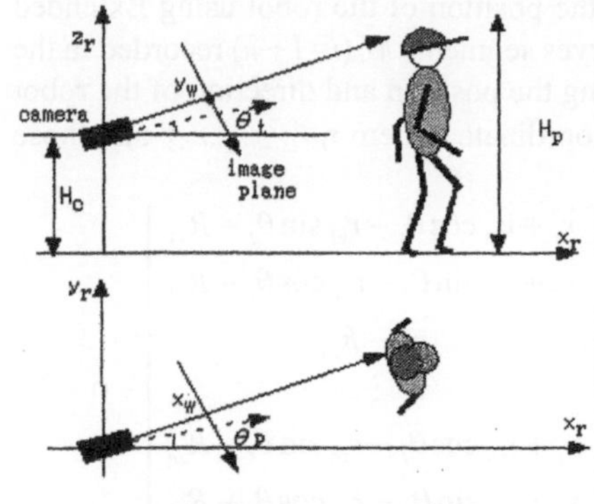

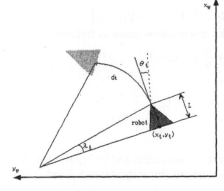

Figure 6. Observation of target

$$x_p = \begin{bmatrix} \dfrac{f + y_\omega \tan\theta_t}{f\tan\theta_t - y_\omega}(H_p - H_c) \\ -\dfrac{f\tan\theta_p + x_\omega}{f - x_\omega \tan\theta_p}\dfrac{f + y_\omega \tan\theta_t}{f\tan\theta_t - y_\omega}(H_p - H_c) \end{bmatrix}, \quad (5)$$

where x_w and y_w are the position in the image; f is the focal length, θ_p and θ_t are the pan and the tilt angle, respectively. Since the height of the cap may change as the target walks, the target position may include uncertainty. The positional uncertainty due to the height change is calculated similarly to the case of stereo uncertainty.

4 UPDATE POSITION OF ROBOT AND OBSTACLES

4.1 Uncertainty of Robot Motion

The position of robot is represented by vector $x_t=[x_t,y_t, \theta_t]^t$. Since the robot motion includes uncertainty, we estimate the position of the robot with uncertainty. Figure 7 shows the relationship between a control value (steering value λ_t and the movement distance d_t) and the movement of the robot. The position after movement is represented by the following equation [1] (for the case $\lambda=0$).

Figure 7. Movement of robot

$$x_{t+1} = f(x_t, d_t) = \begin{bmatrix} x_t + \dfrac{L}{\sin \lambda_t}(\sin(\theta_t + \lambda_t + \dfrac{d_t}{L}\tan \lambda_t) - \sin(\theta_t + \lambda_t)) \\ y_t + \dfrac{L}{\sin \lambda_t}(\cos(\theta_t + \lambda_t) - \cos(\theta_t + \lambda + \dfrac{d_t}{L}\tan \lambda_t)) \\ \theta_t + \dfrac{d_t}{L}\tan \lambda_t \end{bmatrix} (\lambda \neq 0) \quad (6)$$

Since this equation is the non-linear function, we make this function linear as follows.

$$x_{t+1} = f(\hat{x}_{t/t}, \hat{d}_t) + \frac{\partial \hat{f}}{\partial x}(x_t - \hat{x}_{t/t}) + \frac{\partial \hat{f}}{\partial d}(d_t - \hat{d}_t) \quad (7)$$

We estimate the uncertainty of the robot motion by

$$\hat{x}_{t+1/t} = f(\hat{x}_{t/t}, \hat{d}_t) \qquad \Lambda^x_{t+1/t} = F_x \Lambda^x_{t/t} F_x^T + F_d \Lambda^d_t F_d^T$$

$$where \qquad F_x = \frac{\partial \hat{f}}{\partial x}, \ F_d = \frac{\partial \hat{f}}{\partial d} \qquad (8)$$

4.2 Robot Position Estimate by Extended Kalman Filter

The robot records the position of vertical observed segments on the map. Using this segment, we update the position of segments on the map and the position of the robot using Extended Kalman Filter [3]. In Figure 8, suppose the robot observes segments M_i ($i=1 \sim n$) recorded in the map. Equation (9) is represents the relationships among the position and direction of the robot $x_t = [x_t, y_t, \theta_t]^T$, the position of segments in the robot coordinate system $r_i = [r_{xi}, r_{yi}, r_{zi}]^t$ and those in the world coordinate system $R_i = [R_{xi}, R_{yi}, R_{zi}]^t$.

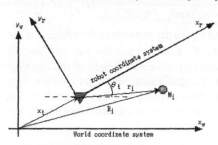

Figure 8. Observing a segment

$$h(X_t, r) = \begin{bmatrix} x_t + r_{x1} \cos \theta_t - r_{y1} \sin \theta_t - R_{x1} \\ y_t + r_{x1} \sin \theta_t - r_{y1} \cos \theta_t - R_{x1} \\ r_{z1} - R_{z1} \\ \vdots \\ x_t + r_{xn} \cos \theta_t - r_{yn} \sin \theta_t - R_{xn} \\ y_t + r_{xn} \sin \theta_t - r_{yn} \cos \theta_t - R_{xn} \\ r_{zn} - R_{zn} \end{bmatrix} = 0$$

$$where \qquad (9)$$

$$X_t = [x_t, R_1, ..., R_n]^t \qquad r = [r_1, ..., r_n]^t$$

Since equation (9) is a non-linear equation, we make this equation linear as follows.

$$h(X_t, r_t) = h(\hat{X}_{t/t-1}, \hat{r}_t) + \frac{\partial \hat{h}}{\partial X_t}(X_t - \hat{X}_{t/t-1}) + \frac{\partial \hat{h}}{\partial r_t}(r_t - \hat{r}_t) = 0 \qquad (10)$$

We rewrite the equation (10) as follows.

$$-h(\hat{X}_{t/t-1}, \hat{r}_t) + H_x \hat{X}_{t/t-1} = H_x X_t + H_r(r_t - \hat{r}_t)$$

$$H_x = \frac{\partial \hat{h}}{\partial X_t}, \ H_r = \frac{\partial \hat{h}}{\partial r_t} \qquad (11)$$

In equation (12), we can consider that the left side is the observation and the right side is the state of the robot. We construct the Kalman Filter based on this linear system as follows.

$$K_t = \Lambda^X_{t/t-1} H^T_X \left[H_X \Lambda^X_{t/t-1} H^T_X + H_r \Lambda'_t H^T_r \right]^{-1}$$

$$\hat{X}_{t/t} = \hat{X}_{t/t-1} - K_t h\left(\hat{X}_{t/t-1}, \hat{r}_t\right) \qquad (12)$$

$$\Lambda^X_{t/t} = \Lambda^X_{t/t-1} - K_t H_X \Lambda^X_{t/t-1}$$

5. MODELING OBSERVATION AMBIGUITY

Since the observation result includes matching error, we cannot always trust the detected obstacles just by one observation. If the robot detects false obstacles and records them on the map, they could be problematic in generating a safe path to the target person. To solve this problem, we take the following strategy.

1. We consider far range data are less reliable than near range data.
2. We would like to reject accidentally-detected range data.

To evaluate the reliability of an observation result, we model the ambiguity of stereo vision statistically. The reliability is estimated by equation (13) and equation (14) based on the Bayes' theorem. Equation (13) applies to the obstacles that the robot succeeded to detect, and equation (14) applies to the region between the robot and the detected obstacle.

$$P_n\left(E(x) \mid O(x)\right) = \frac{P_n(O(x) \mid E(x)) P_{n-1}(E(x))}{P_n(O(x))} \qquad (13)$$

$$P_n\left(E(x) \mid \overline{O}(x)\right) = \frac{P_n(\overline{O}(x) \mid E(x)) P_{n-1}(E(x))}{P_n(\overline{O}(x))} \qquad (14)$$

where $P(E(x))$ represents the probability that the obstacle exists at position x, and its initial value is 0.5. $P(O(x))$ represents the probability that the obstacle is observed at position x.

$P(O(x)|E(x))$ is given as Figure 9; $P(O(x)|E(x))$ changes depending on the distance from robot. Moreover $P(O(x)|E(x))$ is distributed around the detected position using the model of the

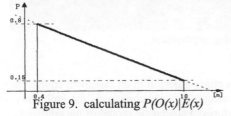

Figure 9. calculating $P(O(x)|\overline{E}(x)$

positional uncertainty of stereo vision. $P(O(x)|\overline{E}(x))$ is set to constant 0.05 at every position. Figure 10 shows the result of a simulation using this method; the existence probability distribution in the environment after the robot made 7 observations are shown. In the figure, the whiter a point is, the higher the existence probability of obstacle at the point is. The robot decides the position of obstacles by thresholding the existence probability.

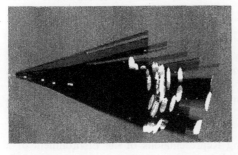

Figure 10. Existence probabilities after 7 observations.

6. PATH GENERATION

Since the robot has a motion uncertainty and the robot is bigger than the target, the path where target has moved is not necessarily safe for the robot movement. The obstacle regions on the map are enlarged by a half of the robot width and the estimated motion uncertainty. We call these enlarged regions dangerous regions as shown in Figure 11. The robot decides the path to avoid this region. To decide the optimal path for

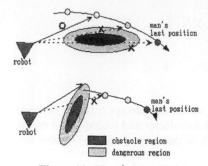

Figure 11. The shortest path

chasing depending on the situation, the robot takes following steps (see Figure 11).

The robot chooses the direct path to the last position of the target if obstacles do not exist on the path. If it is impossible to take such a path, the robot tries to generate a direct path to one of the previous target positions if the path is not obstructed by any obstacles. If the robot cannot take such paths, it chooses the edge of an obstacle which is nearest to the target position as the subgoal for chasing.

7. CHASING EXPERIMENT

The robot repeats the following steps for chasing: (1) detecting target; (2) obtaining range data; (3) updating the map; (4) generating the path; and (5) moving. In this experiment, we used a simplified version of map making, where the vision uncertainty due to quantization error is not considered; this can still generate a safe path because a enough margin is considered in path generation. Figure 12 shows snapshots of a chasing experiment.

Figure 12. Chasing experiment

8. SUMMARY

We have proposed a method that the robot can chase a person in an unknown environment even if the environment information includes uncertainty. To cope with the ambiguity in obstacle detection by stereo vision, we have developed a stochastic model of the ambiguity to evaluate the reliability of observation. Experimental results show the effectiveness of the method.

REFERENCES

1. I. Moon, J. Miura, and Y. Shirai, "On-line Viewpoint and Motion Planning for Efficient Visual Navigation under Uncertainty," *Robotics and Autonomous Systems*, Vol. 28, pp. 237-248, 1999.
2. K. Kidono, J. Miura, and Y. Shirai, "Autonomous Navigation of a Mobile Robot Using a Human-Guided Experience", Proc. Asian Conference on Computer Vision, Vol.1, pp. 449-454, 2000.
3. N. Ayache and O.D. Faugeras, "Maintaining Representaions of the Environment of a Mobile Robot", *IEEE Trans. on Robotics and Automation*, Vol. 5, No. 6, pp. 804-819, 1989.

Human Friendly Mechatronics (ICMA 2000)
E. Arai, T. Arai and M. Takano *(Editors)*

Using desktop OCR and adaptive preprocessing for recognizing text on extruded surfaces

Jani Uusitalo

Tampere University of Technology, Institute of Production Engineering,
P.O. Box 589, Tampere, Finland

This paper proposes three new preprocessing methods for optically recognizing characters on three dimensional extruded surfaces. Presented is a method for recognition with a priori knowledge about object shapes as well as method for recognition without preknown information about the shapes. A background estimation and removal method is presented for OCR system segmentation phase simplification.

1. INTRODUCTION

Most desktop OCR systems on the market today are meant for recognizing images obtained from scanners. Using scanners ensures the fact that the resolution of the images is usually always ideal for text recognition. Because of the scanner operating principle, drum or flatbed, the text to be recognized is always on two-dimensional surface and there are no unexpected geometrical deformations or abrupt variations of lighting. To recognize text on the surfaces of real three dimensional objects, one has to be able to cope with several disturbing factors. These include camera lens aberrations and deformations arising from perspective projection. There are also sudden changes of lighting as the recognition environment is not closed.

In this paper three preprocessing methods are described which enable one to use off-the-shelf OCR programs for text recognition from camera images of 3D objects. The paper suggests methods for measuring the 3D shape of an object visible in the camera image and applying geometrical and tonal corrections in order to rectify 3D surfaces into flat planes for text recognition. After these corrections, the background is removed and only the letters and other shapes to be recognized, such as bar and matrix codes, remain. Separating the shapes readily from the uninteresting background one can be sure the segmentation, and moreover the recognition result, is not affected by the poorly behaving segmentation algorithm of the commercial OCR system.

Section 2 describes a model-based solution for rectifying images where objects with only preknown shapes exist to be recognized. This solution is often efficient enough for example for most mobile robotics applications where only one or two kinds of objects are handled. Section 3 considers the case where no *a priori* knowledge about the periphery to be recognized is available. This is the case when attempting to read text from the surface of a crushed aluminum can in a recycling machine, for example. Section 4 discusses about simple and efficient background estimation and removal method capable of flattening out uninteresting features from the image and this way helping the OCR segmentation phase. Section 5 briefly mentions a couple of interesting application areas where the newly discussed methods could be used and section 6 presents some experimental results we have gained using a prototype system in our laboratory.

2. MODEL-BASED RECTIFICATION

In situations where the object to be recognized is known *a priori* one can use bodel-based rectification methods. The overall solution is simple: The object is first found on the recognition area. Then a model representing the object is fitted over the image acquired from the object. Finally, the image is rectified using the model and its inverse solution to a plane perpendicular to the camera optical axis.

Several methods for finding the object and fitting its model over the image can be found. The application specifies the methods which can and cannot be used. For locating extruded shaped objects lying on the shop floor, one can use, for example, several distance sensors pointed to known directions and then crosscalculating. Another possibility would be using the camera itself as a sensor by triangulating to a laser stripe projected over the image area. Using optimization software and a threshold for the optimization error one can at the same time detect and locate the object in the image. This is the solution we used in our experimental system. An example of the laser positioning system in use can be seen in figure 3.

Before using the inverse object model for rectification, lens aberrations and the decentricity of the optical axis have to be removed. Aberrations deform the camera image so that the acquired image no longer matches the model perfectly. For calculating the implicit camera calibration coefficients several methods have been proposed, of which at least [1] – [3] deserve to be mentioned. In [4] Heikkilä *et al.* present a calibration procedure based on a known 3D calibration object and capable of calculating both implicit and explicit camera calibration parameters. In our experimental system the *model* used for rectification is corrected to match the aberration and decentricity parameters of the camera system.

Figure 1 shows an example of model fitting. Explicit and implicit camera parameters are known in advance and the paper reel has been found using a laser stripe projected on the reel and triangulating the reel center axis position and direction. The reel model, which can be seen

Figure 1. Paper reel and its fitted model.   Figure 2. Reel rectified using the model inverse.

superimposed on the image, includes the camera implicit calibration parameters and has been positioned over the image by the help of the laser stripe triangulation. Figure 2 shows the result of the rectification using the inverse object model. As can clearly be seen, the perspective effect has almost completely been removed. With better OCR systems this image would already produce tolerable results. The lighting, however, changing from scene to scene, can still produce unexpected results and most OCR systems require additional background removal.

3. FREE FORM RECTIFICATION

In contrast to the model-based method presented in the previous section free form rectification can also be used when *no a priori* information about object shapes is at hand. The free form rectification closely links together the measurement principle of the laser stripe triangulation, the imaging geometry and perspective deformations arising from the scene. At the same time an assumption of the object extrusion directions has to be made: The laser stripe used can give distance information over one cross section only and one has to know in which direction the object distance is not changing. In the rest of this section a vertical equidistant direction and a camera with known tilt angle is assumed.

A laser stripe is projected on the image area and can be seen in the camera image, as in figure 3. The laser source is located below the camera and the camera is slightly tilted downwards in respect to the laser. The immediate effect of this is that the closer the object surface is the farther down the laser stripe is in the image. This gives us a hint of the distance from the camera to the object surface. The distance, on the other, can be directly related to the perspective effect and the magnification factor needed to enlargen the object surface to produce a flat result image.

The angle of the object surface normal and the camera optical axis can be estimated by differentiating the distance measures. The faster the distance changes over the image the more the object surface is turned away from the camera. The more the surface is turned away the more it has to be magnified (horizontally) to compensate. The solution is to present an additional coefficient for horizontal magnification and link it directly to the differentiated distance measures.

An example of the free form rectification in action can be seen in figures 3 and 4. The leftmost picture is taken from the corrugated wall of an outdoor container used in transporting goods by trucks. The laser stripe can be seen on the bottom of the picture. The rightmost picture is the result of applying free form rectification based on the position information of the laser stripe. The slight deformations still present in the picture originate from low pass filtering the laser stripe position.

Figure 3. Laser stripe projected on corrugated outdoor container wall.   Figure 4. Free form rectified container wall.

4. BACKGROUND ESTIMATION AND REMOVAL

To further help the segmentation phase of the commercial OCR system the interesting information on the image can be highlighted or the uninteresting information can be hidden. In the most sensitive OCR systems a noise, for example, arising from the imaging system and occurring at a certain frequency can produce untolerable results on the system output. Besides noise, there's also always a possibility of rapid changes of lighting, low intensity textured surfaces beneath the text an so on.

The image can be simplified by hiding low frequency changes. To further validate the results the background should be removed completely. For this, subtracting a separate background image from the currently handled image is sufficiently efficient and simple method. The background subtraction is used to remove the "dc" component of the image to produce a flat background level below and above which the foreground features fluctuate.

When imaging "onetime" scenes where no separate background image can be had, the background can be estimated from the actual source image. The background estimation is based on Pratt's thresholding [5] where the pixel neighbourhood histogram is examined to find a suitable threshold value between the foreground and the background. Here the histogram deviations are checked to interprete the pixel either to belong to the foreground or the background. Small deviations imply low frequency features in the neighbourhood and the pixel belongs to the background. If the histogram deviation is too great there are high frequency features nearby and the pixel belongs to the foreground. By changing the neighbourhood size and shape and by setting different deviation thresholds one can decide which shape features should belong to the foreground and which not. In our case, the estimator was built so that it responded easier to characters of size 5x5 – 20x20 and horizontal rows.

The pixels belonging to the background are immediately put to the background estimate. This way the background estimate is the original image with the difference that there are holes on the spots where foreground used to be. Appropriate pixel values for these spots can be interpolated by using the background pixel values on the edges of the spots. If assuming vertically extruded objects in the image, vertical linear interpolation between the top edge value and the bottom edge value can be used. Other efficient possibilities are horizontal interpolation and nearest value interpolation. Of course, higher order interpolations could be used as well, but they tend to exaggerate the background changes and as such can be thought of not so well suited for the task.

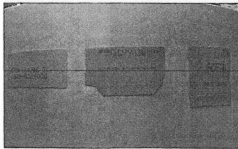 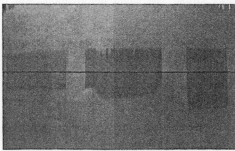

Figure 5. Original image.	Figure 6. Vertically estimated background.

Figure 7. Image flattened using estimation subtraction.

Figure 5 shows an image of paper reel with three labels and text. The background of the image is estimated to figure 6. Neighbourhood size used was 9x9 square and the maximum deviation for a background histogram was 30. Figure 7 shows the result of subtracting the background estimate from the original image. Only the high frequency features have been left in the image and the foreground pixels still have their original values.

5. APPLICATION AREAS

One can find uses for the methods presented in this paper in every system where three dimensional objects are imaged and where text on their surface should be read or characters recognized. The research of these methods began when labels on paper reels were first attempted to read by using commercially available OCR systems. The methods have found their way into the warehouse recognition of the reels as well as the recognition taking place inside the paper mills.

Another use for the methods could be for example inside the aluminum soft drink can recycling machines where the cans should be checked for proper bar code or text label before accepting the cans. The cans are easily deformed and they are usually quite battered when they are brought to the machine. Free form rectification could be used here to first straighten the surface as well as the background texture removal. This way, the bar code or text recognition would be possible in spite of small bruises on the cans.

Third, and most multiform, use for the methods would be using them on a mobile robot navigation system. A navigation system equipped with these methods could read text and markings on natural landmarks. This way the robot could differentiate otherwise equal looking landmarks such as shop labels or name tags besides office doorways.

6. EXPERIMENTAL RESULTS

The methods were tested on a forklift loader used in paper reel handling. The test system was utilized for recognizing the text on paper reel labels visible in front of the loader. The shape of the recognized objects and the changing operating environment of the imaging system made the problem very well suited for this kind of testing. Furthermore, the identification of the reels is currently one of the real problems yet to be solved in automating the cargo warehouses and the logistic chain leading from paper manufacturers to clients.

Currently the system is used for recognizing texts only on cylindrical objects. Model-based rectification and background estimation and removal is used for this mostly. The preprocessing methods mentioned here have increased the recognition propability from below 30% to above 90%. This kind of recognition method of the reels of course demands that several markings can be found from the reel. Commercial possibilities of the system are being studied at the moment.

7. CONCLUSIONS

Three preprocessing methods for commercial OCR systems were presented in this paper. The methods can be utilized in various machine vision applications, but they are most beneficial in camera based OCR applications and in high level robot vision. Methods presented are easily implemented and can be optimized for normal frame rate systems. The model-based preprocessing requires information about the 3D objects to be recognized and the camera has to be calibrated. The free form preprocessing doesn't need any a priori information about shapes or calibration parameters. The cost of easier use is in most applications, however, reduced accuracy. A pilot application has successfully shown that the methods work and the 3D object recognition propability can be radically improved using right kind of preprocessing mentioned here.

ACKNOWLEDGEMENTS

The author would like to thank the Technical Research Centre of Finland, its Machine Automation group as well as the National Technology Agency for making this research possible. The support and guidance of professor Jouko Viitanen and Mr Juha Korpinen is also gratefully acknowledged.

REFERENCES

1. Slama, C. C. (ed.) Manual of Photogrammetry, 4th ed., American Society of Photogrammetry, Falls Church, Virginia (1980).
2. M. Devy, V. Garric and J. J. Orteu. Camera calibration from multiple views of a 2D object, using a global non linear optimization method. Proceedings of IEEE/RSJ Intl. Conference on Intelligent Robots and Systems. Vol. 3, pp. 1583-1589. (1997).
3. S. W. Shih, Y. P. Hung, W. S. Lin. Accurate linear technique for camera calibration considering lens distortion by solving an eigenvalue problem. Optical Engineering 32(1), pp. 138-149. (1993).
4. J. Heikkilä, O. Silvén. A four-step camera calibration procedure with implicit image correction. Proceedings of IEEE Computer Society Conference on Computer Vision and Pattern Recognition. pp. 1106-1112. (1997).
5. W. Pratt. Digital Image Processing. John Wiley & Sons. New York. 750 p. (1978).

Human Friendly Mechatronics (ICMA 2000)
E. Arai, T. Arai and M. Takano *(Editors)*

253

Road Surface Sensor using IR Moisture Sensor

M.Ikegami[a], K.Ikeda[a] Y.Murakami[b], N.Watanabe[b], K.Isodad[b], D.Tsutsumi[c], M.Nami[c]

[a]Hokkaido National Industrial Research Institute,
2-17 Tsukisamu-higashi, Toyohira-ku, Sapporo, Japan, 062-8517

[b]Hokkaido Electric Power Co. Inc.,
2-1 Tsuishikari, Ebetsu-shi, Japan, 067-0033

[c]Hokkaido Industrial Research Institute,
Kita-19 Nishi-11 Kita-ku, Sapporo, Japan, 060-0819

This paper describes IR moisture sensors to detect the road surface using an infrared detector and optical filters as a method for remote sensing of water and snow on roads. Moisture has absorption spectrum in the infrared region at wavelength of 1450[nm] and 1940[nm]. At first, a Ge photo diode combined with two optical band pass filters (1450[nm], 1680[nm]) was tested to detect the moisture (wetness) on target. In order to use outdoors, elimination method for sunshine was developed. For detecting more details of target, for example dry snow, wet snow and sherbet like snow, new classification map for relation between reflection and wetness are developed. These methods were tested outdoors over one month.

1. INTRODUCTION

Many types of water censor are developed based on electric resistance and dielectric constant measured by electric device. Because these sensors need being touched to target, the applications of them are limited. Therefore the remote sensors for moisture detection using light were developed using the relation between regular and diffused reflection[1]. This method, however, can detect moisture but it seems to be difficult to apply this method to detection of ice and snow, because it dose not detect characteristics of moisture essentially. On the other hand moisture sensor using infrared ray (IR) have been developed and are currently used to improve a weak point of connection type sensors[2]. This kind of sensor, however, is affected by obstacle like sunshine and makes errors. In consequence it has been unavailable to use outdoors and to detect long-distance targets.

In this paper an IR moisture sensor is proposed including new technique to eliminate obstacle like sunshine. This technique can apply to the condition that targets are several meters apart from a sensor and to the outdoors where the sun and streetlights shine on the targets. Secondly a new method to classify various conditions is developed. This method can classify water, wet and dry road, wet and dry snow. One application of this sensor is the moisture sensor for a road heating system, which gets rid of snow and ice on the road. A practical test was demonstrated using a test road with a road heating system outdoors for over one month.

2. IR MOISTURE SENSOR

2.1. Moisture detector using an IR sensor and band pass filters

As shown in Fig. 1[3], the absorption spectrum of moisture at near field IR increases gradually with wavelength and has two maximum peaks around 1450[nm] and 1940[nm] and two minimum peaks around 1680[nm] and 2200[nm]. The causes of maximum peaks are vibration energy of OH. Consequently, intensity of IR reflection at absorption spectrum $R1$ from a wet target contains weak IR comparing with other wavelength $R2$, because moisture on a target absorbs IR at these wavelength $R1$. This means that $R1$ can represent wetness of targets but $R1$ is also affected by reflectivity of targets. We assumed that these reflectivity at $R1$ and $R2$ were the same and they depended on the condition of surface and material, it is possible to neglect the reflectivity effect. In response to this, wetness W can be defined as follows:

$$W = R2/R1. \tag{1}$$

Figure 2 shows the block diagram of apparatus to use for moisture detection experiments. They consist of an IR illumination (one tungsten light which contains wide band IR), two Multi-coating band pass filters (their center wavelength are 1450[nm] and 1680[nm], -3dB bandwidth is 30[nm]), an IR sensor (Hamamatsu Photonics K.K., Ge photo-diode, B2614-05, it's sensitivity is from 800[nm] to 1800[nm]) and a filter selector. This apparatus can detect IR at two different wavelengths.

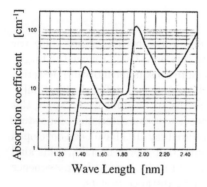

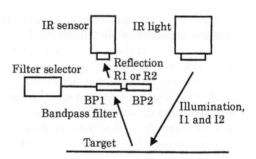

Figure 1. Moisture absorption spectrum at infrared region.

Figure 2. Block diagram of IR moisture sensor using two band pass filters.

In order to verify above equation (1), we made an experiment using above apparatus. A concrete block floor was used as a target and the distance between an IR detector and targets is 1.5 [m]. The floor was covered with water and was shining with water at the beginning of the experiment. Three or four minutes later, the concrete absorbed the water and it seemed to be black color with moisture in comparison with dry area. At the end of experiment, half of the floor became dry and gray color. Figure 3 shows the wetness calculated using equation (1). This result shows that the relation between wetness W and the surface condition had a good correlation.

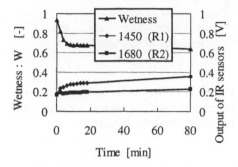

Figure 3. Variation of detected wetness on concrete surface. Wetness decreases in accordance with becoming the surface dry and time passed.

Figure 4. Experimental apparatus and outdoor temporary road with road heating system.

2.2. Elimination of IR included in sunshine

Because the sunshine contains IR which interferes with the IR light and it's reflection from targets and the conventional IR sensors are applied to the targets which are in a blackout curtain to avoid the influence of environment IR, it is difficult to apply conventional IR sensor to the outdoor environment. In order to expand the utilization of moisture IR sensor, this proposed method was developed to apply the IR sensor to outdoors environment.

To eliminate such obstacles, the following method was developed and applied in order to detect moisture outdoors. Figure 2 shows the similar experimental setup for outdoor moisture detection. In addition to Fig. 2, sunshine, $S1$ and $S2$, illuminates a target and $RS1$ and $RS2$ are the reflection of sunshine. The measurement processes are next two steps. First is to record $RS1$ and $RS2$ selecting band pass filters $BP1$ and $BP2$ without an illumination. Second is to record $(R1 + RS1)$ and $(R2 + RS2)$ selecting band pass filter $BP1$ and $BP2$ with an IR illumination. The wetness is defined as follows:

$$W = ((R2 + RS2) - RS2)/((R1 + RS1) - RS1) = R2/R1. \tag{2}$$

Because it is not easy to modulate intensity of a kind of tungsten illumination at high frequency, we switch the illumination simply from *on* to *off* in this experiment. For much reduction of obstacle, the modulated illumination is also available. In order to expand the utilization of moisture IR sensor, the proposed method were tested in outdoor

256

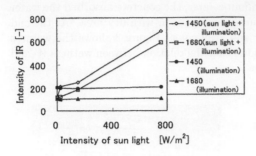

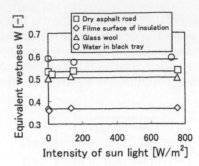

Figure 5. Relation between intensity of sun-shine [x-axis] and detected IR before and af-ter elimination of sunshine [y-axis].

Figure 6. Relation between intensity of sunshine and equivalent wetness using various reflectivity targets.

environment. Four test targets, which have different reflectivity of IR instead of different wetness targets, were used to measure the influence of sunshine intensity on sensitivity of wetness. Figure 5 shows the relation between intensity of sunshine and intensity of reflection, for instance $R1, R2, (R1+RS1), (R2+RS2)$. In this figure, calculated $R1(1450)$ and $R2(1680)$ are uniform approximately. Figure 6 shows the relation between intensity of sunshine and equivalent wetness for various reflectivity targets. In this figure, wetness is independent on intensity of sunshine and in consequence the process to subtraction of sunshine is going well.

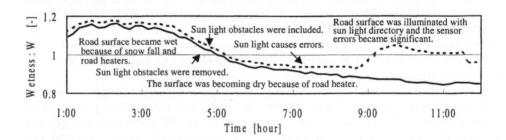

Figure 7. Wetness of road surface, which is becoming dry, with and without elimination process of sunshine.

The aim of next experiments is to demonstrate ability of detecting wetness and the eliminating effect for sunshine obstacles. This experiment was made outdoors using a temporary asphalt road with a road heating system. Figure 7 shows two measured wetness history on the road. Solid line is the results of equation (2) and dotted line is the results of equation (1). It snowed and the surface became wet at 2:00 in the night and stopped snowing at 3:00 and started melting and drying by road heating system. The sun rose at 7:00 and caused sensor error in case of dotted line. Because the road surface was illuminated with sunshine directory from 9:00, the sensor errors became significant in case of dotted line. On the other the solid line kept steady decrease in accordance with road condition. This means that the elimination process for sunshine goes well and wetness also can be measured properly in outdoor environment.

3. DETECTION OF ROAD SURFACE CONDITION

3.1. Classification map for dry and wet condition

According to the section 2, wetness on the road surface can be measured. By using this wetness, it is possible to classify road surface condition into three types. A: flood (there is much water and surface is covered with water), B: wet (there is no water above surface, surface is wet and it's color is black), C: dry (surface is dry and it's color is dark). One of salient characteristics of this method is that the difference of wetness: W at the condition A and B is bigger than that of B and C. In other word, the weak point of this method is low resolution (sensitivity) between B and C condition. One demand on these sensors is moisture detector for a road heating system. In this system it is necessary to detect precisely from B to C condition because there is the threshold between these two condition to control the system.

In order to improve the sensitivity between these two condition, a kind of map is introduced to classify the various surface condition, which is made by the relationship between intensity of reflected IR: $R1$ and wetness: W. Figure 8 shows this map which consist of 104 samples recorded every 10 minutes from 0:00 to 19:50. At the beginning of this period the temperature was below zero and it snowed. Consequently the surface was covered with white and dry snow. As time passed weather became fine and the fallen snow changed into wet snow, snow like sherbet and water and then the water became wet and finally the surface was dry by road heating system. These various condition is record using a CCD camera. As shown in Fig. 8, A, B and C condition are clearly classified and lie around the upper left, the lower left and the lower middle in the map respectively. Consequently it became possible to separate clearly condition B and C comparing with Figs. 3 and 7.

3.2. Classification map for snow condition

In addition to classification of wetness, Fig. 8 can detect dry snow, wet snow and snow like sherbet. Because the moisture in snow absorbs IR illumination and reflectivity of white snow is strong, the wetness W becomes small and intensity of reflection $R1$ become large respectively. Therefore dry snow is classified around the lower right consequently. As the snow melts and changes wet, the reflectivity becomes smaller and wetness W increases slightly, the position on the map move left correspondingly. Anyway snow on the road is distinguished from water.

In order to verify this classification map, long term experiment was conducted from March 3rd to April 6th. $R1, R2, (R1 + RS1), (R2 + RS2)$, temperature and pictures of the target were measured every 10 minutes under condition of IR illumination being *on* and *off* and it took 30 [S] to measure one set of data approximately. Figure 4 shows the

258

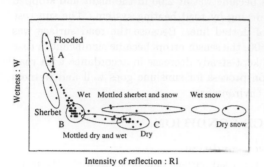

Intensity of reflection : R1

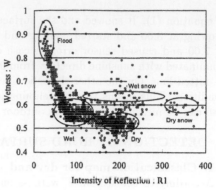

Intensity of Reflection : R1

Figure 8. Relation between intensity of reflection and wetness, which shows several
condition for instance dry, wet, flooded
surface, dry snow and wet snow.

Figure 9. Outdoor experiment to detect
various condition of road surface with road
heating system for one month in winter.

outdoor apparatus. According to Fig. 9 the tendency of classification is same as Fig. 8.
In this figure, the plotted dots spread widely in a belt from flood to dry condition. If the
intensity of sunshine varied by a break in the clouds while measuring one set of data, this
causes error in results. It is better to measure them in a short period comparing with the
variation speed of sunshine intensity.

4. CONCLUSION

We proposed two methods to detect moisture and surface condition on road using an
IR photo diode and two optical band pass filters. First method using a photo diode can
detect averaged wetness on some area and can be applied to outdoor condition eliminating
obstacles like sunshine. Second method, which uses new classification map for relation
between intensity of reflection and wetness, can classify water, wet and dry road, wet and
dry snow. The further investigation is to distinguish freeze from water and is to utilize
these sensors for agriculture field.

REFERENCES

1. K.Okumura, H.Sato, H.Muraki, M.Hiramatsu, Y.Shinomoto, Proceedings of 7th Symposium on Industrial Imaging Sensing Technology.
2. N.Watanabe, Y.Murakami, K.Isoda, D.Tsutsumi, M.Nami, M.Ikegami, Proceedings
   of Cold Reason Technology Conference, 2000.
3. HAMAMATSU Technical letter, TVID1003J03.

Human Friendly Mechatronics (ICMA 2000)
E. Arai, T. Arai and M. Takano (Editors)

Development of a system for measuring structural damping coefficients

H. Yamakawa[a], H. Hosaka[b], H. Osumi[c], K. Itao[b]

[a]Department of Precision Machinery Engineering, The University of Tokyo
7-3-1 Hongo, Bunkyo-ku, Tokyo 113-8656, Japan

[b]Institute of Environmental Studies, The University of Tokyo
7-3-1 Hongo, Bunkyo-ku, Tokyo 113-8656, Japan

[c]Department of Precision Machinery Engineering, Chuo University
1-13-27 Kasuga Bunkyo-ku Tokyo, 112-8551

A simple and accurate system for measuring structural damping coefficients was developed. A new system was developed which uses a microphone as a vibration sensor and a personal computer for signal processing. Compared with the conventional system which utilizes a laser doppler vibrometer and FFT analyzer, the new system is only about 1/10 the cost of conventional system, almost the same measurement accuracy was obtained, and measuring time was reduced to less than 1/10 that of the convensional system. The effectiveness of the system was verified by measuring several types of glass samples.

1. INTRODUCTION

Recently as the miniaturization and performance enhancements of mechanical systems are progressing, requirements to dynamic characteristic of their components are diverging and advancing. Resonant frequency and damping ratio are the basis of dynamic performance and they depend on material property. As far as the damping is concerned, for examples, it is desirable that the damping of the disk part of HDD is as big as possible, because the vibration of it influences the sensitivity of recording and playback signal . And, the Q-value of damping for scanning probe used for the AFM must be set an appropriate value in order to raise the sensitivity and shorten the response time. Therefor, it is important that material's damping characteristic is accurately measured.

Structural damping is the damping generated by the damping force which arises inside the material. It is necessary to hold airflow damping and energy dissipation at supporting point minimum in order to measure the structural damping coefficient accurately. The requested accuracy for the measurement is the order of 10^{-5} at the damping ratio. Until now, the accelerometer and the laser doppler vibrometer were used for measuring vibration, and the FFT analyzer was used for signal processing. However, there are some problems shown as

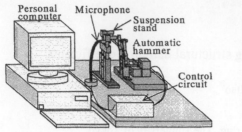
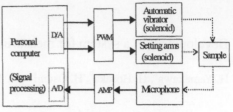

Fig. 1   Whole composition of developed system   Fig. 2   Block diagram of the new system

follows: when the accelerometer is used, damping generated at the contact point causes considerable error , and when the laser doppler vibrometer is used, system becomes expensive and reflection tape must be attached to the object. And, when the FFT analyzer is used for signal processing, the system becomes expensive and it is necessary to measure the exciting force.

In this paper, a new system was described using a microphone as a vibration sensor and a personal computer for signal processing to solve above mentioned problem. Total configuration of the developed system is explained at first. Next, it is shown that the microphone and personal computer is useful for measuring structural damping coefficients. Further, newly designed automatic hammer is explained. Finally, it is shown that the developed system is effective for measuring damping coefficients by comparing experimental results with those by the conventional system for the several types of glass.

2.   TOTAL CONFIGURATION OF THE DEVELOPED SYSTEM

Whole composition of developed system is shown in fig.1, and the flow of the signal is shown in fig. 2. The measuring process is shown as follows: the cylindrical sample is hung in the thread, and it is excited by the automatic hammer, the vibration is measured by the microphone, and the vibrational signal is processed by the personal computer. This system has following features in order to measure simply and accurately.

(1) The cylindrical sample which is 10mm in diameter and about 100mm in length is used in order to reduce the effect of airflow damping. The shape which reduce the effect of airflow damping is adopted, from the viewpoint that the structural damping coefficient is decided by the material and the dependence on the material's shape is small.

(2) The sample is suspended by thin threads at the nodal points of the first resonance in order to reduce the energy dissipation of the supporting point.

(3) For cost reduction of the sensor and simplification of the setup, the vibrational wave is measured by the microphone.

(4) For the cost reduction of the processor, vibrational wave is intermittently sampled, and the signal is processed in the time domain by personal computer.

(5) The sample is vibrated by automatic hammer in order to stabilize the exciting force. In these features, (1) and (2) are shown in [1]-[3]. In this paper, items (3)-(5) and measuring examples are explained.

3. MEASUREMENT OF VIBRATION USING MICROPHONE

The most expensive part in the conventional measuring system is the laser doppler vibrometer. It is possible to lower the system's cost if the laser doppler vibrometer is replaced with the microphone. And, in the measurement by the microphone, the usability is also improved because it does not need to set the reflection tape to the object. On the other hand, temperature, atmospheric pressure, humidity, wind and noise affect measurement accuracy.[4]

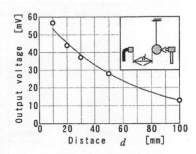

Fig. 3 Relationship between output voltage and distance d

Thus, output voltage of the microphone for the distance from the sample to the microphone and the damping ratio of the vibrational wave were measured in order to examine the amount of these effects and optimum distance from the sample to the microphone. A commercially available microphone for conference was used. Using a glass rod as a sample at first, the output voltage of the microphone from the vibration after 100ms was measured by the oscilloscope. Next, the relationship between the distance from the sample to the microphone and its damping ratio was examined. The mean value and standard deviation was measured. The results are shown in Fig. 3 and Table 1. Except for d =100mm, it is proven that there is not large difference for the value of the damping ratio. In d =100mm, deviation is large because the output voltage is small and noise influences results. In d =10mm, deviation is also increased. Because the swing of the sample following vibration changes sound intensity. From these cases, it is concluded that the optimum distance d is 30~50mm and them the damping is accurately measured at a usual laboratory condition.

4. SIGNAL PROCESSING

The vibrational wave measured by the microphone is converted from an analog signal to a digital signal by an A/D converter whose sampling frequency is 500kHz, conversion accuracy is 12bits. The signal is inputted to a personal computer. The damping ratio is obtained by processing the signal as follows:

① Data smoothing by the moving average method is applied in order to remove noise in the vibrational wave signal and detect the peak point accurately.

② Afterwards, peak detection is carried out using the Savitzky-Golay method of smoothing differentiation .

Table 1 Relationship between damping ratio and distance d

Distance from the sample d (mm)	10	20	30	40	50	100
Mean value of damping ratio ($\times 10^{-4}$)	2.65	2.81	2.49	2.55	2.35	3.52
Standard deviation of damping ratio ($\times 10^{-4}$)	0.22	0.22	0.06	0.11	0.09	0.09

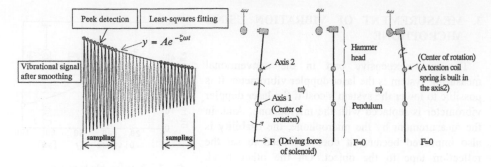

| Fig. 4 | The conceptual scheme of signal processing | Fig. 5 | Motion of the automatic hammer |

③ Using the peek values, exponential approximation by least-squares fitting is carried out according to the theoretical decaying vibration of 1-DOF system,

$$x(t) = Ae^{-\zeta\omega t} \cos\left(\omega\sqrt{1-\zeta^2}t\right) ,$$

where x, t, A, ζ and ω are displacement, time, initial amplitude, damping ratio and natural frequency, respectively.

④ The structural damping coefficient η is obtained by $\eta = 2\zeta$.

Peak detection must be accurately carried out in order to accurately obtain the damping ratio. For this reason, a high sampling frequency is desirable. On the other hand, long-time sampling is effective in order to accurately measure the value of small damping ratios. In this way, high sampling frequency and long-time sampling are necessary for the accurate measurement of the damping ratio. However, the two of them become a contradictory problem from the viewpoint of the effective utilization of memory. Thus, to solve this problem, the intermittent sampling; that is sampling when immediately after vibration and stop sampling temporarily, and resampling after a short period of time has passed; was adapted. The conceptual scheme of signal processing is shown in fig 4.

5. AUTOMATIC HAMMER

Since the measured sample is hung by threads and easily swing, high accuracy is required to give impact force to it. Thus, the automatic hammer was designed in order to perform stable excitation. The hammer is composed of driver, pendulum, and hammer head. The solenoid is used as an actuator for the driver. The vibrational operation is explained as follows: The pendulum is initially rotated with hammer around axis 1 when the iron core of the solenoid is drawn. When the iron core reaches the perfect attraction point, the pendulum stops rotating, but the hammer head continues to rotate around axis 2 by inertial force. However, the hammer quickly returns, due to the torsion coil spring is built in axis 2 in order to apply the reverse force in reaction to the inertial force of the hammer. This motion allows the impulse vibration to be transmitted to the sample (Fig.5). The operating command to the vibrator is output by computer, with the current value changing in a possible 128 steps. In order to shorten measuring time in continuous measurements, another mechanism was added that stops swinging of the sample after measurement. A similar solenoid is used to operate the additional mechanism. A photograph of the automatic vibrator including a stand for hanging the sample and microphone is shown in fig. 6.

6. MEASUREMENT EXPERIMENT

6.1 Comparison with the conventional system

A comparison with the conventional measuring method[3] was performed in order to evaluate the new measuring method . A glass rod was used as a

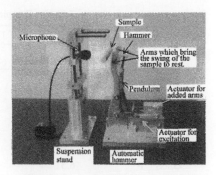

Fig. 6 Photograph of the automatic hammer

sample, and mean value and standard deviation of the damping ratio for 20 measurements were obtained. As well, 10 measurements time was obtained in order to compare measuring time in continuous measurement. The result is shown in fig.7. The results prove that the value determined by new measuring method can be trusted, because the mean value of the damping ratio is close value arrived by the previous measuring method. It is also proven that the new system can obtain equal or higher accuracy in comparison with the previous system, because its standard deviation is considerably smaller than that of the previous system. In addition, a time necessary for 10 continuous measurements has been shortened drastically to 1/10 that of the previous system.

6.2 Measurements using several types of glass

The damping ratios of several types of glass, including SiO_2 glass and 8 other types, were measured using the new system. Anticipated tendencies of structural damping coefficients considering the chemical properties of the samples are shown as follows:
$SiO_2 < \{(A)(B)(C)(D)\} < \{(E)(F)\}, \quad SiO_2 < (G) < \{(E)(F)\}, \quad (H) < (F).$
The experimental results are shown in Table 2. The measurements arrived at by the conventional system are also shown for reference. From this, it is proven that the measured

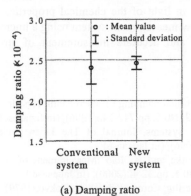

(a) Damping ratio

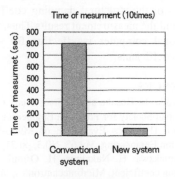

(b) Time of measurement

Fig. 7　Comparison between conventional system and new system for experimental result

Table 2 Damping ratio of several types of glass

Sample	Mean value ($\times 10^{-4}$)		Standard Deviation ($\times 10^{-4}$)	
	Conventional system	New system	Conventional system	New system
SiO$_2$ Glass	0.51	0.87	0.21	0.008
Glass A	1.35	1.54	0.12	0.02
Glass B	1.62	1.98	0.08	0.01
Glass C	2.46	2.18	0.15	0.04
Glass D	2.49	2.23	0.17	0.02
Glass E	4.97	4.84	0.28	0.02
Glass F	6.31	6.08	0.33	0.04
Glass G	3.19	2.83	0.09	0.02
Glass H	3.26	3.43	0.20	0.02

damping ratio in conventional system and new system are close , though there are some differences between them. As well, the tendencies of the measured damping ratio shown by the conventional system and the new system agree in all types of glass. A comparison of the anticipated tendencies of the structural damping coefficients and measurement results show agreement in regard to the tendencies of the structural damping coefficients. In addition, it is proven that the standard deviation of the value of the new system is smaller than that of the value of the conventional system regarding all samples. From the above result, it is confirmed that the new measurement system is valid for the accurate measurement of structural damping coefficients.

7. CONCLUSION

As a simple and accurate measurement system to measure structural damping coefficients, a new system was developed which uses a microphone as a vibration sensor and a personal computer for signal processing. Compared with the conventional system which uses a laser doppler vibrometer and FFT analyzer, the new system shows convenience and economical efficiency. As a result of comparisons with the conventional system, almost the same measurement accuracy was obtained, and measuring time and system cost were reduced to less than 1/10 those of the conventional system. Agreement was obtained in regard to tendencies of the structural damping coefficients in light of the chemical properties of the samples and the measurement results.Thus, this new measurement system using a microphone and personal computer is effective for simple and accurate measurement of structural damping coefficients.

REFERENCES

[1] S. Kuroda, H. Hosaka and K. Itao: Study on vibrational damping of Microcantilevers, Journal of The Japan Society for Precision Engineering, Vol.62, N0.5, pp.742-746(1996), (in Japanese)
[2] K. Itao: Dynamics Evaluation of Micro-Motion Systems, Journal of The Japan Society for Precision Engineering, Vol.63, N0.3, pp.317-322(1996), (in Japanese)
[3] H. Yamakawa, H. Nakajima, H. Osumi, H. Hosaka and K. Itao :Measurement of structural damping coefficient, Micromechatronics, Vol.44, N0.1, pp.32-42(2000), (in Japanese)
[4] S. Nishiyama et al.: Acoustic and Vibration Engineering, Corona Publishing, Tokyo,(1979) (in Japanese)

Robotics

Human Friendly Mechatronics (ICMA 2000)
E. Arai, T. Arai and M. Takano *(Editors)*
© 2001 Elsevier Science B.V. All rights reserved.

Development of an anthropomorphic robotic hand system

Kengo Ohnishi and Yukio Saito

Department of Applied Systems Engineering, Graduate School of Science and Engineering
Tokyo Denki University
Ishizaka, Hatoyama, Hiki, Saitama, 350-0394, JAPAN

This paper describes our development of an anthropomorphic five-digit robotic hand. In addition we describe an cooperative hand -arm control by system. A new model of a mechanical hand is originated from referring to Kapandji's physiological study of the human hand. A 16-D.O.F. robotic hand including a 1-D.O.F. flexible palm was designed as a test bed for evaluating the functionality of our design. The hand is mounted on a robotic arm, Intelligent Arm, to fully perform the dexterous hand function and to present the hand and arm control obtained from categorization and analysis of the relation in the upper limb movements.

1. INTRODUCTION

The research and development of anthropomorphic manipulators is an essential topic in human-friendly machine automation. Research in the area of exercise physiology in the dexterous actions of the digits has had a significant impact on setting a milestone for designing manipulators and upper limb operation devices. Most importantly, creating an appropriate model of the human upper limb is an indispensable issue that affects the design of the human-machine coexisting environment, especially for prosthesis and other assistive devices.

The aim of our research is to propose a new design of an anthropomorphic five-digit robotic hand and a hand-arm cooperation control system based on our hand and arm model. The exercise physiology of a human hand is studied and categorized in order to present a fundamental target for the biomechatronical design of the mechanism and the control system. A human-like five-digit robotic hand with 16 DC micromotors mounted within the hand was designed. We produced the new hand based on our previous research on robotic and prosthetic hands. This new hand is mounted on an 8 D.O.F robotic arm. The control and mechanism of the robotic hand and arm is to be evaluated by experimenting with its performance to demonstrate the effect of our new upper limb model.

1.1. Previous work

Worldly, Human-like robotic hands have been developed based on 20-D.O.F. hand model of Morecki's, which is a sum of the D.O.F. in the digits. In this model, the use of the palm is disregarded and the prehensile function of the hand is extensively limited. The loss of the arch structures in the palm strongly restricts the opposition of the thumb and the fingers, thus reducing the dexterous adjustability in the prehensile configurations of the five-digit structure.

In our research, we modified the functionality of the five-digit mechanical hand. The Intelligent Hand (see Fig. 1.) is a robotic hand with four fingers and thumb developed in our laboratory.[1, 2] This hand has 8 degrees of freedom, seven DC micro motors, and an electromagnetic clutch. Each digit is designed with a 4-bar link mechanism to simultaneously flex and extend the joints in the digit. A universal joint is built in the index finger and the middle finger's MP joint for adduction and abduction movement. The flexion and opposition of the thumb are driven by motors, the electromagnetic clutch switches the driving gear connection mechanism. A distinguishing feature of this hand was in its palm shape design resembling the human hand. The palm is designed with a fixed transverse arch at its MP joint and the fingertips of the ring finger and little fingers are arranged to spread out. This characteristic extended the prehensile capability of the hand at a low degree of freedom. However, the hand had a small drawback of having a fixed curvature palm and a linked flexion and extension digit action. Our next technical challenge was set to propose a new robotic hand model for improving joint and actuator layout that expands the prehensile function with minimum redundancy.

2. MODEL AND DESIGN FOR AN ANTHROBOTIC HAND

A suitable structural model of the human hand and a mechatronical design to furnish the characteristic of this model is proposed. The human hand and arm movements are examined with the relation of the joint and modes of hand usage.[3] The biomechnism of the digit and palm is applied to the robotic hand design.

2.1. Human hand model
The human hand structural model presented in this research was inspired by the Kapandjiís physiological view of the human hand.[4] The model is defined to mechanically compose an active 21 D.O.F. structure, including flexibility in the palm. This model takes in consideration the weak engaged flexion in the fourth and fifth CM joints, which are observed distinctly when creating a tightly closed fist. Our goal is to design and control a dexterous robotic hand with equivalent functions to this human hand model.

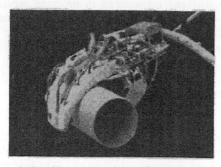 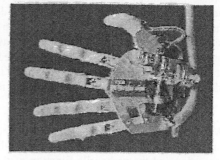

Figure 1. Intelligent Hand: a robotic hand with 7 DC micro motor and an electromagnetic clutch installed on the palm.

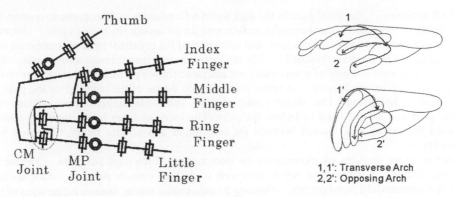

Figure 2. Our proposed structural model of the human hand with the CM joint, left, and the flexible arch structure in the hand which is realized with this joint, right.

2.2. Design Criteria

There is no true necessity nor criteria for designing a mechanical model equal to the human hand model. Reasonably, it is preferred to realize the function while suitably simplifying the mechanism and minimizing the complexity. Therefore, we developed a check form for classifying the movement and poses of the human upper limb and synthesizing the sequential flow for the manipulator. In previous works, classification on the human hand mode was discussed [5], but their emphasis in analyzing the hand has a small drawback in producing a linked joint motion that could be structured with a simple mechanism. Furthermore, the previous classifications are limited to the hand and do not discuss the combined action of the upper limb. The movements of the arm, wrist, thumb, and fingers, along with the modes of prehensile and non-prehensile contact are classified.[3]

The layout of D.O.F., mechanism for the target movement, and the layout of the tactile sensors are determined from the result of classification and mapping the relative items in the combinational usage of the upper limb parts. Analyzing the frequency of usage in the joints shows that the joint in the palm is heavily involved in creating a suitable arch in the hand usages. Therefore, placing a joint in the palm and having control over this enables a smoother performance of the hand.

3. ROBOTIC HAND

A human-like five-digit robotic hand, PSIGHN (Prehension and Symbolization Integrator for General Hand Necessity) is designed and produced. 16 DC micromotors are mounted within the digits and palm. This robotic hand is to be the test bed for evaluating the human hand model and the classification.

3.1. Mechanical structure

The mechanical structure of all digits is similar. The human thumb consists of an equivalent structure, when comparing the first CM joint and MP joint of the thumb to the MP joint and PIP joint of the four fingers. Hence, each digit of this robotic hand composes a 3

D.O.F. structure. The distal joint in the digit forms a fixed angle. We choose to restrain the flexibility of this joint to maximize the surface area for mounting the tactile sensor. Flexion of the mediate joint and proximal joint, and adduction of the proximal joint are composed of a mechanism to be driven individually. The mediate joint is driven through a bevel gear, while the proximal joint consists of a universal joint activated through a link mechanism connected to a linear motion mechanism. A motor placed on the dorsal of the hand drives the palm to compose a hollowing. The thumb, index finger, and little finger curves around the distal-proximal axis of hand to hallow the palm. The inner surfaces of all digits are rotated toward the center axis, located between the ring finger and middle finger, through a link mechanism.

Tactile sensor modules are mounted on the inner surface of the digit and palm. The tactile sensor module is a digital type sensor composed of thirteen contact switches, which are laid out in a continual diamond pattern. Forming a mufti-tactile sensor enables recognition of the holding condition and detection of the slippage with wider sensing range.

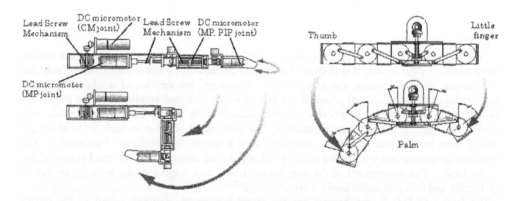

Figure 3. The diagram of the mechanical structure of the digits and palm.

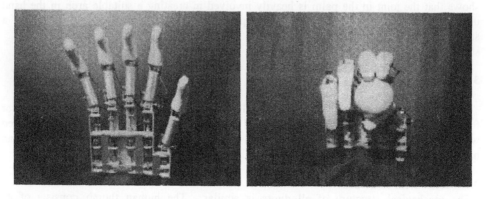

Figure 4. PSIGHN: Anthropomorphic Hand. Hand open (left) and tetra-digit sphere grasp (right).

Table 1. Specification of the Anthropomorphic Hand and the motors in the digits

Length		(mm)	342
Width		(mm)	166
Weight		(kg)	1
Palm	Length	(mm)	131
	DOF		
Thumb	Length	(mm)	160
	DOF		3
Index finger	Length	(mm)	201
	DOF		3
Middle finger	Length	(mm)	211
	DOF		3
Ring Finger	Length	(mm)	207
	DOF		3
Little finger	Length	(mm)	194
	DOF		3

Palm joint		
Output power	(Watt)	1.34
Supply voltage	(DC V)	6
Friction torque	(g-cm)	20
Gear ratio		1/41.53
Metacarpo-phalangeal joint (Flexion)		
Output power	(Watt)	1.34
Supply voltage	(DC V)	6
Friction torque	(g-cm)	20
Gear ratio		1/19.36
Metacarpo-phalangeal joint (Adduction)		
Output power	(Watt)	0.14
Supply voltage	(DC V)	6
Friction torque	(g-cm)	2
Gear ratio		1/32
Interphalangeal joint		
Output power	(Watt)	0.14
Supply voltage	(DC V)	6
Friction torque	(g-cm)	2
Gear ratio		1/209

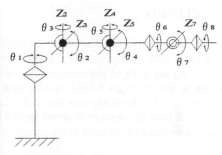

Figure 5. Intelligent Arm: robotic arm with 8 D.O.F. developed for Intelligent Hand is used for the robotic arm for PSIGHN to evaluate the hand and model.

4. HAND-ARM CONTROL

PSIGHN is mounted on the Intelligent Arm, which was developed for the Intelligent Hand. The Intelligent Arm has 8 DOF and consists of base frame, shoulder frame, upper arm frame, and forearm frame. The rotation mechanism is on the base frame for horizontal rotation movement.   For the shoulder and the elbow joint, universal joints are applied for independent longitudinal and transverse rotation.   A coaxial differential gear mechanism is designed for a rotation-flexion-rotation mechanism at the wrist. The length ratio of each frame is designed similarly to a human arm. Seven AC servomotors and one DC motor are mounted to drive the joints.

Studying the relation of the movements of the human upper limb has set a target on designing a control for human-like movement sequence of the multiple D.O.F. in the hand and arm. We arranged a X-Y-Z coordinate system at the shoulder that divides the area around the body, and classified the movements of the arm into 14. This method is also applied to the wrist and digits to classify the movements and setting a reference model. The classified movements are then described by linking the required joint actions and to the modes of hand. The five-digit hand performs a broader range of modes for prehension, but the modes are strongly restricted by the arm and wrist orientation. Therefore, the classification and networking of the hand-arm relation becomes the key in controlling the anthropomorphic hand.

Joint angles are previously computed at grind points set around the orthogonal coordinates. The arm movement will be driven by the differentiation of the present and target wrist position. The palm and elbow are reoriented at target grid point based on the hand mode and object approaching direction, and finely positioned toward the object. This method is implemented to reduce the time delay in response for positioning the arm.

Tactile sensory information is feedback to adjust the digits for performing the hand modes. The multiple sensors laid out in a continual diamond pattern works as a slippage detection sensor by observing the time-series behavior of the sensory response. The classification of the hand mode based on the contact area provides a focused monitoring and digit control for gripping the target object. The hand and arm is controlled to change its posture to overcome the slippage. The composite sensor for tracking the orientation of the arm is in development.

5. CONCLUSION

A new hand model was motivated through our previous work on humanoid hand. A robotic hand with ability to easily adapt to multiple target constraints was designed from reconsidering the work of physiology of the human hand and arm usage. Our 16-DOF hand is design to mount on an 8-DOF robotic arm to fully perform its effect in prehension and manipulation. The hand and arm are to be controlled cooperatively based on the model and the feedback from the sensors mounted on the hand and arm. Our hypothesis of driving the arm as an assistance of the hand mode actions is to be evaluated using this system.

REFERENCES
1. K. Ohnishi, Y. Saito, T. Ihashi, and H. Ueno, Human-type Autonomous Service Robot Arm: HARIS, Proceedings of the 3rd FRANCE-JAPAN CONGRESS & 1st EUROPE-ASIA CONGRESS on MECHATRONICS, 2 (1996) 849-854
2. K. Ohnishi, Y. Saito, D.M. Wilkes, and K. Kawamura, Grasping control for a humanoid hand, Proceedings of Tenth World Congress of Machine Theory and Mechanism, 3 (1999) 1104-1110
3   K. Ohnishi, Y. Saito, Research on a functional hand for electric prosthetic hand, Proceedings of the 7th International Conference on New Actuator & International Exhibition on Smart Actuator and Drive Systems, (2000) 647-652
4   I.A. Kapandji, The Physiology of the Joints, Churchill, Livingstone, 1970
5   C.L. MacKenzie and T. Iberall, The Grasping Hand, NORTH-HOLLAND, 1994

Human Friendly Mechatronics (ICMA 2000)
E. Arai, T. Arai and M. Takano *(Editors)*

Design of a Humanoid Hand for Human Friendly Robotics Applications

Naoki Fukaya, Shigeki Toyama[a] and Tamim Asfour, Rüdiger Dillmann[b]

[a]Department of Mechanical System Engineering, Tokyo University of Agriculture and Technology, Japan. Email: {fukaya,toyama}@cc.tuat.ac.jp

[b]Department of Computer Science, Institute for Control and Robotics, University of Karlsruhe, Germany. Email: {asfour,dillmann}@ira.uka.de

In this paper the mechanism and design of a new, single-motor-driven hand with human-like manipulation abilities is discussed. The new hand (called TUAT/Karlsruhe Humanoid Hand) is designed for the humanoid robot ARMAR that has to work autonomously or interact cooperatively with humans and for an artificial, lightweight arm for handicapped people. The new hand is designed f or anatomical consistency with the human hand. This includes the number of fingers, the placement and motion of the thumb, the proportions of the link lengths and the shape of the palm. It can also perform most part of human grasping types. The hand possesses 21 DOF and is driven by one actuator which can be placed into or around the hand.

1. INTRODUCTION

Humanoid robots are expected to exist and work in a close relationship with human beings in the everyday world and to serve the needs of physically handicapped people. These robots must be able to cope with the wide variety of tasks and objects encountered in dynamic unstructured environments. Humanoid robots for personal use for elderly and disabled people must be safe and easy to use. Therefore, humanoid robots need a lightweight body, high flexibility, many kinds of sensors and high intelligence. The successful introduction of these robots into human environments will rely on the development of human friendly components.

The ideal end-effector for an artificial arm or a humanoid would be able to use the tools and objects that a person uses when working in the same environment. The modeling of a sophisticated hand is one of the challenges in the design of humanoid robots and artificial arms. A lot of research activities have been carried out to develop artificial robot hands with capabilities similar to the human hand. The hands require many actuators to be dexterously moved [6,7]. However, the control system of the humanoid robot becomes more complicated if more actuators are additionally used for the hand design. This is a key aspect for the artificial arm because a handicapped person might not be able to control a complex hand mechanism with many actuators. For this reason, we propose to develop a lightweight hand driven by a single actuator. To this end we adopted a new mechanism for the cooperative movement of finger and palm joints.

2. HUMANOID ROBOT AND ARTIFICIAL ARM

At the Forschungszentrum Informatik Karlsruhe (FZI) we develop the humanoid robot ARMAR, which will be able to assist in workshops or home environments [1]. The research group of the Tokyo University of Agriculture and Technology (TUAT) developed an artificial arm for handicapped people [3]. The humanoid robot has twenty-five mechanical degrees-of-freedom (DOF). It consists of an autonomous mobile wheel-driven platform, a body with 4 DOF, two lightweight anthropomorphic redundant arms each having 7 DOFs, two simple gripper and a head with 3 DOF. Main focus of our research is the programming of the manipulation tasks of ARMAR by a motion mapping between the robot and the person, who demonstrates the task [2]. Since the robot should support a simple and direct cooperation with the human, the physical structure (size, shape and kinematics) of the arm is developed as close as possible to the human arm in terms of segment lengths, axis of rotation and workspace. The mobile platform is equipped with ultrasonic sensors, a planar laser-scanner, and sufficient battery power to allow for autonomous operation.

Figure 1. The humanoid robot ARMAR   Figure 2. The artificial arm using a spherical
                                         utlrasonic motor

We also developed an artificial arm using a spherical ultrasonic motor [3]. The spherical ultrasonic motor has 3 DOF in one unit, which is similar to the human joint [4]. We have also developed a sandwich type ultrasonic motor, it produce twice as high torque output as the same size conventional ultrasonic motor. The arm has one spherical ultrasonic motor for the wrist joint, one sandwich type ultrasonic motors for the elbow and the shoulder joint. These motor are suitable to use for an artifical arm for handicapped people. Consequently, the artificial arm has 7 DOF and represents a good approximation of the human arm. The advantages of the arm are small size, lightweight and high torque at low velocity compared with conventional ones. With a few assignments about output characteristics, it is enough for a handicapped person

3. THE NEW HUMANOID HAND

3.1. Human hand movement

One of the main characteristics of the human hand is the capability of conformably grasping objects with various geometries. Figure 3 clearly shows the typical grasping types of various objects [5].

We do not think about the position of fingers and palm when we grasp objects, because we already know instinctively which position is the best for stable grasp and for saving grasping force. This matter has an important meaning, namely our hand and fingers are automatically moved when we want to stable grasp an object, and the grasping force of each finger becomes even. We simultaneously make adjustments of the position of every finger and of the contact position of the palm. From the above discussion we can conclude that the design requirements for the hand may be satisfied with each finger and the palm acting together in a coordinated group, being independently controlled and making self-adjustment of the position of fingers and palm.

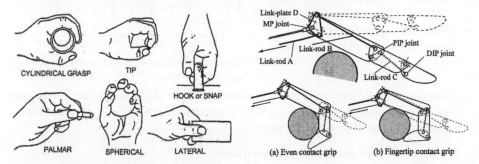

Figure 3. Typical human grasping. Figure 4. Mechanical works on the finger.

3.2. Study of the grip form

The type of human grip figure 3 shows can be classified into 2 categories: a grip achieved by the fingertips and an even contact grip carried out by the whole surface of the hand (i.e. cylindrical or spherical grip). However the even contact grip may change into the fingertip grip. During some actions like squeezing an object or holding heavy objects, we can observe that the fingertip stays in contact with the object surface by flexing the DIP and PIP joints, while the MP joint remains straight and the middle phalanx leaves the object. The even contact grip needs to flex all the joints of a finger around the surface of the object. Specifically it is necessary for the realization of both grip types that the IP and the MP joints move independently. Figure 4 shows the way this link mechanism works to achieve independent movements of each joint. To grasp an object, the link-rod A pulls the link-plate D and the finger moves and keeps its form. If the proximal part touches an object, the following happens: the link-plate D moves independently, while the middle proximal part is moved by the link B. The link C attached to the proximal part pulls the distal proximal part. Finally, the finger curls around the object with a small grasping force.

We studied this structure by using the mechanical analysis simulator A1 MOTION. It is difficult to analyze the contact condition, therefore we decided to place springs in the center of gravity of each block to represent the contact with the object. The rotational moment M on the link-plate D contains the grip force. P1, P2 and P3 are tensions applied on each spring in order to generate contact force with the object. In the analysis model of the fingertip grip we removed the spring on the middle phalanx because the latter has no contact to the object in this case.

276

Figure 5 shows the analysis results performed by A1 MOTION. We can observe remarkable difference at P1. The even contact grip is better than the fingertip grip to grasp breakable objects when the rotational angle is small. Indeed when using the even contact grip, P1 decreases as the rotational angle increases and it becomes the half when the angle is 40 degrees.Therefore the fingers might drop the object if the object is heavy. On the contrary in the case of the fingertip grip, P1 increases as the rotational angle increases, therefore it is possible to firmly hold the object.

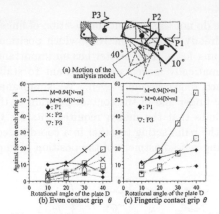

(a) Motion of the analysis model

(b) Even contact grip θ (c) Fingertip contact grip θ

Figure 5. Motion of the analysis model and analysis results

This mechanics has several good characteristics similar to the human being movements and it can automatically change the grip shape in response to the object condition. Therefore it can be very useful to control systems.

3.3. Humanoid hand mechanism description

Based on this research work we constructed an experimental model of the humanoid hand. Its weight is 125 g. We take the size from 27 yeas old Japanese man, his height is 165 cm, and weight is 55 kg. Figure 6(a) shows a clear, rather complex picture of the skeletal structure. We constructed a palm using a ball-joint rod with 2 DOF. The index and middle metacarpal rods are rigid on the base of a plate which is connected to the wrist. The metacarpal rods of the ring and little fingers are free movable like the human palm and are connected with short ball-joint rods like tendons. Thus, each metacarpal joint has 2 DOF (figure 6(b)). The thumb finger has a simple structure: it has only one degree of freedom, which is fixed on the basic plate. The rotational axis is set at an angle of 6.5 degrees to the vertical line of the basic plate. We deem this is sufficient in order to grasp many objects because the hand can already grasp a wide variety of objects if the thumb finger only operates to fix the object position as a fulcrum.

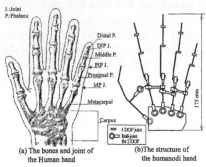

(a) The bones and joint of the Human hand

(b)The structure of the humanodi hand

Figure 6. Structure of the human hand and the humanoid hand

Figure 7 shows the linkage system of the finger motion. The four fingers and the thumb finger are connected with some links by the link-plate E to H and every joint is free movable. By using weak springs or rubber bands, each finger can easily return to its index position. Each link starts working by itself with the strain of the link-rod A when the link-rod J is pulled. Every finger will be moved because link-rods A pull each finger through the link-plate D. The link-plate E starts a rotation centered on a joint point of the link-rod A of the little finger when the little finger touches the object to be grasped.

The link-plate G also starts a rotation centered on a joint point from the link-plate E when the ring finger touches the object, while the index, middle and thumb fingers keep moving. Thus, every link keeps being moved by the link-rod J until every finger touches the object. Therefore, each link moves to keep the balance of the strain force on every link when the contact force of some finger becomes suddenly weak. Finally, the contact force becomes evenly again. The link-plate E affects the palm rods towards the object if the ring and little fingers touch the object (figure 8). Since the fingers and the palm are capable of evenly touching the object, a more stable and safety grasping and holding of the object can be achieved. The link-plate D of the index and little fingers is located at an angle α to the link-rod J (figure 7). Therefore, both fingers can grasp the object with an adduct force and it can be touched with the side surface like as a human being.

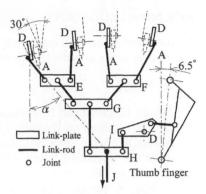

Figure 7. Link mechanism

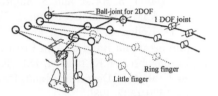

Figure 8. Movement of the palm part

4. RESULTS AND EXPERIMENTS

Figure 9 shows grasping experiments performed by the humanoid hand with the grasping types mentioned above. It is evident that every grasp type is sufficiently well holding. This shows that the link mechanism, the adduction/abduction function of the MP joints and the palm motion ability are effectively working in grasping and holding the objects. We appreciate that extra devices should be employed in order to accomplish the special cases of the Palmar and Lateral grasps. For instance, it is difficult to execute a pick-up operation of a pen and then to change the grip style, because independent movements of every finger are necessary to change the grip style. However, we retain that it is possible to control easily each finger to a certain extent by using simple devices like an electrical brake or an electromagnet pin control on the link-rod of every finger.

5. CONCLUSION

In this paper we presented the mechanical design concept and experimental results of a new humanoid hand for human friendly robotics applications. The humanoid hand is able to grasp and hold objects with the fingers and the palm by adapting the grip to the shape of the object, through a self-adjustment functionality. The hand is driven by only one actuator. Therefore, it is not necessary to use feedback sensors in the hand because the griping force adjusts the grasp position and the posture of the five fingers. This is greatly useful to simplify the hand control system. As a further result, we confirmed that with the humanoid hand, it is possible to fulfill the demands for typical manipulation tasks of humanoid robots in the human everyday world such as offices, homes or hospitals.

6. ACKNOWLEDGMENT

The presented research is supported by the Japan Foundation for Aging and Health. Special thanks to the president Takeshi Nagano.

REFERENCES

1. T. Asfour, K. Berns and R. Dillmann, The Humanoid Robot ARMAR, Proc. of the 2nd Int. Symposium on Humanoid Robots (HURO'99), pp. 174-180, 1999.
2. T. Asfour, K. Berns, J. Schelling and R. Dillmann, Programming of Manipulation Tasks of the Humanoid Robot ARMAR, the 9th Int. Conf. on Advanced Robotics (ICAR'99), pp. 107-112, 1999.
3. N. Fukaya, S. Toyama and T. Seki, Development of an Artificial Arm by Use of Spherical Ultrasonic Motor, The 29th Int. Sympo. Robotics, England, 74, 1998.
4. S. Toyama, S. Hatae and S. Sugitani, Multi-Degree of Spherical Ultrasonic Motor, Proc. ASME Japan/USA Sympo. Flexible Automation, 169, 1992.
5. A. D. Keller, C. L. Taylor and V. Zahm, Studies to Determine the Functional Requirements for Hand & Arm Prostheses", Dept. of Engr., UCLA., CA, 1947.
6. S.C.Jacobsen, E.K. Iversen, D.F. Knutti, R.T. Johnson and K.B. Biggers, Design of the Utah/MIT Dexterous Hand, Proc. IEEE Int. Conf. On Robotics and Automation, pp. 1520-1532, 1986.
7. M. Rakic, Mutifingerd Robot Hand with Selfadaptability, Robotics and Computer-Integrated Manufacturing, Vol. 5, No. 2/3, pp. 269-276, 1989
8. H. Ito, *Tsukamitenodousakinoukaiseki*, Biomechanism, Vol.3, pp. 145-154, 1975 (*in Japanese*).

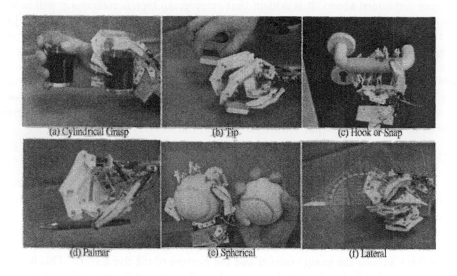

Figure 9. Grasping examination by the TUAT/Karlsruhe humanoid hand

Human Friendly Mechatronics (ICMA 2000)
E. Arai, T. Arai and M. Takano *(Editors)*
279

Humanoid robot "DB"

Shin'ya Kotosaka [1], Tomohiro Shibata [1], Stefan Schaal [1,2]

1 Kawato Dynamic Brain Project (ERATO/JST), 2-2 Hikari-dai, Seika-cho, Soraku-gun, Kyoto 619-02, Japan
2 Computer Science and Neuroscience, University of Southern California, Los Angeles, CA 90089-2520

Abstract: In this paper, we discuss what kinds of capabilities are needed for a humanoid robot for brain science, introduce our humanoid robot DB, and outline our current research topics concerning this experimental setup.

1. Introduction

Brain science has made remarkable advances in the twentieth century and has discovered numerous new insights into the functional and structural organization of nervous systems. It is important that not only neuro-biological research is needed for brain science, but also theoretical research, and, moreover, research that systematically combines both approaches in order to gain a system's level understanding of complex processes such as language, perceptuo-motor control, and learning. Based on these premises, our project, the Kawato Dynamic Brain Project (1), JST, investigates higher-level brain function using theoretical approaches with neural models, psychophysical experiments with human subjects, and a humanoid robot. Our research with the robot has two aspects. One is to provide a test bed for the validation of hypotheses of computational theories derived from computational neuroscience. The other aspect is to develop new algorithms for motor learning, trajectory generation, and perceptuo-motor control with the goal to explore physical, mathematical, and statistical limitations that are also relevant for brain science.

We have developed a hydraulic anthropomorphic robot based on our research concept above. In this paper, we will introduce this humanoid robot, called "DB" (2), and its peripheral equipment and demonstrate our current research with the humanoid robot.

2. A humanoid robot for the brain research
2.1 Parallels between brain research and robotics

Robotics has assisted brain research in various situations. For example, a robot can be employed for laboratory automation, such as moving samples, manipulating an electro-probe for neuronal recording, and robotic surgery (3)(4). Another example of the usage of a robot in brain science is as a manipulandum (5)(6). These "manipulandum robots" have usually one or two degrees-of-freedom (DOFs), use impedance control, and are used for psychophysical experiment with force perturbations. Subjects grasp the tip of the end-effector of the manipulandum and move their arms. At certain times, the manipulandum generates a perturbation in response to the subject's arm movement. From the change of the subject's arm movement in response to the perturbation, it possible to investigate biomechanical and neurophysiological issues, such as muscle stiffness, muscle viscosity, spinal reflexes, cerebral learning capacity in altered physical environments, and trajectory planning and execution. More spectacular research even uses neuronal firing in the brain to directly control the movement of a robotic arm. For instance, Chapin (7) and his colleague developed an artificial neural network to control the robot's movement from recorded neural signals of the motor area in a rat's brain. They connected it to a one DOF robotic arm. After some practice, the rat could operate the robotic arm solely based on controlling its brain activation. This research is based on result by A. V. Lukashin and A. Georgopoulos (8) whose results showed how movement targets for arm movement are represented in the brain.

2.2 Humanoid robot research for the practical verification of motor learning

Recently, several researcher groups have started to investigate "Humanoid" robots from various aspects. One of the recent study is concerned with trajectory generation for walking, dancing or jumping movements (9)(10). A full body humanoid robot was built to generate several kinds of human-like motion. The P3 (11) by Honda Corp. is a self-contained type humanoid robot. It can fully autonomously walk on moderately irregular terrain. Humanoid robots are also investigated as communication devices. Another novel study is the Cog project (12) at MIT and the Humanoid robot project (13) at the Waseda Univ. Both research groups focus on investigating cognitive behavior with their humanoid robots. The collaboration with psychology, neuroscience and robotics in such projects seems to be very productive from both a technological and biological point of view (14).

One of the most interesting research areas for neuroscience and robotics research is the theory of motor learning in humans, and humanoid robots can effectively be used to validate research hypotheses. What kinds of capability are needed for a humanoid robot in such a research area? One of the most important properties is that kinematics and dynamics are similar to humans, e.g., that weight, size, position of the center of the mass, and hopefully the viscoelastic properties of the joints are human-like. Equally important is the availability of sensory information to mimic human proprioception and that joint torques can be produced to realize human levels of performance, but also human limitations.

3. Overview of the humanoid robot "DB"
3.1 Hardware specifications

Based on the insight mentioned above, we have developed a hydraulic anthropomorphic robot "DB". The robot has two arms, two legs, a torso, and a head with a binocular vision system, resulting in a total of 30 hydraulically activated degrees of freedom (DOF). Figure 1 illustrates the appearance of the humanoid robot DB. The height of the robot is approximately 1.85 [m] and its weight is around 80 [kg]. The robot was designed and built by the SARCOS Research Corp., USA. Figure 2 shows that the configuration of DB's DOFs. The robot's neck has three DOFs. The robot head has a binocular vision system (see Fig. 3). Each eye has a pan and tilt DOF. The arms have seven DOFs like human arms. Legs can only perform planar movements with three DOFS per leg. The robot's trunk has three DOFs. Every DOF is equipped with a position sensor and a load sensor except for the DOFs of the camera systems, which have only position sensors. Linear and rotary hydraulic actuators control the joints. Due to the control out of a torque loop based on the load cells in each joint, all DOFs of the robot are very compliant (except for the eyes). The robot runs out of a hydraulic pressure of approximately 0.45 [MPa]. A four-bar linkage pivot stands it attached to the robot's pelvis and prevents the robot from falling over laterally. A counter weight is equipped at the opposite side of the four bar linkages to cancel some of the weight of the robot.

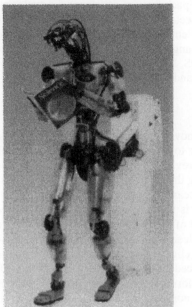

Figure 1 Humanoid robot DB

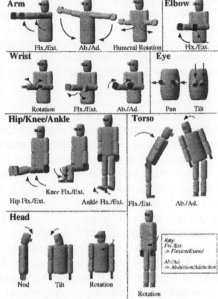

Figure 2 Joint configuration of DB

3.2 Binocular vision system

Each eye of the robot's oculomotor system consists of two cameras, a wide angle (100 degrees view-angle horizontally) color camera for peripheral vision, and second camera for foveal vision, providing an narrow-viewed (24 degrees view-angle horizontally) color image. This setup mimics the foveated retinal structure of primates (see Fig.3).The weight of each eye is only around 160 [g] without cables. This   feature is essentially required to implement saccades, a very fast mechanism of gaze control (1000 [deg/s] in monkeys and 700 [deg/s] in humans) to change overt visual attention, because saccadic motion with heavy cameras can perturb other body parts such as the head. For visual processing, we are using Fujitsu tracking vision boards in a VxWork-Tornado/VME system, Hitachi image processing boards in the a PC/Linux system, and a QuickMag, a commercial stereo color vision system.

3.3 Robot controller and software development environment

To control the many degrees of freedom of the robot, we are using a parallel processing system based on VME technology. The robot's control system is composed of 6 CPU cards (PowerPC G3 266MHz or 366MHz, 64MRAM) in a VME bus, and D/A, A/D, and AJC (Advanced Joint Controller) cards. AJC cards are special purpose analog signal conditioning cards that drive the hydraulic valves, collect sensory information, and communicate with the VME bus. All CPU cards of the robot manage different tasks, e.g., feedback control, signal I/O, feedforward control, movement planning, etc. The CPU cards run the real-time operating system VxWorks (Wind River Corp.). PD or PID feedback is available on the system, and also feedforward control is available with inverse dynamics controllers. Servo control loops run at 420Hz.

A full body dynamics simulation software package was developed for exploring algorithms for robot control such as learning and trajectory generation. The simulation environment has full compatibility with the real execution environment and allows switching between the simulation and the real robot without any change of code. The simulation environment works on several kinds of architecture, e.g., SunOS, Digital UNIX on Alpha Station, IRIX on SGI, Linux on PC with OpenGL graphics library, and MacOS 9.

4 Robot experiments
4.1 Biomimetic adaptive gaze stabilization

Since the oculomotor system resides in a moving body and the foveal vision has a narrow view, it is essential to stabilize gaze in order to obtain stable visual input. In biological systems, the stabilization of visual sensation is provided by the phylogenetically oldest oculomotor reflexes, called the VOR-OKR, involving the vestibulocerebellum to be adaptive for self-calibration through life (14). Inspired by research in the vestibulocerebellum, we have developed an adaptive VOR-OKR circuit which is applicable to any other oculomotor systems without any major modifications. The core of the learning system is derived from the biologically inspired principle of feedback-error learning (15) combined with a state-of-the-art nonparametric statistical learning network and eligibility traces, a concept from biology, to compensate for unknown delays in the sensory feedback pathway (16). With this circuitry, our humanoid robot is able to acquire high performance visual stabilization reflexes after about 40 seconds of learning despite significant nonlinearities and processing delays in the system (See Fig.4, 5).

4.2 Trajectory generation by neural oscillator network

The coordination of movement with external sensory signals is found commonly in the daily behaviors of human. In our work on humanoid robots, we are interested in equipping autonomous robots with similar sensorimotor skills since sensorimotor integration and coordination are important issues in both biological and biomimetic robot systems. We investigate a ubiquitous case of sensorimotor coordination, the synchronization of rhythmic movement with an external rhythmic sound signal, a signal that however, can change in frequency. The core of our techniques is a network of nonlinear oscillators that can rapidly synchronize with the external sound, and that can generate smooth movement plans for our humanoid robot system. In order to allow for both frequency synchronization and phase control for a large

Figure 3 The DB's head

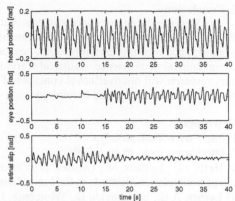

Figure 4 The monitor output of four cameras: The bottom half images show he peripheral vision output, and the upper half images present the foveal vision output

Figure 5 A VOR learning result: the head was perturbed according to top plot. Without a priori knowledge of the oculomotor system, our adaptive VOR-OKR circuit can learn the proper oculomotor controller producing the trajectory in the middle plot, minimizing the image slip on the CCD (the bottom plot shows the learning time course).

range of external frequencies, we developed an automated method of adapting the parameters of the oscillators. The applicability of our methods is exemplified in a drumming task. Our robot can achieve synchronization with an external drummer for a wide range of frequencies with minimal time delays when frequency shifts occur. Figure 6 shows the this capability of the system for drumming sounds generated by a human drummer. The upper graph indicates a sound signal of the human. The bottom graph shows the robot drum beats measured by a vibration sensor on the drum, identical to the case above. Although the human drummer changes the drumming period, the robot is able to follow the drumming sounds without any problems (17).

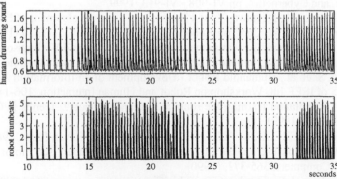

Figure 6. The robot drums in synchrony with external sounds: (top) the average magnitude of the sound the robot hears; (bottom) the robot drumbeats measured by a vibration sensor on the drum.

5. Conclusion

We introduced our concept of humanoid robot research for brain science and demonstrated the humanoid robot DB, an experimental platform that is about one year old. This humanoid robot system is currently rapidly expanding through the use by various researchers in computational neuroscience, statistical learning, and robotics. We hope that our interdisciplinary research will discover novel concepts of perceptuo-motor control and learning, applicable to both neuroscience and technology.

References

1. http://www.erato.atr.co.jp/
2. http://www.erato.atr.co.jp/DB/
3. Hefti, J.L., Epitaux, M., Glauser, D., Fankhauser, H.: Robotic three-dimensional positioning of a stimulation electrode in the brain, Computer Aided Surgery, 3(1), 1-10 (1998).
4) Lerner, A.G., Stoianovici, D., Whitcomb, L.L., Kavoussi, L.R.: A passive positioning and supporting device for surgical robots and instrumentation, Proceddins of Medical Image Computing and Compuer-Assisted Intervention (MICCAI'99) Second International Conference, 1052-1061 (1999).
5. Gomi, H, Kawato, M: Equilibrum-point control hypothesis examined by measured arm stiffness druing multijoint movement, Science, 272, 117-120 (1996).
6. Mussa-Ivaldi, F.A., Patton, J.L.: Robots can teach people how to move their arm, Proc. of the 2000 IEEE Int. Conf. on Robotics & Automation, 300-305 (2000).
7. Chapin, J.K., Moxon, K.A., Markowitz, R.S., Nicolelis, A.L.: Real-time control of a robot arm using simultaneously recorded neurons in the mortor cortex, Nature neuroscience, 2(7), 664-670 (1999).
8. Lukashin, A. V.; Amirikian, B.R.; Georgopoulos, A.P.: A simulated actuator driven by motor cortical signals, NEUROREPORT, 7 (15-17), 2597-2601 (1996).
9 Yamaguchi, J., Soga, E., Inoue, S., Takanishi, A., Development of a Bipedal Humanoid Robot - Control Method of Whole Body Cooperative Dynamic Biped Walking -, Proc. of IEEE Int. Conf. on Robotics and Automation, 368-374 (1999).
10. Ken'ichirou Nagasaka, Masayuki Inaba, Hirochika Inoue, Walking Pattern Generation for a Humanoid Robot Based on Optimal Gradient Method, Proc. of 1999 IEEE Int. Conf. on Systems, Man, and Cybernetics, pp. VI-908 - VI-913, 1999.
11. Hirai, K., Hirose, M., Haikawa, Y. and Takenaka, T. The development of Honda humanoid robot, Proc. of IEEE Int. Conf. on Robotics and Automation, 1321-1326 (1998)
12. Brooks, R. A., Breazeal, C., Marjanovic, M., Scassellati, B. and Williamson, M. (in press) The Cog Project: Building a Humanoid Robot, Lecture Notes in Computer Science: Springer.
13. Hashimoto, S. et. al., Humanoid Robot - Development of an Information Assistant Robot Hadaly -, Proc. of IEEE Int. Workshop on Robot and Human Interaction, 106-111 (1997).
14. Schaal, S.: Is imitation learning the route to humanoid robots?, Trends in Cognitive Sciences, 3(6), 233-242 (1999).
15. Ito,M., THE CEREBELLUM AND NEURAL CONTROL, Raven Press, 1984.
16. Kawato, M., Feedback-error-learning neural network for supervised motor learning, Advanced Neural Computers, 365-372, (1990).
17. Shibata, T., Schaal, S.: Fast learning of biomimetic oculomotor control with nonparametric regression networks, Proc. of IEEE Conf. on Robotics and Automation, 3847-3854 (2000).
18. Kotosaka, S., Schaal, S., Synchronized robot drumming by neural oscillator, International Symposium on Adaptive Motion of Animals and Machines, (2000).

Human Friendly Mechatronics (ICMA 2000)
E. Arai, T. Arai and M. Takano *(Editors)*
© 2001 Elsevier Science B.V. All rights reserved.

Teleoperation master arm system with gripping operation devices

Hitoshi Hasunuma[a] Hiroaki Kagaya, Masataka Koyama, Jun Fujimori,
Fumisato Mifune, Hisashi Moriyama, Masami Kobayashi, Toshiyuki Itoko
and Susumu Tachi[b]

[a]Kawasaki Heavy Industries, Ltd. 118 Futatsuzuka Noda, Chiba, 278-8585 JAPAN

[b]Department of Mathematical Engineering and Information Physics,
The University of Tokyo
7-3-1 Hongo Bunkyo-ku, Tokyo 113-8654 JAPAN

This paper describes a master arm system with gripping operation devices in a teleoperation platform system which we developed in the national project;"Humanoid and Human Friendly Robotics System", organized by AIST(Agency of Industrial Science and Technology) which belongs to MITI(the Ministry of International Trade and Industry).

The teleoperation platform system is connected to a humanoid robot with communication links. An operator can remotely operate the robot by using the platform system with reality of existence at the remote site. The master arm system in the teleoperation platform system can provide the operator with reacting force sensation of the robot arms. The gripping operation devices, which are installed on each master arm, can also provide the oeprator with gripping sensation.

1. Introduction

A humanoid robot has substantial advantages when working in environments where human beings live. The main advantage is that a humanoid robot can act as human beings in such an environment without any previous adjustment for the robot. On the other hand, human friendly and functional machinery become more neccesary as robots are used closer to human beings to care.

Based on the needs above mentioned, since 1998 fiscal year, AIST, which belongs to MITI, has promoted the reserch and development project of "Humanoid and Human Friendly Robotics System" as a part of the Industrial Science and Technology Frontier Program(ISTF).

In the first term, from 1998 to 1999 fiscal year, platform systems as a common base of the research and development has been developed. In the second term, from 2000 to 2002 fiscal year, various kinds of element technologies as for applications, in which humanoid and human friendly robots are expected to be used, will be developed by using the platform systems developed.

A teleoperation platform system, which is one of the platform systems, consists of a humanoid robot and a remote control cockpit system to operate the robot.(Figure 1) The

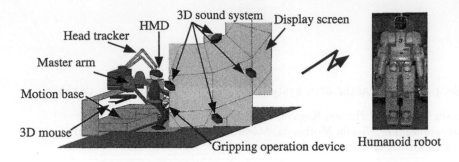

Figure 1. Teleoperation platform system

cockpit system communicates with the humanoid robot which exists at a remote site. The communication utilizes wireless or optical fibers LAN. The remote control cockpit system makes an operator possible to operate the humanoid robot with sense of high reality. The system consists of a master arm system to operate a humanoid robot arm with reacting force sensation, an audio-visual display system to provide realistic information as for robot's views and surrounding sounds, and a motion-base system to provide the operator with motion sense of the humanoid robot. We have developed the master arm system and the motion-base system[1]. In the following chapters, the features of the master arm system and the several experimental results will be shown.

2. System concept and design

There are many kinds of master arm system[2]. In this project, we have developed a master arm system with reality of presence to manipulate the arm of the humanoid robot, or a "slave arm" on the sequel. The sphere of the master arm motion widely covers the sphere of the operator arm's. Functions as for force feedback and control for a redundant degree of freedom(d.o.f.) are implemented on the control system.

We have developed a gripping operation device which makes an operator possible to manipulate a slave hand, which is on the tip of the slave arm. The gripping operation device is installed on the grip part of the master arm. The features of the master-arm and the gripping operation device will be described in the following sections.

2.1. Master arm

A humanoid robot has usually seven d.o.f. as for an arm and can execute not only a position tracking task with a desired position and orientation of the end point, but also an obstacle avoidance task by using redundant d.o.f. of the arm. Therefore, to control such a slave arm, we need to constrain the redundant d.o.f. besides commanding a desired position and orientation of the end point. We designed the master arm system as a seven d.o.f. serial link arm, and defined "elbow angle" as an index of the redundant d.o.f. to utilize it as a command input to the slave arm.

In general, joint arrangement of human beings is described as a simple model with two ball joints at a shoulder and a wrist, and one revolute joint at an elbow. One of the most popular ways to avoid interference of a master-arm with a human operator is to design a master arm as an exoskeleton type with joint arrangement similar to human beings. We adopted the same design. The arrangement of joints and links was carefully examined to avoid interference with a human operator. This designing concept resulted in a compact mechanisim and large sphere of the master arm motion.

Since this master arm system has joint arrangement similar to human beings, it can rotate its elbow around a line segment between the wrist joint and the shoulder joint, keeping a position and orientation of the wrist joint fixed. We defined an angle, called "elbow angle", to depict this rotational motion of arm, and utilized it as a constraint condition of the master arm. By controlling elbow angle of the master arm according to operator's elbow motion, the master arm can avoid interference with a human operator. At the same time, by using elbow angle with a position and orientation of the end point, we can generate an input command for a redundant slave arm.

Elbow angle is calculated from a moving direction of an operator's arm which is measured by several optical sensors located on the lower link of the master arm. By using information from the sensors, elbow angle is controlled so that it keeps roughly a relative distance between an operator's arm and the master arm.

We implemented seven actuators for seven joints of the master arm for force feedback. We control six of them to perform force feedback control to an operator, and the other to adjust elbow angle.

Figure 2 shows a schematic block diagram of the master-slave control system[3]. In this control system, elbow angle is calculated by "elbow angle generator" as described above. And also a position and orientation of the end point of the master-arm is calculated by "end point generator". These outputs, elbow angle and end point's position and orientation, are put into "joint motion generator" in order to generate input commands for each joint actuators. To realize force feedback function, force information of the master-arm and the slave-arm is put into "end point generator". "Slave command generator" generates input commands for a slave arm.

The joint arrangement of the master-arm is shown in Figure 3. Actuators are located on each joint, and a six-axis force/torque sensor is put on the wrist joint to measure force and torque generated by an operator. The specifications of the master-arm system are shown in Table1.

Table 1
Specifications of the master arm

Interface type	Exoskeleton dual master-arm
Degree of freedom	7(each arm)
Output force	10N(Maximum at a moment)
Max speed at end point	100mm/sec
Weight	75kg(each arm)
Joint angle range	JT1: -60 — 80 [deg] JT2: -15 — 50 [deg] JT3: -30 — 90 [deg] JT4: -50 — 55 [deg] JT5,6,7: -70 — 70 [deg]

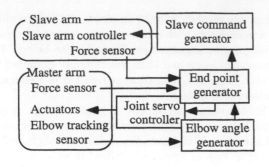

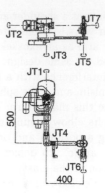

Figure 2. Schematic block diagram of master-slave control system

Figure 3. Joint arrangement

2.2. Gripping operation devices

If complex mechanisms, by which an operator can operate with five fingers, were put on the grip of the master arm, it would cause undesirable increasing weight of master arm and loss of manipulability. Then, we have focused on the gripping operation by using two fingers, and have developed a gripping operation device with which an operator can easily operate open-close motion to grip by his /her thumb and index finger, feeling gripping force of a slave hand. Maximum displayed force is 5N at a moment.

We employed a wire tension mechanism to display gripping force in small size and light weight. There are two actuators to independently display forces of a thumb and an index finger. The mechanism for thumb has another passive joint to allow thumb's radial abduction and ulnar adduction to keep work space wide. As for gripping operation, commands to a slave hand are generated according to lengths of wires which are pulled without any looseness by keeping wire tensions constant.

3. Experimental results

3.1. Performance test for the master arm system

First of all, we have examined performances of the master arm system developed. We confirmed its available operation area, the function to follow an operator's elbow motion without interfering with the operator, force feedback function, and stable operation speed at the end point.

We have developed a two d.o.f. slave hand system with two fingers to test the performance of the gripping operation device. The slave hand system can measure gripping force and control position of the finger. We confirmed that an operator could manipulate the slave hand and feel gripping force of the slave hand through the gripping operation device. Figure 4 shows a scene of an experiment for the gripping operation device.

3.2. Task operation test for the master-slave control system

Next, we have examined performances of the master-slave control system. An industrial robot, FS-06, made by Kawasaki Heavy Industries, Ltd., was used as a slave arm in this experiment. six-axis force/torque sensor is installed on the wrist of the slave-arm.

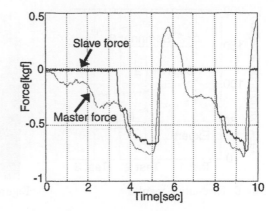

Figure 4. Photo of experiment for gripping operation task

Figure 5. An example of experimental results

In this experiment, an operator operated the slave arm through the master-arm to push a rigid wall with constant force. We confirmed that the operator could feel fedback force from the master arm and perform a task stably. Figure 5 shows trend data of the master arm's and the slave arm's force and position; the master-arm's force follows the slave-arm's after contact with the wall.

3.3. Performance test of operation task

Finally, we have examined performances as for remote operation tasks using a humanoid robot which has developed in this project as a slave robot. In order to verify the manipulability of the whole system, we have executed several remote operation tasks: lifting a stuffed toy clipping with dual-arms, open-closing the sliding door to pick up a can, pushing a shopping cart, performing the tower of Hani, and toy blocks assembly.

"Toy blocks assembly" is proposed as a benchmark task for the purpose of maneuverability evaluation[4]. Thus, as the fifth experiment, the toy blocks, "Mega Block(MAXI)", are used to evaluate maneuverability of this system. Several operators manipulated the slave robot to assemble the toy blocks into the desired form through the master arm.

Four operators tried the test; three of them were beginners in operating remote control system, and the other was an expert. Before the experiment, each operator practiced the task once. The task completion time of the assembly task was measured. The results of this experiment are shown in Figure 3.3; the task completion time of the expert operator is faster than that of beginners. According to the operator's comments on the experimental results, the difference in the task completion time could be due to their experience of distance recognition through the visual display system.

However, even a beginner for remote operation could accomplish such a skillful task without failure. This shows that this master arm system is an easy-to-operate system for a human operator. In the future, we hope to improve the system to execute more skillful tasks and evaluate the maneuverability using a benchmark task mentioned above.

290

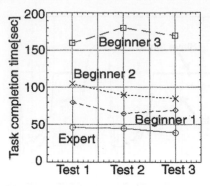

Figure 6. Evaluation results

Figure 7. A photo of assembly task

4. Conclusion

We have developed a master arm system which provides an operator to control a humanoid robot with sense of reality. The master arm has wide sphere of motion, a tracking function for an operator's elbow motion with non-contact sensors, and force feedback function. A gripping operation device, which makes an operator possible to control a slave hand, is installed on the grip of the master arm. We have evaluated the performance of the system with several experiments. As the results of these experiments, an operator could execute skillful tasks with force feedback and gripping operation. Our next step is to improve the control system and the mechanisms to realize faster response and higher stability of the whole system.

The master arm system will be used to investigate various applications as for humanoid and human friendly robot in the second term of the project. We expect that the system can serve realistic tele-presence when teleoperating a humanoid and human friendly robot in the near future.

5. Acknoledgement

The authers would like to thank MSTC and NEDO for their entrusting development of the project "Humanoid and Human Friendly Robotics System" and the members cooperating in the project for their constructive support. Also the authers would like to thank Prof. Yokokohji, Kyoto University for his instructive advice.

REFERENCES

1. H.Hasunuma et al., Development of Teleoperation Master System with a Kinethetic Sensation of Presence, ICAT'99 (1999).
2. G.C.Burdea, Force and Touch Feedbak for Virtual Reality, J.Wiley&Sons,Inc., 1996.
3. T.Miyazaki, S.Hagihara, Parallel Control Method for a Bilateral Master-Slave Manipulator, JRSJ Vol.7 No.5 (1988).
4. Y.Yokokohji et al., Maneuverability Evaluation of a Unified Hand-Arm Master-Slave System, 17th Annual Conference of RSJ (1999).

Human Friendly Mechatronics (ICMA 2000)
E. Arai, T. Arai and M. Takano *(Editors)*
© 2001 Elsevier Science B.V. All rights reserved.

Stable Grasping and Posture Control for a Pair of Robot Fingers with Soft Tips

K.Tahara[a], M.Yamaguchi[a], P.T.A.Nguyen[a], H.−Y.Han[a] and S.Arimoto[a]

[a]Department of Robotics, Ritsumeikan Univercity, Kusatsu, Shiga, 525-8577 Japan

This paper derives and analyzes non-linear dynamics of grasping a rigid object by a pair of robot fingers (1 D.O.F. and 2 D.O.F.) with soft and deformable tips. A method of synthesizing a feedback control signal for stable grasping and posture control of the object is presented, which is based on passivity analysis of the dynamics including extra terms of Lagrange's multipliers arising from holonomic constraints of tight area contacts between soft finger-tips and surfaces of the object. Results of computer simulation based on the derived non-linear differencial equations with geometric constraints are presented. Finally, usefulness of the proposed control method is discussed from the practical viewpoint.

1. Introduction

Human easily pinches and operates an object using thumb and other fingers, but it is very difficult and complicated to let robots to do the same motion. One of the reasons may come from the fact that such a physical motion in ordinary tasks that we do in our everyday life is hardly analyzed and understood from the logical viewpoint.

This paper aims at analyzing dynamics of a pair of robot fingers with soft and deformable finger-tips grasping a rigid object. It is shown theoretically that dynamic stable grasping is realized by means of sensory feedbacks based on the mesurement of changes of each center of two contact areas between finger-tips and surfaces of the rigid object. It is assumed in this paper as in previous papers[1~3] that motion of the overall system is confined to a horizontal plane and the effect of gravity is ignored. At the same time it is assumed that the shape of finger-tips is hemispherical and the distributed pressure generated by deformation of each finger-tip is lumped-parameterized into a single representative reproducing force that has a direction to the center of curvature of the finger-tip from the center of contact area between the finger-tip and the surface of the object.

In section 3, it is shown that the dynamics satisfy passivity when the output is taken as a vector of velocity variables though they are quite complicated and subject to holonomic constraints that arise from tight area-contacts. Then, we propose a method of systhesizing feedback control signals. One of these signals aims at stable grasping and the other aims at posture control of the rigid object. In section 4, results of computer simulation by using the derived non-linear differencial equations with geometric constraints are presented. Then, usefullness of this control methodology is discussed from the practical viewpoint.

2. Dynamics of Pinch Motion

Firstly we derive the dynamics of pinch motion made by using two robot fingers with 1 D.O.F. and 2 D.O.F. whose tips are made of soft material. All symbols and coordinates in the overall system are defined in Fig.1. There are four equations of geometric constraint as shown:

$$Y_1 = c_1 - r_1\varphi_1 = c_1 - r_1(\pi + \theta - q_{11} - q_{12}) \tag{1}$$
$$Y_2 = c_2 - r_2\varphi_2 = c_2 - r_2(\pi - \theta - q_{21} - q_0) \tag{2}$$
$$\tau = x_1 + \frac{l}{2}\cos\theta - Y_1\sin\theta = x_2 - \frac{l}{2}\cos\theta - Y_2\sin\theta \tag{3}$$
$$y = y_1 - \frac{l}{2}\sin\theta - Y_1\cos\theta = y_2 + \frac{l}{2}\sin\theta - Y_2\cos\theta \tag{4}$$

The eqns.(1) and (2) are generated from tight area-contacts between each surface of finger-tips and the surface of the rigid object(see Fig.2). Here, Y_1 and Y_2 in eqns.(1) and (2) should be expressed using position variables $z = (x, y, \theta)^T$, $q_1 = (q_{11}, q_{12})^T$, and $q_2 = q_{21}$ (q_0 is a fixed angle of the 1 D.O.F finger) as follows:

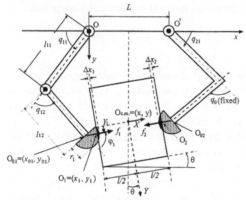

Figure 1. A setup of two fingers(1 D.O.F. and 2 D.O.F.) with soft tips pinching a rigid object

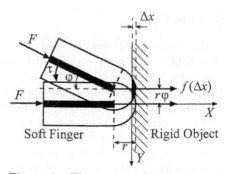

Figure 2. The center of contact area moves on the object surface by inclining the last link against the object.

$$Y_1 = (x_{01} - x)\sin\theta + (y_{01} - y)\cos\theta \tag{5}$$
$$Y_2 = (x_{02} - x)\sin\theta + (y_{02} - y)\cos\theta \tag{6}$$

where

$$x_{01} = -l_{11}\cos q_{11} - l_{12}\cos(q_{11} + q_{12}) \tag{7}$$
$$y_{01} = l_{11}\sin q_{11} + l_{12}\sin(q_{11} + q_{12}) \tag{8}$$
$$x_{02} = L + l_{21}\cos q_{21} + l_{22}\cos(q_{21} + q_0) \tag{9}$$
$$y_{02} = l_{21}\sin q_{21} + l_{22}\sin(q_{21} + q_0) \tag{10}$$

The eqns.(3) and (4) mean that the loop starting from the orgin O of the first joint center on the left finger and return to it through the $O_{01}, O_1, O_{c.m.}, O_2, O_{02}, O'$ is closed, as shown in Fig.1. The Lagrangian of the overall system is given as

$$L = K - P + S \qquad (11)$$

where K is the kinetic energy, P is the potential energy and S is related to constraints. These are defined as

$$K = \sum_{i=1,2} \frac{1}{2} \dot{q}_i^T H_i(q_i) \dot{q}_i + \frac{1}{2} \dot{z}^T H \dot{z} \qquad (12)$$

$$P = \sum_{i=1,2} \int_0^{\Delta x_i} f_i(\xi) d\xi \qquad (13)$$

$$S = \sum_{i=1,2} \lambda_i (Y_i - c_i + r_i \varphi_i) \qquad (14)$$

where $H = \text{diag}(M, M, I)$, M and I denote the mass and inertia moment of the rigid object. Δx_i ($i = 1, 2$) denotes the maximum desplacement of deformation and λ_i ($i = 1, 2$) are Lagrenge's multipliers. Here, by applying Hamilton's principle for the variational form

$$\int_{t_1}^{t_2} \left\{ \delta(K - P + S) + u_1^T \delta q_1 + u_2^T \delta q_2 \right\} dt = 0 \qquad (15)$$

the Lagrange eqnation of the overall system is obtained, which is described as follows:

$$\left\{ H_i(q_i) \frac{d}{dt} + \frac{1}{2} \dot{H}_i(q_i) \right\} \dot{q}_i + S_i(q_i, \dot{q}_i) \dot{q}_i + \frac{\partial \Delta x_i}{\partial q_i} f_i - \lambda_i \frac{\partial \Phi_i}{\partial q_i} = u_i, \quad i = 1, 2 \qquad (16)$$

$$H \ddot{z} + \sum_{i=1,2} \left(\frac{\partial \Delta x_i}{\partial z} f_i - \lambda_i \frac{\partial \Phi_i}{\partial z} \right) = 0, \quad \Phi_i = Y_i - c_i + r_i \varphi, \quad i = 1, 2 \qquad (17)$$

where eqn.(16) denotes dynamics of each of the two robot fingers. The eqn.(17) expresses dynamics of the rigid object where $\Phi_i = 0$ for $i = 1, 2$ express tight area-contact constraints. Here, it is possible to show that the overall dynamics of eqns.(16) and (17) satisfy passivity, i.e.,

$$\dot{q}_1^T u_1 + \dot{q}_2^T u_2 = \frac{d}{dt} \left\{ \sum_{i=1,2} \frac{1}{2} \dot{q}_i^T H_i(q_i) \dot{q}_i + \frac{1}{2} \dot{z}^T H \dot{z} + \sum_{i=1,2} \int_0^{\Delta x_i} f_i(\zeta) d\zeta \right\}$$

$$= \frac{d}{dt} (K + P) = \frac{d}{dt} V(t) \qquad (18)$$

whose integral over $[0, t]$ leads to

$$\int_0^t (\dot{q}_1^T u_1 + \dot{q}_2^T u_2) d\tau = V(t) - V(0) \geq -V(0) \qquad (19)$$

Clearly the input-output pair (u_1, u_2) and $(\dot{q}_1, \dot{q}_2)$ satisfies passivity, where V signifies the total energy of the overall system, that is,

$$V = K + P \qquad (20)$$

3. Dynamic Stable Grasping and Posture Control

We propose the feedback control signals for stable grasping and posture control based on passivity analysis of the overall dynamics in the following may:

$$
u_{fi} = -k_{v_i}\dot{q}_i + \frac{\partial \Delta x_i}{\partial q_i} f_d + (-1)^i f_d \Big\{ \frac{r_i}{r_1 + r_2}(Y_1 - Y_2)e_i \Big\}
$$

$$
+ (-1)^i \gamma \Big\{ \frac{\partial \Phi_i}{\partial q_i} \Big(\frac{\dot{Y}_1 - \dot{Y}_2}{l - \Delta x_1 - \Delta x_2} \Big) + \frac{r_i}{r_1 + r_2}(\dot{Y}_1 - \dot{Y}_2)e_i \Big\} \tag{21}
$$

$$
u_{\theta i} = (-1)^i \frac{\partial \Phi_i}{\partial q_i} \Big(\frac{\beta \Delta \theta}{l - \Delta x_1 - \Delta x_2} + \alpha \dot{\theta} \Big) \tag{22}
$$

$$
u_i = u_{fi} + u_{\theta i} + U_i, \quad i = 1, 2 \tag{23}
$$

where $\Delta\theta = \theta - \theta_d$, $\alpha > 0$, $\beta > 0$, $\gamma > 0$, $e_1 = (1,1)^{\mathrm{T}}$, $e_2 = 1$, $l - \Delta x_1 - \Delta x_2 > 0$. The inner product between (U_1, U_2) and $(\dot{q}_1, \dot{q}_2)$ of the overall closed-loop system in which this control signal u_i is substituted into eqn.(16) for $i = 1, 2$ is reduced to:

$$
\int_0^t (\dot{q}_1^{\mathrm{T}} U_1 + \dot{q}_2^{\mathrm{T}} U_2) \mathrm{d}\tau = W(t) - W(0) + \int_0^t \sum_{i=1,2} k_{v_i} \|\dot{q}_i\|^2 \mathrm{d}\tau + \int_0^t \alpha(l - \Delta x_1 - \Delta x_2)\dot{\theta}^2 \mathrm{d}\tau
$$

$$
+ \int_0^t \frac{\gamma}{r_1 + r_2}(\dot{Y}_1 - \dot{Y}_2)^2 \mathrm{d}\tau \geq -W(0) \tag{24}
$$

where

$$
W(t) = K + \Delta P + \frac{f_d}{2(r_1 + r_2)}(Y_1 - Y_2)^2 + \frac{\beta}{2}\Delta\theta^2 \tag{25}
$$

$W(t)$ is positive definite in all state values under algebraic constraints of eqns.(1) and (2). Then, the closed-loop system including this control signal satisfies passivity.

Next, we prove that the closed-loop system with $U_1 = 0$ and $U_2 = 0$ satisfies asymptotic stability. In fact, it follows that

$$
\frac{\mathrm{d}}{\mathrm{d}t} W(t) = -\sum_{i=1,2} k_{v_i} \|\dot{q}_i\|^2 - \alpha(l - \Delta x_1 - \Delta x_2)\dot{\theta}^2 - \frac{\gamma}{r_1 + r_2}(\dot{Y}_1 - \dot{Y}_2)^2 \tag{26}
$$

Hence, $W(t)$ becomes a Lyapunov function under constraint conditions, since $\dot{W}$ satisfies

$$
\dot{W} \leq 0 \tag{27}
$$

Thus, according to LaSalle's invariance theorem, all $\ddot{q}_i$, $\dot{q}_i$ (for $i = 1, 2$), $\ddot{z}$, and $\dot{z}$ converge to zero as $t \to \infty$ and hence each sum of remaining terms of the closed-loop system except the inertia and Coriollis-centripetal terms tend to vanish as $t \to \infty$. This leads to the conclusion that the solution trajectory of the closed-loop system converges to the single equilibrium expressed as $f_i = f_d$ $(i = 1, 2)$, $Y_1 - Y_2 = 0$, $\Delta\theta = 0$ together with constraint eqns.(1) and (2). In this paper the details of the proof is omitted but more precisely it proceeds similarly to the proof given in future papers[4,5]. It is important to remark that the novelty of this result lies in finding a feedback control scheme that realizes stable grasping in a dynamic sense differently from the classical concept of stable grasping from the static viewpoint[6].

4. Computer Simulation of Non-linear Differencial Equation with Geometric Constraints

The initial conditions of robot-fingers and rigid object are shown in Fig.3 and parameters chosen in the design of control inputs are shown in Table.1. The result of computer simulaion based on the use of a Constraint Stabilization Method[2,7] are shown in Figs.4~6. In this paper we show two simulation patterns when the desired reproducing force f_d is set at 0.5[N] and 1[N]. It is assumed that the relation between the reproducing force f and the maximum deformation displacement Δx_i is described[1] as:

$$f = K(\Delta x_i)^2, \quad i = 1, 2 \tag{28}$$

where K is a constant value.

Table 1
Parameters of control inputs

f_d[N]	θ_d[deg]	K[N/mm^2]	k_{v_i}	α	β	γ
0.5/1.0	-5.0	1.0	0.09	0.5	0.3	9.5

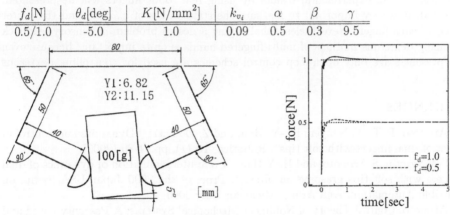

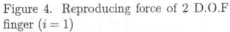

Figure 3. Initial condition of robot-fingers and a rigid object

Figure 4. Reproducing force of 2 D.O.F finger ($i = 1$)

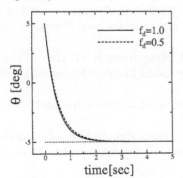

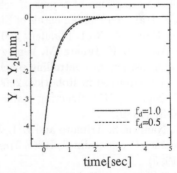

Figure 5. Posture of the rigid object

Figure 6. Vanishing phenomena of the couple of force exerted on the object

In Fig.4, both the reproducing forces f_i $(i = 1, 2)$ converge to the desired force and the settling time of convergence is about 1.5 second. In Figs.5~6, $Y_1 - Y_2$ also converges to 0 and the posture of the rigid object converges to the desired angle. These settling times are about 2 second. Thus, both stable grasping and posture control of a rigid object are simultaneously realized by using the proposed design of feedback control signals.

5. Conclusion

In this paper, we derived and analyzed dynamics of a pair of robot fingers with soft and deformable tips pinching a rigid object. Then, a design method of feedback control signals for stable grasping and posture control of the object was proposed. We simulated non-linear differencial equations with geometric constraints and showed performances of the proposed feedback scheme. Through these simulations, we found that a linear superposition of feedback signals realizes stable grasping and posture control of the object at the same time.

In a near future, experimental results by using real robot fingers will be presented. To this end, it is very important to measure the state values of q_i, θ, Y_i and the center point O_i of each finger. Nevertheless, challenging a design problem of sensory feedback for realizing dexterous motions of multi-fingered hands is quite important, because even at the present stage only open-loop control schemes are used for controlling dexterous hands[8].

REFERENCES

1. S. Arimoto, P. T. A. Nguyen, H. -Y. Han, and Z. Doulgeri, "Dynamics and control of a set of dual fingers with soft tips", Robotica, **18** (1), pp.71-80 (2000).
2. P.T.A.Nguyen, S.Arimoto, and H.-Y.Han, "Computer simulation of dynamics of dual fingers with soft-tips grasping an object", Proc. of the 2000 Japan-USA Symp. on Flexible Automation, Ann Arbor, Michigan, July 2000.
3. S. Arimoto, Control Theory of Nonlinear Mechanical Systems: A Passivity-based and Circuit-theoretic Approach, Oxford University Press, Oxford, UK, 1996.
4. S. Arimoto and P. T. A Nguyen, "Principle of superposition for realizing dexterous pinching motions of a pair of robot fingers with soft tips", to be published in IEICE Trans. on Fundamentals of Electronics, Communications and Computer Sciences, Vol.**E84 − A**, No.1, (January 2001).
5. S. Arimoto, K. Tahara, M. Yamaguchi, P. T. A. Nguyen and H. -Y. Han, "Principle of superposition for controlling pinch motions by means of robot fingers with soft tips" ,to be published in Robotica, 2000.
6. V. Nguyen, "Constructing stable grasps", Int. J. of Robotics Research, **8** (1), pp. 26-37 (1989).
7. T. Naniwa, S. Arimoto and L. L. Whitcomb, "Learning control for robot tasks under geometric constraints", IEEE Trans.on Robotics and Automation, **11** (3), pp.432-441 (1995).
8. K.B.Shimoga, "Robot grasp synthesis algorithms: A survey", Int. J. of Robotics Research, **15** (3), pp.230-266 (1996).

Human Friendly Mechatronics (ICMA 2000)
E. Arai, T. Arai and M. Takano *(Editors)*

An Implementation of Chopsticks Handling Task by a Multi-fingered Robotic Hand

Hajime Sugiuchi[a] and Hirofumi Yagi[b]

[a]The department of mechanical engineering, Yokohama National University,
79-5 Tokiwadai, Hodogaya-ku, Yokohama, Japan 240-8501

[b]United System Engineers Inc.,
1589—1 Kitaono, Shiojiri, Nagano, Japan 399-0651

A human mimetic robotic hand and its control system are developed. This hand has five fingers and 17 D.O.F. (3 for each finger, 4 for thumb and 1 on palm). The distributed touch sensor that has more than 600 measuring points is equipped. The developed system can control the position, orientation, velocity and force of multiple points on the hand simultaneously. The event driven task execution system is also developed. This system watches events that signal the change of constrained state of hand and switches the action of hand dynamically according to the detected event. The effectiveness of this system is shown by the experiment in which our robotic hand holds chopsticks, pinchs a soft object as human successfully. We found that the palm joint is very important for the dexterity of the hand through this experiment. It should be impossible to handle chopsticks without the palm joint.

1. INTRODUCTION

In the future, many robots will work at home for house keeping aids and so on. Such robots should be able to perform various kinds of tasks. The working environment is optimized for human hands. So, the human mimetic general-purpose hand should be suitable. Such hand can perform each task without replacing hand and reduce the complexity and the size of total system. On the other hand, the complexity of the hand itself and the control system are disadvantages.

In order to realize various dexterous hand tasks by robotic hand, we should write some robot programs. But this is quite difficult work because, in dexterous hand tasks, we use multiple fingers in cooperation with each others, each finger may contact with handling object on multiple points and position, orientation, velocity and force of these points should be controlled simultaneously. The distributed touch sensor should be necessary for force control because the location of contact point is not limited on the fingertip and it should be detected by the sensor. In addition, the locations and the number of those contact points will change while the task proceeds. We propose a framework for the hand task description. The Task Description System manages the reference of each controlled point. A control system that supports simultaneous multiple points control is

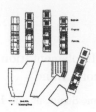

Figure 1. The posture used palm joint.

Figure 2. The photo. of developed hand(pinching a sponge cube by chopsticks).

Figure 3. The measuring points arrangement of hand.

also developed.

Some robotic hands have been developed [1],[2],[3] but their configuration are different a little from human's because of cost or mechanical reason. In the case of such hands, we may be forced to find deferent way from human's to achieve some tasks. Many kinematic analyses have been made for the hand works[4], [5]. These should be effective to find the way of performing the task. But we consider the human mimic robot hand and the human mimic task implementation approach should be suitable. The effectiveness of our approach is shown by the chopsticks handling experiment. Through this experiment, we found that human has at least one degree of freedom on his hand and add one joint on the palm of our robot hand.

2. FEATURES OF DEVELOPED ROBOTIC HAND

Our developed hand has four 3 D.O.F. fingers and a 4 D.O.F thumb. Each finger has 4 joints but the last joint is coupled to the next one as human. In addition, it has one extra joint on the palm to realize the palm motion shown in figure 1. So, totally it has 17 D.O.F. The photo. of this hand is shown in figure 2.

All active joints are driven by RC servo unit. The size of hand is approximately twice of human hand and it's weight is about 2.5kg. The thumb, forefinger, middle and ring finger and a part of the palm are covered with distributed touch sensor[6] which can detect the intensity of pressure force and it's location. The arrangement of measuring points are shown in figure 3. All measuring points are scanned within 20ms.

The reachable area of each fingertip are shown in figure 4 and 5 to illustrate the effect of palm joint. The light gray areas are the union of reachable areas of all fingers. The dark gray areas are the intersection of two unions. One is the union of reachable areas of thumb, forefinger and middle finger and the other is the union of reachable areas of ring and little finger. Thumb, forefinger and middle finger are more important than the others. These three fingers usually play a main role in the hand tasks and the others assist them. If the size of dark gray area is small, it is difficult for ring and little fingers to collaborate with the main role fingers. So, the size of this area will indicate the dexterity of hand in some meaning. The case without palm joint is shown in figure 4 and the other is shown

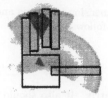

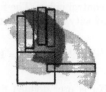

Figure 4. The working area hand without palm joint.

Figure 5. The working area hand with palm joint.

in figure 5. These figures show that the palm joint enlarge the intersection area and very effective to the dexterity of the hand. In fact, the palm joint is necessary for handling chopsticks because the ring finger cannot reach to the lower chopstick without this joint.

3. CONTROL SYSTEM

In our hand control system, the position and force references of control points are transformed to the velocity references of corresponding control points. The velocity references of all control points are transformed to the joint angular velocity reference vector by using the Jacobean matrix.

3.1. Control of one point on the hand

The velocity control is a basic control law in our control system. The relation between the i-th control point velocity reference vector v_{ri} in Cartesian space and the joint angular velocity reference vector $\dot{\theta}_r$ is as follows. Here, $\mathbf{J_i}(\theta)$ is the Jacobean matrix of i-th contact point. We can get $\dot{\theta}_r$ by solving equation 1 for given v_{ri}.

$$v_{ri} = \mathbf{J_i}(\theta)\dot{\theta}_r \qquad (1)$$

In the case of Position control, the reference position p_{ri} and reference orientation $\mathbf{R_{ri}}$ in rotational matrix form are given and the actual position and orientation of the i-th control point are p_i and $\mathbf{R_i}$ respectively. The positional error vector Δx_i in 6 dimension is defined and by using Δx_i, the velocity reference of contact point v_{ri} is given as follows.

$$\Delta p_i = p_{ri} - p_i, \mathbf{R_{ri}R_i}^{-1} \approx \begin{pmatrix} 1 & -\Delta\gamma & \Delta\beta \\ \Delta\gamma & 1 & -\Delta\alpha \\ -\Delta\beta & \Delta\alpha & 1 \end{pmatrix}, \Delta x_i = \begin{pmatrix} \Delta p_i \\ \Delta\alpha \\ \Delta\beta \\ \Delta\gamma \end{pmatrix}, v_{ri} = \mathbf{K_{pi}}\Delta x_i \qquad (2)$$

Here, $\mathbf{K_{pi}}$ is user defined 6 by 6 position control gain matrix. It should be chosen not to violate the system stability. Thus the position control is delegated to velocity controller.

In the case of force control, the reference force f_{ri} is given and the actual force on the i-th contact point f_i is detected by touch sensor. Both values are scalar because our distributed touch sensor can detect only normal force on the surface. z_{ni} is the normal vector at control point. The force error vector Δf_i in 6 dimension is defined and by using Δf_i, the velocity reference of contact point v_{ri} is given as follows.

$$\Delta f_i = \left(\begin{array}{c} (f_{ri} - f_i) z_{ni} \\ 0 \end{array} \right), v_{ri} = \mathbf{K_{fi}} \Delta f_i \tag{3}$$

Here, $\mathbf{K_{fi}}$ is user defined 6 by 6 force control gain matrix. It also should be chosen not to violate the system stability. Thus the force control is also delegated to velocity controller.

3.2. Motion integration of all contact points in one action

Generally, the robotic hand contacts with the handling object at several points and each point has a position, velocity or force reference. All reference values are converted to corresponding velocity references as mentioned above and the equation 1 is available for all control points. We pile up all equations and build up an equation as follows. Here, n is the number of control points.

$$V_r = \mathbf{J}(\theta)\dot{\theta}_r, \, V_r = \left(\begin{array}{c} v_{r1} \\ \vdots \\ v_{rn} \end{array} \right), \mathbf{J}(\theta) = \left(\begin{array}{c} \mathbf{J}_1(\theta) \\ \vdots \\ \mathbf{J}_n(\theta) \end{array} \right) \tag{4}$$

The integrated velocity reference vector V_r is a quite large 6n dimension vector and $\mathbf{J}(\theta)$ is the 6n by 22 integrated Jacobean matrix. By solving the equation 4, we will be able to obtain the joint angular velocity reference vector $\dot{\theta}_r$. But, in the case that the system has many control points, the equation 4 falls into over constrained situation and is impossible to solve. Even in such case, the control system can minimize the velocity error.

V is the integrated velocity vector which can be realized by the joint angular velocity vector $\dot{\theta}$. The integrated and weighted velocity error E is defined and the optimal joint angular velocity reference vector $\dot{\theta}_{ropt}$ that minimizes E is obtained as follows. Here, $\mathbf{C}$ is a user defined weight matrix and $^+$ means the Moore Penrose's generalized inverse matrix[7].

$$E = (V_r - V)^t \mathbf{C}^t \mathbf{C}(V_r - V), \dot{\theta}_{ropt} = (\mathbf{J}(\theta)^t \mathbf{C}^t \mathbf{C} \mathbf{J}(\theta))^+ \mathbf{J}(\theta)^t \mathbf{C}^t \mathbf{C} V_r \tag{5}$$

The Moore Penrose's generalized inverse matrix is unique for any matrix and minimizes not only E but also the norm of $\dot{\theta}_{ropt}$. So, the equation 5 is applicable to all cases including redundant case. Thus the motion of each control point is integrated to the action of hand.

4. TASK DESCRIPTION

We adopted human mimic approach to implement dexterous tasks with short time. Some system which can describe dexterous tasks easily and efficiently should be necessary for this approach. In order to inherit the implemented tasks to the successive new hardware, it is preferable that the task description has physically recognizable structure. So, we developed the contact position based task description system.

The easy error recovery description should be considered because robot works in uncertain real world. And the new functions should be implemented easily. The system should start from the minimal function set and gradually grow by implementing various tasks one by one because it is quite difficult to define the necessary and sufficient function set.

In our system, the dexterous hand work task is divided into several Actions. The robot hand contacts with the handling object(s) and/or environment at several points. We call this set of contact conditions as State of hand. The State restricts the possible action of hand and, on the contrary, the State will be changed by the Action and so the Action should be associated with the State. The change of State means the change of contact condition of the hand. When the contact condition changes, some signal will be detected from the sensor. We call this signal as Event. Thus, the hand task will be executed as follows. The hand starts from initial State and execute the specified Action. The State of the hand is changed by it's own Action and this is detected by the Event. The Event directs the next State to the system.

The task is described by the network that is constructed from the States, Actions and Events. At first, the task is broken down to the Actions. Each Action is combined to the State and each State is connected by the Event. We call this network as State Action Event Network (SAEN). All SAEN components are implemented as the class library in C++ programming language and the task program is also described in C++ by using those components. It is compiled and linked with other control library.

5. EXPERIMENT

The effectiveness of our system is demonstrated by the chopsticks handling experiment. Our hand system holds chopsticks stably and open/close them successfully. It takes only around one week to implement this task including control parameter tuning. The human mimic approach brings us this short time implementation. The program flow of chopsticks handling task is shown in figure 6. The task is divided into 6 States(Actions). The task starts from the initial state and goes on along to the state number. The position references of 6 locations on the hand are specified for preparation from state 1 to 3 as shown in figure 7. During these states, the operator inserts chopstick one by one according to the posture of the hand.

The 4th through 6th states have the same control points setting as shown in figure 8. The hand holds chopsticks in the 4th state. The index, middle and ring fingers form a proper posture by using position control and The thumb holds both chopsticks by using force control. In the 5th state, the hand opens chopsticks by the position control specification. In the 6th state, the positional references are reversed and the hand close chopsticks while watching the force exerted on the fingertip of index finger. The 5th and 6th state configure the loop and the hand repeats open and close actions. By watching the arising force on the forefinger fingertip, the system can detect that a soft sponge cube is pinched on the end of chopsticks.

6. CONCLUSION

We have developed multi-finger robotic hand system. The developed hand has five fingers and the surface of hand is covered with the distributed touch sensor. A control

302

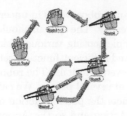 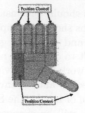 

Figure 6. Task description of chopsticks handling experiment.

Figure 7. The control points setting of State 1 to 3.

Figure 8. The control points setting of State 4 to 6.

system is proposed. This system is able to control the position, velocity or force of multiple points on the hand surface simultaneously. A task description system is also developed. The system is able to detect the change of constrained condition of the hand and change the control points setting dynamically according to the detected event. It is demonstrated by the experiment that the proposed control system has enough ability to execute dexterous hand work. It is also shown by the rapid implementation that our task description system is very effective for dexterous hand task programming and our human mimic approach is suitable for the task realization.

Acknowledgments

The authors wish to gratefully acknowledge Toshihiro Sano, Toshiaki Ishihara and Takeharu Kitagawa for the development of joint level control and the force sensing system.

REFERENCES

1. J. K. Salisbury, Design and Control of an Articulated Hand, Int'l. Symp. on Design and Synthesis, Tokyo, July 1984.
2. André van der Ham, A Dexterous Teleoperator for Hazardous Environments, the Universiteitsdrukkerij Delft, 1997.
3. T. Oomichi et al, Development of Working Mulitfinger Hand Manipulator, IROS'90, IEEE, pp.873-880, 1990.
4. A. Bicchi et al, On the Mobility and Manipulability of General Multiple Limb Robots, IEEE Trans. Robot. and Automat., vol. 11, no. 2, Apr. 1995.
5. P. R. Sinha et al, A Contact Stress Model for Multifingered Grasps of Rough Objects, IEEE Trans. Robot. and Automat., vol. 8, no. 1, Feb. 1992.
6. M. Shimojo et al, A Flexible High Resolution Tactile Imager with Video Signal Output, IEEE Int. Conf. Robotics and Automation, pp. 384-391, 1991.4.
7. R. Penrose, A generalized inverse for matrices, Proc. Cambridge Philos. Soc. 51, 406-413, 1955.

Human Friendly Mechatronics (ICMA 2000)
E. Arai, T. Arai and M. Takano (Editors)
© 2001 Elsevier Science B.V. All rights reserved.

Stability Map for Graspless Manipulation Planning

Yasumichi AIYAMA[a]and Tamio ARAI[b]

[a]Inst. Engg. Mech. & Sys., Univ. of Tsukuba, Tsukuba, Ibaraki 305-8573, Japan

[b]Dept. Precision Engg., Graduate School of Engg., the Univ. of Tokyo,
Hongo 7-3-1, Bunkyo-ku, Tokyo 113-8656, Japan

Graspless manipulation is a process in which objects are manipulated by using environment as a support. In this paper, we propose a planning method for graspless manipulation with consideration of stability of manipulated object. Our proposed stability is calculated from fingers position and the optimal contacting force. We make a 'stability map' against object configuration and finger configuration to plan a path in the map from start object configuration to goal configuration. 2-dimensional example is simulated.

1. Introduction

We have proposed graspless manipulation which uses environment around as a support of an object[2]. With that method a manipulator can manipulate an object without grasping There exist some previous researches which we can call as graspless manipulation. Figure 1 shows some methods. Mason showed pushing[5] which a finger pushes an object on a flat table. Sawasaki et al. showed tumbling[7] which two fingers tumble an object on a table. Aiyama et al. showed pivoting[1] which fingers tilt an object to stand on a floor with only one corner and rotate it around the corner.

Graspless manipulation is efficient in the field of assembly. Figure 2 shows an example. When a robot tries to insert a peg into a hole at the edge of a wall, it cannot insert

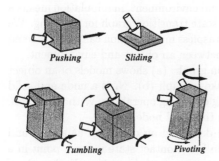

Figure 1. Graspless manipulation

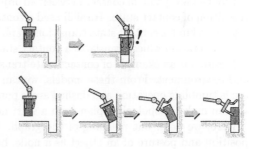

Figure 2. Graspless manipulation for insertion task

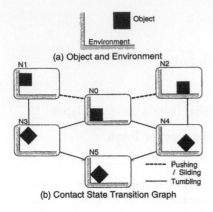

(a) Object and Environment

(b) Contact State Transition Graph

Figure 3. Contact state transition graph

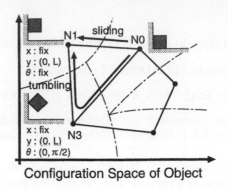

Configuration Space of Object

Figure 4. Contact state in C-space

by pick-and-place operation because of collision with the wall. But by using graspless manipulation, it can easily insert the peg into the hole. Some fingers can be removed when the peg contacts at the top of the hole. But how does the robot know whether it can remove some fingers? It is difficult problem for graspless manipulation planning.

In this paper, we propose a planning method for graspless manipulation. It directs to the most stable finger position for manipulation. The stability depends on a contact state between an object and environment and on finger position and its applying force.

In Section 2, we introduce a contact state transition graph. In Section 3, we define stability index of graspless-manipulated object. With this stability, we make stability map of an object for each contact state and finger position in Section 4 and propose the planning method. Numerical planning simulation is shown in Section 5.

2. Contact State Transition Graph

From a view poing of contact between an object and environment, manipulation means a transition of contact state. Hirai[3] used a contact state transition graph for planning. We also use a kind of contact state transition graph. Graspless manipulation can be considered as an operation which translates a contact state between an object and environment.

Figure 3 is an example of contact state transition graph. (a) shows models of an object and environment. From these models, we can make a graph (b). N_i is a node and solid arc is tumbling operation and dashed arc is pushing or sliding operation. With this graph, we can make a topological plan from a start node to a goal node.

Here, we point out meaning of the graph. We described an operation as an arc and position and posture of an object as a node, but in fact, contact state is not a point in a configuration space. Figure 4 shows the configuration space of the object. Each node of the graph may have some parameters; for example, node N_1 has a range of position y and N_3 has two parameters. So, it is a sub-space of the C-space. In the C-space, an operation is described as transition in a pair of neighbored sub-spaces. In the later sections, we consider these kind of operations included in a node region.

3. Stability Index of Object for Graspless Manipulation

In the field of multi-fingered hands, stability of a grasped object has been often discussed. Two concepts "form closure" and "force closure" are mainly used. Nakamura extended these theories to an object contacting with its environment, and defined "work closure"[6]. We defined "gravity closure", assuming the gravity as a virtual finger[2]. Using this concept, we define stability in graspless manipulation.

Let us assume that both fingers and the environment (virtual finger) grasp a rigid object at n contact points. The forces applied by many contact points are indeterminate. So with quasi-static analysis, we use an algorithm for the optimal internal force[4].

Graspless manipulation has many risks of failure during manipulation. In order to prevent object from moving in undesired direction, it is necessary, for all contact points, to prevent the robot fingers from slipping or removing. That is, the force applied from every contact point must be positive and contained inside the relevant friction cone. In addition, it is necessary to take into account the limit for the force.

These conditions are expressed by applied force vector $\boldsymbol{f}_i$ at the contact point i as shown in Figure 6.

$$\begin{cases} d_{i1} = \boldsymbol{f}_{iz} > 0 \\ d_{i2} = |\boldsymbol{f}_{iMax}| - |\boldsymbol{f}_{iz}| > 0 \\ d_{i3} = \mu_i \boldsymbol{f}_{iz} - \sqrt{\boldsymbol{f}_{ix}^2 + \boldsymbol{f}_{iy}^2} > 0 \end{cases}$$

When we discuss about pushing or sliding, however, some contact points slip on an object surface. In such contact points, we have to take into account only d_{i1} and d_{i2}.

We define stability of a manipulated object S as follows. S depends on friction coefficient in contact points, contact point position, and posture of the object.

$$S = \max_{\boldsymbol{f}} \min_{i,j} (d_{ij})$$

S expresses maximum permissible error of the norm of force vector. The larger S becomes, the more reliable an manipulation becomes. Therefore, we must keep S positive during an operation. But we should keep $S > S_{safe}$, a threshold for safely manipulation against unknown disturbant external force.

4. Stability Map and Planning of Graspless Manipulation

With this stability S, we make a map which describes stability against each object configuration and finger configuration. Here, contact points with environment are included in object configuration.

We introduce a planning method with this stability map according to Figure 5.

(a) First, we make a stability map for given object and fingers configuration. We can see safety area in the configuration space. And in the map, we draw lines (hyper-planes) $\boldsymbol{q}_{object} = \boldsymbol{q}_{start}$ (const.) and $\boldsymbol{q}_{object} = \boldsymbol{q}_{goal}$ (const.).

(b) Next, we check collisions between fingers and environment. Then we make finger inhibit regions on the map. We cannot select a configuration in the finger inhibit regions.

(c) We select a start finger configuration which is on the $\boldsymbol{q}_{object} = \boldsymbol{q}_{start}$, within the safety area and not in the finger inhibit regions.

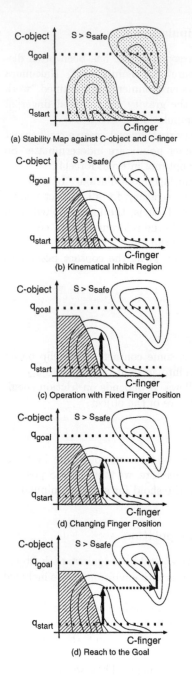

(a) Stability Map against C-object and C-finger

(b) Kinematical Inhibit Region

(c) Operation with Fixed Finger Position

(d) Changing Finger Position

(d) Reach to the Goal

Figure 5. Planning with stability map

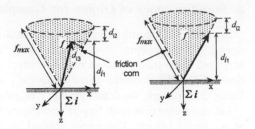

Figure 6. Restriction of force

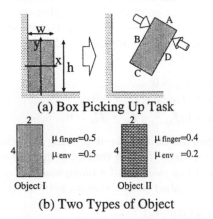

(a) Box Picking Up Task

μ finger=0.5
μ env =0.5

Object I

μ finger=0.4
μ env =0.2

Object II

(b) Two Types of Object

Figure 7. Task and model for simulation

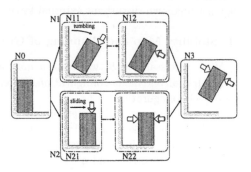

Figure 8. Possible operations to pick up

Here, we assume that no fingers slip on an object surface during manipulation. With this assumption, graspless manipulation is expressed as a line with $q_{finger} = const..$

(d)   Here, we cannot draw a line from q_{start} to q_{goal} in the safety area. Then we change finger position. Since object configuration does not change during changing finger, it is expressed as a horizontal line in the map. Using another finger for the new position, the line can be in the out side of the safety area.

(e)   Then, we can draw a vertical line which reaches to the goal configuration q_{goal}.

5. Numerical Example of Planning

Consider a rectangular box picking up task as shown in Figure 7. A box contacts with environment at two faces. A finger can approach only to top side (Face A) and right side (Face D). Manipulator has two fingers. Here we have two kind of objects. Set their weight $Mg = 1$. Each object has different friction coefficient μ_{finger} between fingers and objects, and μ_{env} between objects and environment.

The size of a finger is approximated as a circle with diameter of 1. Maximum tolerance force for each finger is approximated as $f_{Max} = 5$. And we set S_{safe} as 0.05.

In the box picking up task, we have some possible operations from the contact state graph. Here, we discuss two appropriate manipulations as described in Figure 8. One is through the state **N1** using tumbling. The other is through the state **N2** using sliding.

First, we analyse tumbling in the state **N1**. Finger motion for each object is generated as follows:

Object I (Figure 9):   Finger 1 contacts face **A** at $x = 0.4$ and tumbles the object from $\theta = 0$ to 43[deg]. Then, finger 2 support face **D** at $y = 0$. Then, finger 1 can approach to face **B**. During this series of operation, minimum stability $S = 0.14$.

Object II (Figure 10):   Finger 2 contacts face **A** at $x = 0.7$ and tumbles the object from $\theta = 0$ to 20[deg]. Next, finger 1 contacts face **A** at $x = -0.3$ and tumbles it again from $\theta = 20$ to 35[deg]. Then finger 2 support the object at face **D**. At last finger 1 can contact to face **B**.

Next, we analyse sliding in the state **N2**. In the feasible manipulation region, the best value of the minimum stability during the operation is:

Object I: $S = 0$ (impossible to manipulate by sliding).

Object II: $S = 0.26$ at $x = -0.5$.

Here, if objects are different, different manipulation are planned. And the optimal operations that make the manipulation most reliable are planned.

6. Conclusion and Future Works

In this paper, we show a planning method in order to select the optimal sequence of graspless manipulation. In this method, we also optimize the finger position and force. This method provides robust and dexterous manipulation, which correspond to objects in different shapes in diverse environment.

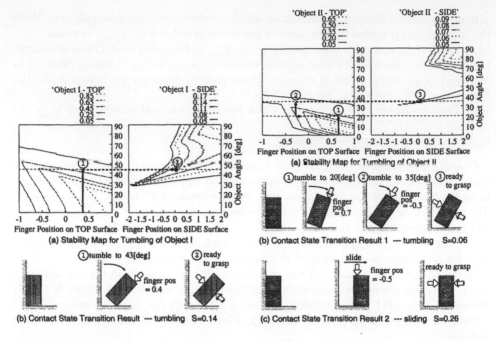

Figure 9. Map and operations for **Obj. I**   Figure 10. Map and operations for **Obj. II**

We only simulated in simple 2D examples. So it is required to discuss more complex situations in 3D, and to experiment using actual manipulator.

REFERENCES

1. Y.Aiyama, M.Inaba and H.Inoue, "Pivoting: A New Method of Graspless Manipulation of an Object by Robot Fingers," *Proc. 1993 IEEE/RSJ Int. Conf. on Intelligent Robots and Systems (IROS 93)*, pp.136–143, 1993.
2. Y.Aiyama and T.Arai, "Graspless Manipulation with Sensor Feedback," *Proc. IEEE Int. Symp. Assembly and Task Planning (ISATP'97)*, pp.78–83, 1997.
3. S.Hirai and H.Asada, "Kinematics and Statics of Manipulation Using the Theory of Polyhedral Convex Cones," *Int. J. of Robotics Research*, **12**(5), pp.434–447, 1993.
4. J.Kerr, B.Roth, "Analysis of Multifingered Hands," *Int. J. of Robotics Research*, **4**(4), pp.3–17, 1986.
5. M.T.Mason, "Mechanics and Planning of Manipulator Pushing Operation," *Int. J. of Robotics Research*, **5**(3), pp.53–71, 1986.
6. Y.Nakamura, "Grasping and Manipulation", *SICE Measurement and Control*, **29**(3), pp.206–212 *(in Japanese)*, 1990.
7. N.Sawasaki, M.Inaba and H.Inoue, "Tumbling Objects Using a Multi-Fingered Robot," *Proc. 20th Intl. Symp. on Industrial Robots (ISIR)*, pp.609–616, 1989.

Human Friendly Mechatronics (ICMA 2000)
E. Arai, T. Arai and M. Takano (Editors)

Motion planning and control method of dexterous manipulation utilizing a simulation system for a multi-fingered robotic hand

Chang-Soon Hwang[a], Terunao Hirota[b] and Ken Sasaki[b]

[a]Control and Systems Engineering, Tokyo Institute of Technology,
2-12-1 Ookayama, Meguro-ku, Tokyo 152-8552, Japan

[b]Environmental Studies, University of Tokyo,
7-3-1 Hongo, Bunkyo-ku, Tokyo 113-8656, Japan

This paper presents a motion planning and control method for the dexterous manipulation of a robotic hand. For a given trajectory of an object, a simulation system calculates the necessary joint displacements and contact forces at the fingertips. These joint displacements and contact forces are the reference inputs to the control loops of the robotic fingers. A task is decomposed into a set of primitive motions, and each primitive motion is executed using the output of the simulation system as the reference. Force sensors and dynamic tactile sensors are used to adapt to uncertainties encountered during manipulation. Several experimental results are presented.

1. INTRODUCTION

Control of a multi-fingered robotic hand is usually based on the theoretical analysis of the kinematics and kinetics of the fingers and of the object. However, application of such analyses to a robotic hand is difficult because of modeling errors and uncertainties in the real world. Moreover, the complexity of multi-finger manipulation makes the programming difficult, even for a simple motion. Speeter [1] described a hardware and software hierarchy to control manipulation with the Utah/MIT dexterous hand, and addressed an abstraction of dexterous manipulation and a control scheme for the Utah/MIT dexterous hand with a set of just over 50 primitives. Michelman [2] showed how the primitive manipulations can be combined into complex tasks. Shirai, Kaneko and Tsuji [3] found that the grasp patterns should also be changed according to the surface friction and the geometry of a cross-section of an object in addition to the scale, and verified the grasp strategies by experiments.

In our previous works, we have classified kinematics and kinetics problems [4]. Furthermore, analytical solutions of contact between a finger surface and simple-shaped objects have been derived. Solutions for contact with an object represented by a polyhedron [5] and the B-splines [6] have also been made.

This paper presents a control method for the motion planning of dexterous manipulation using a simulation system for kinematics and kinetics. The aims of this paper are (1) to define a set of primitive motions, and (2) to provide a control method of manipulation based on the use of primitive motions. In this paper, we consider the manipulation of only an object on a table using a two-fingered robotic hand that is vertically suspended

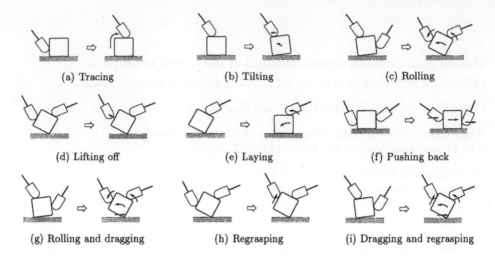

(a) Tracing   (b) Tilting   (c) Rolling

(d) Lifting off   (e) Laying   (f) Pushing back

(g) Rolling and dragging   (h) Regrasping   (i) Dragging and regrasping

Figure 1. Primitive motions for manipulating a block on a table.

from above. We assume that the geometry and dimensions of the fingers and of the object are given. We also assume that contact between the object and the table is maintained during manipulation.

2. MOTION PLANNING AND CONTROL PROGRAM

Generally, a task can be decomposed into a sequence of primitive motions. Various tasks of manipulating a block on a table can be achieved by combining the nine primitive motions shown in Figure 1. Each primitive motion is defined by the trajectory of the object being moved and the motions of the fingers. Variations on these basic primitive motions can be derived by scaling the magnitude of each motion. The use of the predetermined primitive motions shown in Figure 1 simplifies the processes of motion planning and control program. Each manipulation task has several associated processes. The procedure of motion planning and control program is as follows. First, the simulation system calculates the joint angles and contact forces necessary to move the object in a desired way [4] [7]. These necessary joint angles and contact forces will be used as references in the control loops of the fingers. Second, the manipulation task is analyzed, and primitive motions are manually selected and organized in order to produce the desired motion of the object. Third, the control program is executed.

Normally, the transitions of contact states during the manipulation do not necessarily follow the predetermined sequence. Uncertainties can be attributed to many factors, including unmodeled surface properties, friction, and errors in position or orientation. Control programs for error recovery were implemented by incorporating feedback from sensors such as the dynamic tactile sensor.

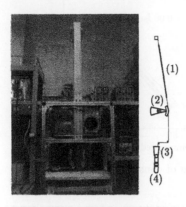

(1) Rubber bands for pretensioning
(2) Four voice-coil motors per finger
(3) Robotic finger
(4) Force sensor and tactile sensor

Figure 2. Experimental system.

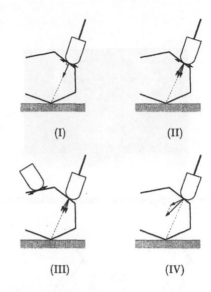

Figure 3. Algorithms for estimating the kinematic constraint.

3. EXPERIMENTAL SYSTEM

The experimental system has a two-fingered robotic hand suspended vertically for manipulation in the vertical plane, as shown in Figure 2. The fingers are driven by wires directly connected to voice-coil motors. The fingers are equipped with three-axis force sensors and with dynamic tactile sensors that detect slippage between the object and the finger surfaces [8]. The joint configuration and driving mechanism are similar to those of the Stanford-JPL hand [9].

4. EXPERIMENTAL RESULTS

4.1. Primitive Motion: Estimating the Kinematic Constraint

During the manipulation of a block on a table, there are many situations in which the fingers must hold or roll the block while the block is standing on edge. This motion involves estimating the position of the contacting edge of the block on the table. We have developed four active sensing algorithms for estimating the contact position: position-based algorithms (I) and (II), and force-based algorithms (III) and (IV), as shown in Figure 3.

(I) Calculate the projected motion of the object from the joint angle displacements of the fingers. The direction of constraint is perpendicular to the trajectory of the contact point on the object.

(II) Increase the contact force in the estimated direction of constraint and slightly change the direction of the force. The direction in which the object does not move is the direction of constraint.

312

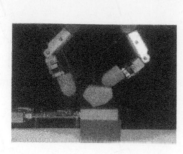

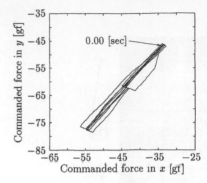

Figure 4. Experimental result of estimating the direction of constraint.

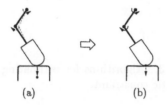

(a)   (b)

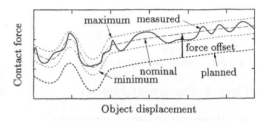

Object displacement

Figure 5. Control of the joint angles.   Figure 6. Control of the contact forces.

(III) Slightly vary the fingertip force direction in one finger in the estimated direction of the contact normal and observe the change in contact force of the other finger. The direction that gives the minimum force change in the other finger is the direction of constraint.

(IV) Vary the fingertip force in one finger. The change in the contact force vector gives the direction of constraint, because the change is in equilibrium with the reaction from the contact point on the table.

Figure 4 illustrates the experimental result of the combination of two algorithms (II) and (IV). A force variation of 30.0 [gf] was repeated four times in 4.0 [sec]. Repeated increases and decreases of 15.0 [gf] were exerted on the object using algorithm (IV). The constraint direction was adjusted at every increase and decrease of 15.0 [gf] using algorithm (II). The primitive motion of lifting a finger off the block was carried out by using these algorithms. Other primitive motions were realized as well.

4.2. Applied Motion: Rolling a Block with Partial Sliding

Basically, finger joints are position-controlled, and feed-forward torques, which correspond to the contact forces, are added. These contact forces, however, are the minimum forces corresponding to a nominal coefficient of friction. In order to increase the margin for variation of the friction, additional torque is applied in each joint. This offset of torque is equivalent to shifting the desired fingertip position slightly to the inside of the object,

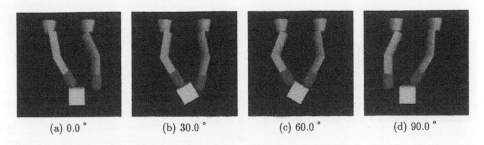

(a) 0.0 °   (b) 30.0 °   (c) 60.0 °   (d) 90.0 °

Figure 7. Procedure of rolling a block with partial sliding.

as shown in Figure 5. The actual force was always kept above the minimum force, as shown in Figure 6.

Figure 7 shows the sequence of rolling a block with partial sliding. Contact conditions, such as pure rolling and slide rolling, are controlled by using the dynamic tactile sensors, which are capable of detecting slippage.

Figures 8 and 9 illustrate the experimental results of rolling the block with partial sliding, where $q_1 = (q_{11}, q_{12}, q_{13})^T$, $q_2 = (q_{21}, q_{22}, q_{23})^T$ are the joint angles of the right and left fingers and F_1, F_2 are the magnitude of contact force of the right and left fingers. For a tilting or rolling motion, monitoring the direction of contact normal force proved to be more reliable than using the trajectory variation of the fingers. Complex manipulations, such as tumbling a block on a table, rolling a cylinder by pure rolling contact, rolling a cylinder with partial sliding, and rolling and dragging a cylinder and a block, were also realized using the proposed control method.

5. CONCLUSION

Manipulation of a block and a cylinder on a table with a two-fingered robotic hand was presented. Nine types of primitive motions were defined for this manipulation. The joint trajectories and contact forces were generated by the simulation system that we developed in the previous work. For the control of primitive motions involving estimation of the direction of the kinematic constraint, four active sensing methods were developed. The change in the contact force vector was used to estimate the contact location on the table, and the small variation in the fingertip position was used to refine the accuracy of the estimate. Experimental results of various manipulative motions have proved that complex manipulations can be realized as sequences of primitive motions.

ACKNOWLEDGEMENT

This research was supported in part by Grants-in-Aid for Scientific Research (C), No.11650418, 2000.

REFERENCES

1. T.H. Speeter, Control of the Utah/MIT dextrous hand: hardware and software hierarchy, J. Robotic Systems 7(5) (1990), 759–790.

314

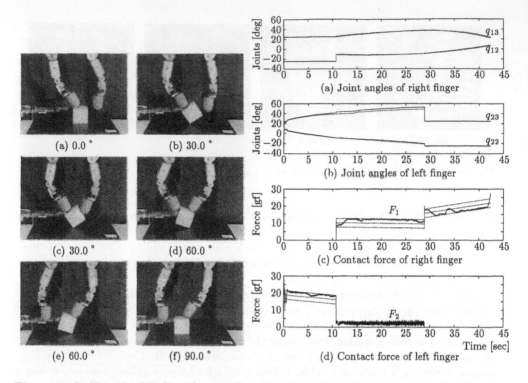

(a) 0.0 °

(b) 30.0 °

(c) 30.0 °

(d) 60.0 °

(e) 60.0 °

(f) 90.0 °

(a) Joint angles of right finger

(b) Joint angles of left finger

(c) Contact force of right finger

(d) Contact force of left finger

Figure 8. Rolling the block with partial sliding.

Figure 9. Experimental result of rolling the block with partial sliding.

2. P. Michelman, Precision object manipulation with a multifingered robot hand, IEEE Trans. Robot. Automat. 14(1) (1998), 105–113.
3. T. Shirai, M. Kaneko and T. Tsuji, Scale-dependent grasp, J. Robot. Soc. Japan 17(4) (1999), 567–576.
4. T. Nagashima, H. Seki and M. Takano, Analysis and simulation of grasping/manipulation by multi-fingersurface, Mech. Mach. Theory 32(2) (1997), 175–191.
5. H. Takahashi, M. Takano, K. Sasaki and H. Seki, Grasping/manipulation of an object with any shape by multi-fingersurfaces, Proc. 2nd ECPD Int. Conf. Advanced Robot., Intelligent Automat. and Active Syst., Vienna, Austria (1996), 498–503.
6. C.-S. Hwang, M. Takano and K. Sasaki, Kinematics of grasping and manipulation of a B-spline surface object by a multifingered robot hand, J. Robotic System 16(8) (1999), 445–460.
7. H. Takao, H. Seki, M. Takano and K. Sasaki, Analysis and simulation of grasping/manipulation dynamics by multi-finger surfaces, Proc. 9th World Congress on the Theory of Machines and Mechanisms, Milano, Italy (1995), 2272–2276.
8. K. Sasaki, T. Hirota, Y. Fujikake and H. Nakaki, Slip sensing tactile sensor for robots, Proc. China-Japan Bilateral Symposium on Advanced Manufacturing Engineering, Yellow Mountain City, China (1998), 60–65.
9. J.K. Salisbury and J.J. Craig, Articulated hands: force control and kinematic issues, Int. J. Robot. Res. 1(1) (1982), 4–17.

Human Friendly Mechatronics (ICMA 2000)
E. Arai, T. Arai and M. Takano *(Editors)*
© 2001 Elsevier Science B.V. All rights reserved.

Distributed Control of Hyper-redundant Manipulator with Expansion and Contraction Motion for Obstacle Avoidance

Kenji Inoue[a], Akinobu Okuda[b], Hiroyuki Tani[a], Tatsuo Arai[a] and Yasushi Mae[a]

[a]Department of Systems and Human Science,
Graduate School of Engineering Science, Osaka University,
1-3 Machikaneyama, Toyonaka, Osaka 560-8531, Japan

[b]Matsushita Electric Industrial Co., Ltd.,
Kadoma, Osaka 571-8501, Japan

A distributed control method for hyper-redundant manipulators, which is applicable to obstacle avoidance in complicated, unknown and varying environment, is proposed. A manipulator consists of serially connected joint units. Each unit is controlled by one controller, using local information from its proximity sensors and its neighboring units. Both the end-effector and several units can have their own desired positions and can converge stably. Hence, only if a unit which detects an obstacle with its sensors sets its desired position a certain distance away from the obstacle, the manipulator can move along the obstacle while keeping the distance. The manipulator basically makes expansion and contraction motion; it can move into/out of long and narrow space like pipes.

1. Introduction

Hyper-redundant manipulators consisting of a large number of serially connected short links are useful in complicated and narrow environment such as disaster area, nuclear power or chemical plants and space stations, because they have high ability in obstacle avoidance. Most of such environments are unknown to the manipulators and may suddenly change. Hence the manipulator is required to observe the surroundings using many proximity sensors attached to its links. Furthermore, the links near obstacles must immediately respond to sudden and unexpected change in the environment and avoid the obstacles, like reflex actions of human beings. For these problems, distributed control is more effective than centralized control. In the former method, each link or unit consisting of a few links controlled by one controller detects obstacles using the proximity sensors attached to itself and autonomously determines its own motion; thus it can immediately avoid the obstacles. In the latter method, one controller gathers and integrates the information from many sensors, then it plans and controls the whole motion of the manipulator. This process is more difficult when the environment is more complicated; that makes real-time control impossible. Some of the previous studies on hyper-redundant manipulators discuss centralized control methods[1-4]. Others propose distributed control methods[4,5], but they are not fully distributed because each link requires the information from far links. Some of these methods use smooth curves defined by a few parameters for generating the motion of the manipulator; they are not applicable to complicated or narrow environment, since the substantial degrees of freedom is reduced.

In the present study, a distributed control method for hyper-redundant manipulators, which is applicable to obstacle avoidance in complicated, unknown and varying environment, is proposed; this is extension of our previous method[6]. A manipulator consists of serially connected joint units. Each unit has the same degrees of freedom as the dimension of task space and can change the relative displacement between its both ends. The proposed method is highly distributed: each unit is controlled by one controller, using local information from its proximity sensors and its neighboring units; only neighboring units communicate with each other. Both the end-effector and several units can have their own desired positions and can converge stably. These features allow local and autonomous obstacle avoidance of units in unknown environment: only if a unit which detects an obstacle with its sensors sets its desired position a certain distance away from the obstacle, the manipulator can move along the obstacle while keeping the distance. The manipulator basically makes expansion and contraction motion; it can move into/out of long and narrow space like pipes. The proposed method is applied to an experimental planar hyper-redundant manipulator. Each unit is controlled by one personal computer, and the computers of neighboring units communicate through shared memory. Each unit has infrared sensors for detecting obstacles. The effectiveness and usefulness of the method are ascertained by computer simulations and experiments.

2. Multiple Point Control Method

2.1. Model of Manipulator

As shown in **Fig.1**, a hyper-redundant manipulator consists of serially connected joint units U_i ($i = 1, \cdots, n$), where n is total number of units. The dimension of task space is denoted by m, and the global coordinate system fixed to this space is by Σ. The base of the manipulator is fixed to the origin of Σ. Each U_i has m degrees of freedom, and its m-dimensional displacement between its both ends expressed on Σ is called "unit displacement" $q_i \in R^m$. Notice that unit displacement is not the set of joint variables of the unit. For example, unit displacement of planar manipulators is a 2-dimensional translational displacement vector. Most of the combinations of two joint mechanisms can be joint units, since they can generate 2-dimensional translational displacement. The degrees of freedom of the manipulator is $m \times n$. Letting $r_i \in R^m$ be the position of the tip of U_i expressed on Σ, the kinematics of the manipulator is given by

$$r_i = \sum_{k=1}^{i} q_k \quad (i = 1, \cdots, n) \tag{1}$$

r_n is the position of the end-effector. This study proposes a distributed control method where each joint unit U_i controls its own unit displacement q_i with one controller.

2.2. Control Law

The distributed method of multiple point control—positioning both the end-effector and the tips of several joint units on their own desired positions—is described[6].

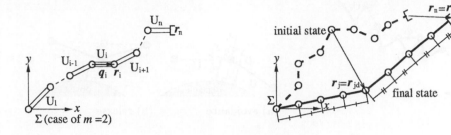

Figure 1. Model of hyper-redundant manipulator   Figure 2. Multiple point control

Letting $\boldsymbol{r}_{nd} \in R^m$ and $\boldsymbol{r}_{jd} \in R^m$ be the desired position of the end-effector and that of the tip of a unit U_j, we have proposed the following control law for each unit:

$$\dot{\boldsymbol{q}}_1 = C(\boldsymbol{q}_2 - \boldsymbol{q}_1) \tag{2}$$

$$\dot{\boldsymbol{q}}_i = C(\boldsymbol{q}_{i-1} - \boldsymbol{q}_i) + C(\boldsymbol{q}_{i+1} - \boldsymbol{q}_i) \quad (1 < i < j, j+1 < i < n) \tag{3}$$

$$\dot{\boldsymbol{q}}_j = C(\boldsymbol{q}_{j-1} - \boldsymbol{q}_j) + K_j(\boldsymbol{r}_{jd} - \boldsymbol{r}_j) \tag{4}$$

$$\dot{\boldsymbol{q}}_{j+1} = C(\boldsymbol{q}_{j+2} - \boldsymbol{q}_{j+1}) \tag{5}$$

$$\dot{\boldsymbol{q}}_n = C(\boldsymbol{q}_{n-1} - \boldsymbol{q}_n) + K_n(\boldsymbol{r}_{nd} - \boldsymbol{r}_n) \tag{6}$$

where $C \in R^{m \times m}$, $K_j \in R^{m \times m}$ and $K_n \in R^{m \times m}$ are positive-definite matrices. The above equations mean that each unit gradually modifies its unit displacement so that it may be equal to the displacements of its neighboring units. The second term in the right side of Eq.(6) plays a role of moving the end-effector toward its desired position. Eq.(4) means that the tip of U_j is regarded as a virtual end-effector.

2.3. Features
2.3.1. Highly Distributed Control
As is obvious from Eqs.(2)-(6), each joint unit uses only unit displacements of its own and its neighboring units; it does not have to obtain information from far units. In unknown environment, the tip unit U_n will directly measure the desired position of the end-effector relative to the current one, $\boldsymbol{r}_{nd} - \boldsymbol{r}_n$, using its external sensors. Thus the second term in the right side of Eq.(6) can be calculated by U_n. The same is said of Eq.(4). Accordingly, the proposed method requires only communication between neighboring units—it is highly distributed.

2.3.2. Stability and Convergence
We have proved using Lyapunov Stability Theory that both the end-effector and the tips of several units can converge to their own desired positions stably. In addition, the unit displacements of all units converge to "equal displacement state" where they are equal. As shown in Fig.2, when the desired positions $\boldsymbol{r}_{nd}$ and $\boldsymbol{r}_{jd}$ of the end-effector and the tip of a unit U_j are constant, the above theorem is expressed by

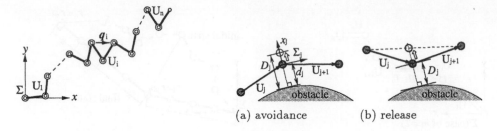

(a) avoidance   (b) release

Figure 3. Application to planar hyper-
redundant manipulator

Figure 4. Obstacle avoidance algorithm

$$\begin{cases} \boldsymbol{r}_{jd} = \text{const.} \\ \boldsymbol{r}_{nd} = \text{const.} \end{cases} \Rightarrow \begin{cases} \boldsymbol{r}_j(\infty) = \boldsymbol{r}_{jd}, \ \boldsymbol{q}_1(\infty) = \cdots = \boldsymbol{q}_j(\infty) = \boldsymbol{r}_{jd}/j, \\ \boldsymbol{r}_n(\infty) = \boldsymbol{r}_{nd}, \ \boldsymbol{q}_{j+1}(\infty) = \cdots = \boldsymbol{q}_n(\infty) = (\boldsymbol{r}_{nd} - \boldsymbol{r}_{jd})/(n-j) \end{cases} \quad (7)$$

2.3.3. Expansion and Contraction Motion

Convergence to equal displacement state causes the manipulator to be straight-line form (See Fig.2). As a result, the manipulator makes expansion and contraction motion.

2.4. Application to Planar Hyper-redundant Manipulator

The proposed method can be applied to a planar hyper-redundant manipulator shown in **Fig.3**, consisting of serially connected rotational joints. One joint unit has two joints and two links. The unit displacement $\boldsymbol{q}_i$ of the unit U_i is a translational vector connecting its both ends. Each U_i converts $\boldsymbol{q}_i$ to the reference of its two joint angles by solving its inverse kinematics, then it drives each joint by local position feedback control. Refer to our previous work[6] for details.

3. Obstacle Avoidance

3.1. Algorithm

It is assumed that each joint unit can measure the position of an obstacle in the sur-roundings relative to its tip position, using its own proximity sensors. When the tip of a unit avoids an obstacle, the tip must keep a certain distance away from the obstacle in the normal direction to the obstacle for collision avoidance, while it should not be constrained in the tangential direction for moving along the obstacle freely. We, therefore, apply the same law as Eq.(4) to the normal direction by setting the desired position away from the obstacle and apply the same law as Eq.(3) to the tangential direction for free motion without desired position. The algorithm for a unit U_j of avoiding an obstacle is described.

1) In "normal mode" when no obstacle exist near the tip of U_j, U_j and its upper unit U_{j+1} are controlled by Eq.(3) (where $i = j$ or $i = j + 1$).

2) If the measured distance d_j between the obstacle and the tip is shorter than the allowable distance D_j, U_j and U_{j+1} enter "obstacle avoidance mode". We consider the local coordinate system Σ_j attached to the tip, whose x axis, x_j, is the normal direction to the obstacle (**Fig.4**(a)). The rotation matrix from Σ_j to the global coordinate system Σ is denoted by $R_j \in R^{m \times m}$. Then the control laws for U_j and U_{j+1} of keeping the tip

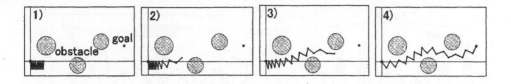

Figure 5. Computer simulation of obstacle avoidance in unknown environment

away from the obstacle in x_j direction and of moving it freely in the other directions are

$$\dot{q}_j = C(q_{j-1} - q_j) + (R_j \hat{C}_j R_j^T)(q_{j+1} - q_j) + (R_j \hat{K}_j)x_j \tag{8}$$

$$\dot{q}_{j+1} = (R_j \hat{C}_j R_j^T)(q_j - q_{j+1}) + C(q_{j+2} - q_{j+1}) \tag{9}$$

where

$$x_j = \begin{bmatrix} D_j - d_j \\ \mathbf{o} \end{bmatrix} \in R^m, \quad \hat{C}_j = \begin{bmatrix} 0 & \mathbf{o}^T \\ \mathbf{o} & \hat{C} \end{bmatrix} \in R^{m \times m}, \quad \hat{K}_j = \begin{bmatrix} k_j & \mathbf{o}^T \\ \mathbf{o} & O \end{bmatrix} \in R^{m \times m} \tag{10}$$

$\mathbf{o} \in R^{(m-1)}$: zero vector, $O \in R^{(m-1) \times (m-1)}$: zero matrix, k_j: positive constant, $\hat{C} \in R^{(m-1) \times (m-1)}$: positive definite matrix.

3) If the arrangement of the obstacle, U_j and U_{j+1} becomes like Fig.4(b), U_j and U_{j+1} return to normal mode. Then the tip moves away from the obstacle.

3.2. Features

The proposed method is applicable to unknown environment because each joint unit autonomously avoid obstacles using only local information from its sensors and its neighboring units. The method is highly distributed because the unit and its upper one cope with obstacle avoidance locally. Thus the manipulator can move its end-effector to the desired position while avoiding obstacles. Furthermore, the manipulator basically makes expansion and contraction motion along the shape of the environment; That enables the manipulator to move into/out of long and narrow space like pipes.

4. Computer Simulation

Fig.5 shows the simulation result when a planar hyper-redundant manipulator is given the desired position of its end-effector in unknown environment with three circular obstacles. The manipulator consists of 10 joint units (20 rotational joints), and all link lengths are the same.

While the units which detect the obstacles autonomously avoid them, the end-effector can reach its desired position. The manipulator expands like an accordion along the shape of the obstacles, thus entering narrow space.

5. Experiment

Fig.6 shows the experimental planar hyper-redundant manipulator consisting of 5 joint units. The link length is 80[mm]. Each joint is driven by a DC servo motor with a

320

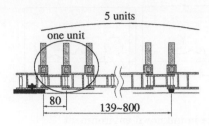

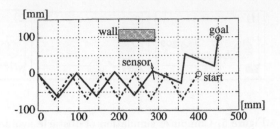

Figure 6. Experimental manipulator   Figure 7. Experimental result

reduction gear and a rotary encoder. Each unit is controlled by one personal computer (Pentium processor 100[MHz], 133[MHz] or 166[MHz]), and the computers of neighboring units communicate through shared memory. The sampling period including control and communication is 2.0[ms]. A infrared sensor with about 80[mm] detecting distance is attached to the tip of the unit U_3. As shown in the experimental result **Fig.7**, the end-effector can reach its desired position while the manipulator avoids the wall.

6. Conclusion

A distributed control method for hyper-redundant manipulators, which is applicable to obstacle avoidance in complicated, unknown and varying environment, is proposed. Only if a joint unit which detects an obstacle with its sensors sets its desired position a certain distance away from the obstacle, the manipulator can move along the obstacle while keeping the distance. The manipulator basically makes expansion and contraction motion; it can move into/out of long and narrow space like pipes. The effectiveness and usefulness of the method are ascertained by computer simulations and experiments.

REFERENCES

1. G. S. Chirikjian, J. W. Burdick : "An Obstacle Avoidance Algorithm for Hyper-redundant Manipulators", Proc. 1990 ICRA, pp.625–631, 1990.
2. N. Takanashi, H. Choset, J. W. Burdick : "Simulated and Experimental Results of Dual Resolution Sensor Based Planning for Hyper-redundant Manipulators", Proc. IROS'93, pp.636–643, 1993.
3. S. Ma, M. Konno : "An Obstacle Avoidance Scheme for Planar Hyper-redundant Manipulators", JRSJ, Vol.15, No.7, pp.1019–1024, 1997.
4. H. Mochiyama, E. Shimemura, H. Kobayashi : "Control of Serial Rigid Link Manipulators with Hyper Degrees of Freedom (Shape Control by a Homogeneously Decentralized Scheme)", JRSJ, Vol.15, No.1, pp.109–117, 1997.
5. H. Kobayashi, E. Shimemura, K. Suzuki : "An Ultra-Multi Link Manipulator", Proc. IROS'91, pp.173–178, 1991.
6. K. Inoue, T. Arai, Y. Mae : "Distributed Control of Hyper-redundant Manipulator with Expansion and Contraction Motion (Position Control)", Proc. TITech CO-E/Super Mechano-Systems Symposium 2000, pp.314–323, 2000.

Human Friendly Mechatronics (ICMA 2000)
E. Arai, T. Arai and M. Takano (Editors)
© 2001 Elsevier Science B.V. All rights reserved.

Inverse kinematics analysis of a parallel redundant manipulator by means of differential evolution

Wu Huapeng and Heikki Handroos

Machine Automation Laboratory, Lappeenranta University of Technology, 53851 LPR Finland

This paper presents a novel redundant parallel manipulator. The inverse kinematics model is studied, which is based on matrices and geometric algorithm. The static stiffness of manipulator is discussed in this paper. To achieve a minimum deflection in inverse kinematics problem the differential evolution method is used. In inverse kinematics solution the links motion of manipulator to avoid collision and joints limit are considered.

1.Introduction

The inverse kinematics analysis is very important for robot control, which is to find the variables of actuator for giving position and orientation of the end effector. To find the solution of inverse kinematics problem some optimization methods have been implemented: A criterion to minimize the integral of kinetic energy was given by Whitney [1]; Hamza D, Li-shan and S.F.P.saramago[2]. Some methods of optimization are available in the literatures to minimize total travel time with position error, velocity, and acceleration constraint [3] [4]. A.C.Nearchu and N.A. Aspragathos used genetic algorithm to solve optimization problem of point-to-point motion of redundant manipulator [5]. Most studies ware based on serial type and rigid manipulator.

Since most manipulators are flexibility, the error cased by the flexibility should be considered in high precision task. Some researchers have contributed their efforts to study the compensation methods for the error [6][7]. But it is every difficult to find the solution for those manipulators which have multi-degree freedom and redundant structure, some times it is nearly impossible to find the solution. Parallel manipulator has high stiffness and high precision characteristic, but this does not mean that there are not any problems caused by the finite stiffness.

This paper presents a redundant parallel manipulator that consists of seven actuators (hydraulic cylinders). This manipulator is designed for special application such as rock drilling.

An optimization model is built, which is to optimize the motion of actuators in the inverse kinematics solution of the manipulator to achieve the minimum deflection at a giving situation of end effector.

In optimization solution differential evolution algorithm (DE) is employed. The DE algorithm was first introduced by Storn and Price [8] a few years ago. It is one kind of evolution Strategies (ES) algorithm, however it is an exceptionally simple ES that promises to make fast and robust numerical optimization accessible to everyone.

2. Redundant Parallel Manipulator

The redundant parallel manipulator is showed in Fig. 1, which is designed for such as rock drilling tasks. This manipulator consists of 7 drivers (six hydraulic cylinders and one revolute hydraulic driver) based on two 3DOF parallel mechanisms. In the first stage the telescopic beam is connected to basement by a universe joint while it is fixed with mid plate. In the second stage the telescopic beam is fixed with mid plate and connected the end plate by a universe joint. So the first stage similar to an arm and can be extended, the second stage looks like a wrist and can be extended too. Universe joints are employed in the manipulator for all the hydraulic cylinder connections.

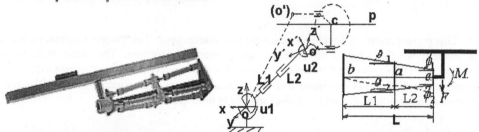

Fig.1 Parallel manipulator   Fig.2 kinematics model   Fig.3 static model

3. Optimization Model

3.1 Inverse kinematics model

The inverse kinematics analysis of the manipulator is based on the homogenous transformation matrices and geometric identity methods. The base frame (coordinate system) is fixed with the center of base universe joint by which the telescopic beam is connected to the base plate, while the wrist frame is fixed on the center of universe joint which connects the telescopic beam to the end plate (see Fig.2). For giving the vector $\mathbf{p}$ of end effector, the vector $\mathbf{q}$ of actuators is

$$q = \lambda(\mathbf{p}) \tag{1}$$

Where, $\mathbf{p} = (x, y, z, \alpha, \beta)'$, α is the angle about the x axis, β is the angle about y axis with respect to the base frame; $q = (l_{11}, l_{12}, l_{13}, l_{21}, l_{22}, l_{23}, \gamma)'$, l_{ij} is the length of cylinder and γ is the angle of the revolute actuator with respect to z'.

Equation (1) is not easy to be solved since it has two redundant degrees, we can simplify the structure of manipulator as equivalent structure (illustrated in Fig .2). Then the model of manipulator can easier be studied. The inverse problem (1) can be denoted as

$$q = \mu(\mathbf{r}) \tag{2}$$
$$r = h(\mathbf{p}) \tag{3}$$

Where, $r = (L1, L2, \zeta, \psi, \varphi, \varpi, \gamma)'$, ψ and ζ are the angles of universe joint $\mathbf{u2}$ about X' and Y' axis; φ and ϖ are the angles of universe joints $\mathbf{u1}$ about X and Y; L1, L2 are the length of two telescopic beams.

In the equation (3) the vector $\mathbf{r}$ has two parameters more than the vector $\mathbf{p}$, which means those two parameters must be determined by other methods. Fig.2 illustrates the possible trajectory of the point o'. Once an arbitrary position B $(x_1, y_1, z_1, \alpha_1, \beta_1)$ is given the vector **r** has multi-solution since the structure of manipulator is redundant. The different location on the circle the deflection of manipulator is different. In order to determine all the parameters of the vector **r** optimization method is used, which achieves the minimum deflection for the manipulator at position B. Once the vector **r** is solved the vector **q** can be got from the Equation (2) that can be solved from two parallel mechanisms and inverse kinematics problem is solved.

3.2 Minimum deflection

The overall compliance of the manipulator contains two parts: firstly, structure compliance; secondly, actuator compliance. The actuator compliance is highly determined by the control system. To improve the stiffness of actuator feedback control method and suitable control law should be considered. In this paper we only focus on the structure compliance. The deflection of the structure mainly includes three parts: the bending deflection of the first stage, the bending deflection of the second stage and shearing deflection of the beam. The deflection model of manipulator is built by static analysis shown in Fig.3. The total deflection is given by

$$
\begin{aligned}
f(\mathbf{r}) = &\frac{F_0 S_2^2 L_{20}}{3EI} + \frac{F_1 L_{10}^3}{3EI_1} - \frac{F_2 S_1^2}{2EI_1}(L_{10} - \frac{S_1}{3}) + S_2(\frac{F_1 L_{10}^2}{3EI_1} - \frac{F_2 S_1^2}{2EI_1}) + \sum_{i=1}^{2} \frac{\rho_i}{8EI_i}(L_{10}^4 + S_i^4) \\
&+ \frac{F'_2 l_2^2 L_{20}}{3EI} + \frac{F'_1 L_{10}^3}{3EI_1} - \frac{F'_2 S_1^2}{2EI_1}(L_{10} - \frac{l_1}{3}) + l_2(\frac{F'_1 L_{10}^2}{3EI_1} - \frac{F'_2 l_1^2}{2EI_1}) + d\sum_{i=1}^{2} \frac{TL_i^2}{4A_i^2 G} \cdot \frac{g_i}{t_i}
\end{aligned}
\tag{4}
$$

Where,

F, M are the force and moment act on the beam-end;

L_{10}, L_{20} are the length of the inside and the outside telescopic beams;

L1, L2 are the lengths of telescope of the first stage and the second stage;

L =L1+L2 is the total length of telescope;

S_1=L1- L_{20} ; $S_2 = L_{10} + L_{20}$ -L1; S_3=L1- L_{10} ;

l_1=L2- L_{20} ; $l_2 = L_{10} + L_{20}$ -L2; l_3=L2- L_{10} ;L2=L-L1;

E is the module of elasticity of telescopic beam;

G is the module of elasticity in shear of telescopic beam;

I, I_1 are the inertia of the inside and the outside telescopes;

T is the torque on the end platform, T= $F \cdot d \cdot \sin(\gamma)$;

ρ_i Is the linear mass density of the beams;

A $_i$ is the square of telescopic beam; g $_i$ and t $_i$ are the parameters of the beams.

In formulation (4) the forces $F_0, F_1, F_2, F', F'_1, F'_2$ are defined by following equations

$$
F_0 = [\frac{\cos(\theta_2)\cos(\theta_2)eF}{\cos(\theta_2)\sin(\theta_2) + \sin(\theta_2)\cos(\theta_1)} - M + F(L - L1)]/(L1 + \frac{\cos(\theta_1)\cos(\theta_2)e}{\sin(\theta_2)\cos(\theta_2) + \sin(\theta_2)\cos(\theta_1)})
\tag{5}
$$

Where, θ_1, θ_2, e are the structure parameters of manipulator.

$$
F_1 = \frac{L_{20}}{S_3} F_0 ; F_2 = \frac{S_2}{S_3} F_0 ; F' = F - \frac{M}{e}(tg(\phi_1) + tg(\phi_2))
\tag{6}
$$

Where, $\phi_{,}$, $\phi_{,}$ are the structure parameters of manipulator.

$$F'_1 = \frac{L_{20}}{l_3} F' \; ; \; F'_2 = \frac{l_2}{l_3} F' \tag{7}$$

The formulations (4), (5), (5), (6), and (7) are high nonlinear and complex within variables of beam length and position. Function (4) is also the object function for optimization.

3.3 Collision and joints limit considering

As the structure of end effector is very long, the collision will happen between the end effector and cylinders during the motion. It is important to consider the collision in inverse kinematics solution.

3.3.1 Collision considering

There are two categories of collision in the motion: the first one is the end effector collides with the cylinders, and the second one is the end effector collides with the telescopic beam. When the angle γ is in $(-\delta, +\delta)$ or $(120-\delta, 120+\delta)$ or $(-120-\delta, -120+\delta)$ area, the first collision may happen, here δ is the limit angle to avoid collision between end effector and cylinders, it is determined by the size of the end effector and cylinders. It can be calculated by

$$\delta = \arcsin \frac{n}{2d} + \arcsin \frac{k}{2d} \tag{8}$$

Where, n is the diameter of cylinder, k is the width of end effector, and d is length of routable beam.

If γ is in the above area and the angle ψ satisfies the following equation the first collision will happen.

$$\frac{(d - D.tg\psi)\cos\psi - \frac{2}{\sqrt{3}} e}{\frac{2}{\sqrt{3}} b - \frac{2}{\sqrt{3}} e} = \frac{\frac{D}{\cos\psi} - L - \frac{2}{\sqrt{3}} e \cos\psi + d \sin\psi - D \cos\psi}{L - \frac{2}{\sqrt{3}} e \cos\psi} \tag{9}$$

Where, D is the distance from the collision point to the mid of end effector,

When γ is out of the above area, the end effector may collide with telescopic beams, the second collision condition is

$$\psi \le \ arctg \frac{D}{d - \frac{w}{2}} \tag{10}$$

Where, w is the width of the telescopic beam.

3.3.2 Actuator limit

All the cylinders should satisfy the inequity in the inverse kinematics solution.

$$\min 1 \le 1_{ij} \le \max 1 \ (i=1,2; j=1,2,3) \tag{11}$$

3.3.3 Joints limit

The angles of universe joints contacting beam should also be limited, the condition is defined as

$$\min \varphi \le \varphi \le \max \varphi \; ; \; \min \psi \le \psi \le \max \psi \; ; \; \min \zeta \le \zeta \le \max \zeta \; ; \; \min \varpi \le \varpi \le \max \varpi \tag{12}$$

The optimization problem is defined as: minimum defection function (4) and subject constraint function (8) (9) (10) (11) and (12).

4. Differential evolution strategy

Differential evolution algorithm resembles that of most other population-based searches, which is a parallel direct search method, which utilizes NP D-dimensional vector as a population for each generation [9]. DE scheme generally contains Population initialize, Mutating with vector differential, Recombination and selection. The optimization program is shown in Fig.4. In the DE the vector **r** in equation (2) was selected as population, and deflection equation (4) was used to calculate the cost.

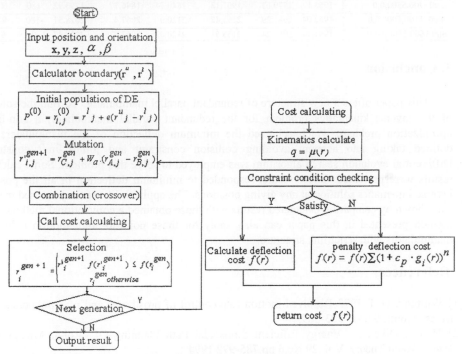

Fig.4. Optimization strategy

5. Calculate results

In the DE strategy the search parameters are defined as: n_{pw}=50, W_a=0.85 and C_a=1. The forces act on beam F=10000N, and moment M=5000NM. In table 1.1 there are two groups of calculating results, one is calculated by Differential evolution algorithm, the other one is calculated by using "L1=L2" method. In the "L1=L2" method the angle γ always is $0°$ or $\pm180°$, in this situation the arm of revolute actuator coincides with the direction of gravity, so that there is no shearing deflection. But once the position is far from origin point or in order to avoid the collision γ should be changed to an angle, then the deflection should take into account shearing deflection and the total deflection increase. Compare with this method DE has no this kind problem and deflection is less than this method.

Table 1.1 Optimal solution

Position and Orientation	The value of actuator (mm) and minimum deflection(mm) Differential evolution							
$(x, y, z, \alpha^{o}, \beta^{o})$	l_{11}	l_{12}	l_{13}	l_{21}	l_{22}	l_{23}	γ^{o}	δf
600, 700, 5500, 0, 0	1999.94	2085.41	2005.32	1684.36	1610.72	1651.92	0.45	3.882
-600, -700, 5500, 0, 0	2016.94	1930.09	2011.07 1	620.06	1694.03	1652.26	179.55	3.882
-800,1100,5900,0,0	2127.56	2272.73	2274.54	1929.59	1845.14	1781.03	-127.45	7.185
600,1600,5500,0,0 10,10	2056.65	2209.67	2035.83	1854.60	1651.54	1710.80	89.63	7.12
L1=L2 method								
600,700,5500,0,0	1907.31	1881.39	1748.934 1	820.05	1804.18	1867.20	0	6.123
-600,-700,5500,0,0	1753.17	1779.03	1901.18	1840.74	1846.77	1793.92	180	6.123
-800 ,1100,5900,0,0	1962.04	2081.29	2155.42	2120.0	2067.4	2028.51	-180	9.33
600,1600,5500,10,10	1961.94	2081.24	2155.51	2120.036	2067.43	2028.47	0	9.323

5. Conclusion

This paper offers a new structure of redundant parallel manipulator. To find the solution of the inverse kinematics problem for the redundant parallel manipulator, a non-linear optimization problem, which achieved the minimum deflection of the manipulator, was defined taking into account avoiding collision constraint and joints limit constraint. Differential evolution search algorithm was employed to solve this problem. Some calculate results ware presented. Those results responded to minimum deflection in all the possible inverse kinematics solution at any giving positions. The applicability of the proposed method was shown by comparing calculated results with those obtained a conventional method. The method presented in this paper can also apply for those point-to-point motion redundant manipulators with serial link structure.

References

[1] Whitney, D .E 1972 Resolved motion rate control of prostheses and manipulators J.Dys, Meas., Contr. 94(4) : 303-309.

[2] Hamza Diken ,   Energy Efficient Sinusoidal Path Planning Of Robot Manipulators Mech. Mech Theory Vol .29 No 6 pp 785-972 1994

[3] Shinobu Sasaki Feasiblity Study Manipulator Inverse Kinematics Problems With Applicatons of Optimization Principle. Mech. Mech. Theory vol. 28 No. 5. Pp685-697, 1993

[4]. J.Y.S. Luh,M.W.Walker, Minimum time along the path for s mechanical arm. Proceeding of the IEEE Conference on Decision and Control, New Orlean, LA1977,pp.755-759.

[5]   A.C.Nearchou, N.A.Aspragathos, Application of genetic algorithms to point-to-point motion of redundant manipulator Mech. Mach. Theory Vol.3, No.3, pp. 261-270, 1996

[6] Surdilovic,D.,Vukobratovic, M. Deflection Compensation for Large Flexible Manipulatoes, Mechanism and Machine, theory Vol.31,No.3.1996,pp.317-329.

[7] Asko Rouvinen and Heikki Handroos Deflection Compensation of A Flexible Hydraulic Manipulator Utilizing Neural Networks; Mechatronics Vol.7 No.4,pp355-368,1997.

[8] Storn,Rainer and Price,Kenneth. Differential Evolution – A simple evolution strategy for fast optimization. Dr.Dobb's Journal ,April 97,pp.18-24 and p.28.

[9] Lampinen.J, Zelinka.I, Mixed Variable nonlinear optimization by Differential evolution, 2nd International prediction conference,pp45-55 1999.

Human Friendly Mechatronics (ICMA 2000)
E. Arai, T. Arai and M. Takano (Editors)
© 2001 Elsevier Science B.V. All rights reserved.

Polishing Robot with Human Friendly Joystick Teaching System

Fusaomi Nagata[a] and Keigo Watanabe[b]

[a]Interior Design Research Institute, Fukuoka Industrial Technology Center,
Agemaki-405-3, Ohkawa, Fukuoka 831-0031, JAPAN

[b]Department of Advanced Systems Control Engineering, Graduate School of Science and
Engineering, Saga University, Honjomachi-1, Saga 840-8502, JAPAN

In this paper, a teaching system for a polishing robot using a game joystick is proposed. The teaching process is as follows: first, a zigzag path to be considered according to both sizes of each object and polishing tool is prepared; next, the polishing robot, which applies an impedance model following force control, profiles the object□ssurface using the zigzag path. The operator has only to control the orientation of the polishing tool using a game joystick. Since the impedance model following force controller keeps the contact force to be a desired value, the operator has to give no attention to a sudden over-load or non-contact state. The desired trajectory is automatically obtained as the data including the continuous information on both the position and orientation along the zigzag path on the object s surface. The robot can achieve the polishing task without any assists of the operator by referring the obtained trajectory.

1. INTRODUCTION

Industrial robots have been improved remarkably and applied to several tasks such as handling, painting, welding, sealing and so on. In such tasks, it is important to precisely control the trajectory of the end-effector. However, when the robots are applied to sanding or polishing task [1-4], it is necessary to achieve compliant force control without damaging the object as the skilled workers are carrying out, and further the polishing tool attached to the robot arm has to be controlled along the surface keeping contact with the object from normal direction with a desired contact force. Such a control scheme to realize above requirements is called profiling control. The profiling control is a basic and essential strategy for polishing or sanding task using the robots, and which is achieved with both a force control and a position/orientation control.

Generally, the desired trajectory that is indispensably required for the position/orientation control, is obtained through a well-known teaching process using a teaching pendant. However, it is complicated and time consuming to implement. Especially, in teaching for the polishing robot, the object with curved surface demands a large number of teaching points. To reduce the load of such teachings, we have already proposed a position compensator based on cutter location (CL) data [5, 6]. The CL data generated by the CAD/CAM system have information of points along the curved surface and their normal vectors. If the object is manufactured with the multi-axis numerical control (NC) machine tool, the CL data can be used as not only desired trajectory in the direction of position/orientation control but also feedforward information in the direction of force control. This technique allows us to generate the desired trajectory without teaching and to improve the performance of profiling control. However, it must be considered how the polishing robot can be applied to such an object as is manufactured without CAD/CAM system.

In this paper, to overcome this problem, a teaching system of force control using a game joystick is proposed. The teaching process is as follows: first, a zigzag path to be considered according to both sizes of each object and polishing tool is prepared; next, the polishing robot, which uses the impedance model following force control, profiles the object⑤ surface using the above zigzag path. The operator has only to control the orientation of the polishing tool using the game joystick. Since the impedance model following force controller keeps the contact force in the direction of the tool coordinate system to be a desired value, the operator has to give no attention to a sudden overload or non-contact state. Through the above process, the desired trajectory is automatically obtained as the data including the continuous information on both the position and orientation of the polishing tool along the zigzag path. After all, the robot can achieve the polishing task without any assists of the operator by referring the obtained trajectory. The effectiveness of the proposed method has been proved through some experiments using an industrial robot JS-10 with a PC based controller.

2. IMPEDANCE MODEL FOLLOWING FORCE CONTROL

Regarding the force control, we use the impedance model following force control that can be easily applied to industrial robots with an open architecture controller. The desired impedance equation for Cartesian-based control of a robot manipulator is designed by

$$M_d \left(\ddot{x} - \ddot{x}_d \right) + B_d \left(\dot{x} - \dot{x}_d \right) + SK_d \left(x - x_d \right) = SF + (E - S) K_f \left(F - F_d \right) \tag{1}$$

where $x \in \Re^6$, $\dot{x} \in \Re^6$ and $\ddot{x} \in \Re^6$ are the position, velocity, and acceleration vectors, respectively. $M_d \in \Re^{6 \times 6}$, $B_d \in \Re^{6 \times 6}$ and $K_d \in \Re^{6 \times 6}$ are the coefficient matrices of the desired inertia, damping and stiffness, respectively. $F \in \Re^6$ is the force-moment vector acting between the end-effector and its environment defined by $F^T = [f^T \ n^T]$, where $f \in \Re^6$ and $n \in \Re^6$ are the force and moment vectors, respectively. $K_f \in \Re^{6 \times 6}$ is the force feedback gain matrix. x_d, $\dot{x}_d$, $\ddot{x}_d$ and $F_d^T = [f_d^T \ n_d^T]$ are the desired position, velocity, acceleration and force-moment vectors; S and E are the switch matrix and identity matrix. It is assumed that M_d, B_d, K_d and K_f are positive-definite diagonal matrices. Note that if $S = E$, Eq. (1) becomes an impedance control system in all directions; whereas if $S = 0$, it becomes a force control system in all directions. If the force control is used in all directions, $X = \dot{x} - \dot{x}_d$ gives

$$\dot{X} = -M_d^{-1} B_d X + M_d^{-1} K_f \left(F - F_d \right) \tag{2}$$

In general, (2) is solved as

$$X = \exp\left(-M_d^{-1} B_d t\right) X(0) + \int_0^t \exp\left\{-M_d^{-1} B_d (t - \tau)\right\} M_d^{-1} K_f \left(F - F_d \right) d\tau \tag{3}$$

In the following, we will consider the form in the discrete time k using a sampling time Δt. It is assumed that M_d, B_d, K_f, F and F_d are constant at $\Delta t (k-1) \le t < \Delta tk$. Defining $X(k) = X(t)|_{t = \Delta tk}$, it follows that

$$X(k) = \exp\left(-M_d^{-1} B_d \Delta t\right) X(k-1) - \left\{\exp\left(-M_d^{-1} B_d \Delta t\right) - E\right\} B_d^{-1} K_f \left\{F(k) - F_d\right\} \tag{4}$$

Remembering $X = \dot{x} - \dot{x}_d$ and setting $\dot{x}_d = 0$ in the direction of force control, the recursive equation of velocity command in the Cartesian space is derived by

$$\dot{x}(k) = \exp\left(-M_d^{-1} B_d \Delta t\right) \dot{x}(k-1) - \left\{\exp\left(-M_d^{-1} B_d \Delta t\right) - E\right\} B_d^{-1} K_f \left\{F(k) - F_d\right\} \tag{5}$$

where $x(k)$ is composed of position vector $[x(k) \ y(k) \ z(k)]^T$ and orientation vector $[\phi(k) \ \theta(k) \ \psi(k)]^T$ expressed by Z-Y-Z Euler angles. Profiling control is the basic strategy for polishing or sanding and it is performed by both force control and position/orientation control. However, it is so difficult to realize stable profiling control under such environments that have unknown dynamics or shape [7]. Undesirable oscillations and non-contact state tend to occur. To reduce such undesirable influences, an integral action is added to Eq. (5), which yields

$$\dot{x}(k) = \exp\left(-M_d^{-1}B_d \Delta t\right)\dot{x}(k-1) - \left\{\exp\left(-M_d^{-1}B_d \Delta t\right)-E\right\} B_d^{-1}K_f\left\{F(k)-F_d\right\} + K_i \sum_{n=1}^{k}\left\{F(n)-F_d\right\} \quad (6)$$

where $K_i = \mathrm{diag}\left(K_{i1},...,K_{i6}\right)$ is the integral gain. The control input $\dot{x}(k)$ given by Eq. (6) is substituted into the reference of servo system every sampling time so that the contact force F can be controlled to be F_d.

3. TEACHING SYSTEM OF FORCE CONTROL USING A JOYSTICK

It has been already described on how to apply the impedance controlled polishing robot to the model which is designed and manufactured by CAM/CAM system [5, 6]. In this section, a novel teaching method using a game joystick is proposed. Using this teaching system, it is expected that the proposed polishing robot can be applied to any models whose shapes are unknown as numerical data.

3.1. Experimental setup

The polishing robot based on an industrial robot JS-10 is shown in Photo 1. On this system, several API (Application Programming Interface) functions, such as kinematics, coordinate transformation for force sensor, and servo control with position command, can be used. In manufacturing industry of furniture, skilled workers usually use air driven vibrational tools as shown in Photo 2. In this experiment, a

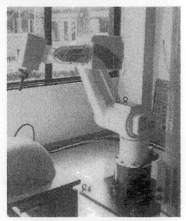

Photo 1 Overview of the JS-10

similar tool is selected and mounted on the top of the robot arm via the 6-DOF force sensor. The diameter of the tool is 50 [mm] and paper roughness is #220.

In teaching, the joystick is only used to control the orientation of the attached polishing tool. Consequently, the desired trajectory including the continuous data of both position and orientation is easily obtained. The block diagram of polishing robot in teaching mode is shown in Figure 1. Here, each switch matrix S_p, S_o, S_f changes the direction of position control, orientation control or force control to be active, respectively. When one value is presented, each controller is in effect, whereas when zero value is presented, each controller is without effect. The interpretation of the block diagram shown in Figure 1 is as follows: in the direction of force control, Eq. (6) generates a velocity command $\dot{x}(k)$, which includes the quantity from the integral controller. On the other hand, in the direction of position control, the zigzag path generator generates desired position $x_p(k)$ based on the profiling velocity $v(k)=[v_x(k)\ v_y(k)\ v_z(k)\ 0\ 0\ 0]$. The error between $x_p(k)$ and $x(k)$ yields a velocity command $v_p(k)$ with a velocity transformation gain $K_v=\mathrm{diag}(K_{v1},...,K_{v6})$. $x_p(k)$ is computed by

Photo 2 Rotational tool used in the experiment

Photo 3 Experimental scene of proposed force controlled joystick teaching

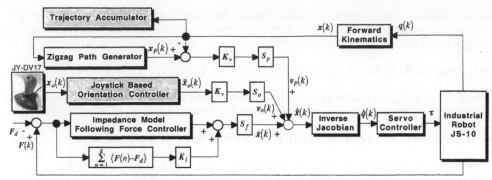

Figure 1 Block diagram of the polishing robot in teaching mode

$$x_p(k) = x_p(k-1) + \Delta t v(k) \tag{7}$$

In the direction of orientation control, multiplying $\tilde{x}_o(k)$ by K_v generates another velocity command $v_o(k)$. $\tilde{x}_o(k)$ is the corrected output of joystick, which is defined as follows:

$$\tilde{x}_o(k) = K_p x_o(k) + \hat{K}_p \hat{x}_o(k) \tag{8}$$

where

$$\hat{x}_o(k) = \begin{cases} \hat{x}_o(k-1)L^{-1} & \text{if } x_o(k) = 0 \tag{9} \\ \hat{x}_o(k-1) + x_o(k) & \text{if } x_o(k) \neq 0 \tag{10} \end{cases}$$

here, $x_o(k)$ and L ($L > 1$) is the raw data from the joystick and the down scaler, respectively. The each directional velocity command is added each other, which yields $\dot{x}(k)$. $\dot{x}(k)$ is transformed into the joint angle velocity $\dot{q}(k)$ with the inverse Jacobian, then $\dot{q}(k)$ is given to the reference of servo system installed in the open architecture controller.

The teaching process is as follows: the polishing tool is moved on the object's surface along the path generated from the zigzag path generator. In this case, since the contact force in tool coordinate system is controlled to be desired value by Eq. (6), the operator has only to manipulate the joystick to keep the orientation of the tool being desirable, for examples, from normal direction. In teaching, the data of both position and orientation are continuously stored into the trajectory accumulator shown in Figure 1. The 3-DOF joystick used in the system is the model JY-DV17 provided by SANWA SUPPLY.

3.2. Teaching result

In the teaching experiment shown in Photo 3, a zigzag path shown in Figure 2 was used. The control parameters are shown in Table 1. Figures 3, 4 show the taught orientation of the tool, which represent the x- and z-directional components of the tool vector, respectively. The tool vector is represented by normalized vector in robot base coordinate system. Note that the joystick teaching was carried out with the x-directional profiling velocity 10 [mm/s]. It has been confirmed that the desired trajectory of the tool can be easily obtained, even if the object is not

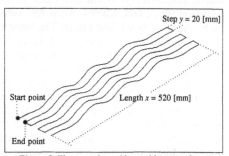

Figure 2 Zigzag path used in teaching experiment

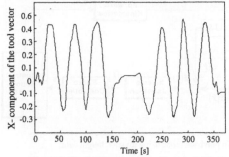

Figure 3 X-component of the tool vector obtained by proposed teaching method

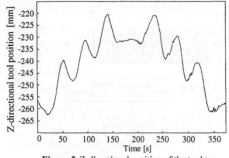

Figure 5 Z-directional position of the tool top controlled by Eq. (6)

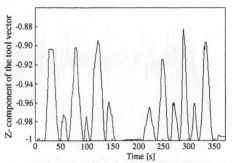

Figure 4 Z-component of the tool vector obtained by proposed teaching method

Table 1 Parameters used in experiments

Desired contact force $\sqrt{(f_{dx})^2 + (f_{dz})^2}$	1 [kgf]
Profiling velocity $\sqrt{(v_x)^2 + (v_z)^2}$	10 [mm/s]
Desired inertia coefficient M_{d1}, M_{d3}	0.01 [kgf·s²/mm]
Desired damping coefficient B_{d1}, B_{d3}	20 [kgf·s/mm]
Force feedback gain K_{f1}, K_{f3}	1
Integral control gain K_{i1}, K_{i3}	0.0001
Velocity transformation gain K_{v1}, K_{v2}, K_{v3}	0.8
Velocity transformation gain K_{v5}	0.2
Gain for joystick control K_{p5}	1
Gain for joystick control $\hat{K}_{p5}$	0.003
Down scaler L	1.2
Sampling width Δt	10 [msec]

manufactured with the CAD/CAM system.

4. PROFILING CONTROL USING ACQUIRED TRAJECTORY

After the proposed teaching, the trajectory accumulator shown in Figure 1 has acquired a skillful trajectory for automatic control. Figure 6 shows the block diagram of the polishing robot in playback mode using the position/orientation compensator based on the joystick taught data. $x_a(k)$ is the output of the compensator in the discrete time k and it is also the k-th step of the joystick taught data. In the direction of force control, a velocity command $\{x_a(k) - x_a(k-1)\}\Delta t^{-1}$ is added to the velocity command generated from Eq. (6), which generates

$$\dot{x}(k) = \exp\left(-M_d^{-1}B_d\,\Delta t\right)\dot{x}(k-1) - \left\{\exp\left(-M_d^{-1}B_d\,\Delta t\right) - E\right\}B_d^{-1}K_f\left\{F(k) - F_d\right\}$$

$$+ K_i \sum_{n=1}^{k}\left\{F(n) - F_d\right\} + \frac{x_a(k) - x_a(k-1)}{\Delta t} \tag{11}$$

In the direction of position/orientation control, $x_a(k)$ is used as a desired trajectory. The error between $x_a(k)$ and $x(k)$ yields a velocity command $v_a(k)$ with K_v. Each directional command is added, which composes $\dot{x}(k)$.

Finally, the performance of the proposed scheme shown in Figure 6 was examined through a polishing task. In this experiment, the polishing tool was made to contact with a work and profile its curved surface with a desired contact force 1 [kgf]. Figure 7 shows the experimental result. From this figure, although some spikes appear when the profiling direction changes from x-direction to y-

332

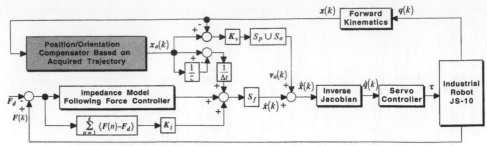

Figure 6 Block diagram of the polishing robot in playback mode using joystick taught data

direction, the contact force was desirably controlled around the reference. Surface accuracy of the polished work was so good condition as well as polished by skilled workers.

5. CONCLUSION

In this study, a user-friendly joystick controlled teaching system has been proposed to obtain the skillful trajectory of each polishing tool based on operator's will, and the usefulness has been examined through some experiments on teaching and polishing using an industrial robot JS-10 with

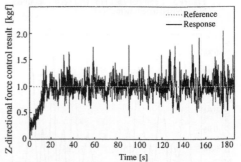

Figure 7 Profiling control result using joystick taught data

an open architecture controller. Consequently, it has been recognized that due to the joystick teaching of force control, the proposed polishing robot can be easily and safely applied to such an object as is manufactured without CAD/CAM system.

REFERENCES

1. F. Ozaki, M. Jinno, T. Yoshimi, K. Tatsuno, M. Takahashi, M. Kanda, Y. Tamada and S. Nagataki, A Force Controlled Finishing Robot System with a Task-Directed Robot Language, *Journal of Robotics and Mechatronics*, Vol. 7, No. 5, (1995) 383.
2. M. Jinno, F. Ozaki, T. Yoshida and K. Tatsuno, Development of a Force Controlled Robot for Grinding, Chamfering and Polishing, *Procs. of the IEEE International Conference on Robotics and Automation*, (1995) 1455.
3. F. Pfeiffer, H. Bremer and J. Figueiredo, Surface Polishing with Flexible Link Manipulators, *European Journal of Mechanics*, A/Solids, Vol. 15, No. 1, (1996) 137.
4. F. Nagata and K. Watanabe, An Experiment on Sanding Task Using Impedance Controlled Manipulator with Vibrational Type Tool, *Pcocs. of the Third Asian Control Conference*, (2000) 2989.
5. F. Nagata, K. Watanabe and K. Izumi, An Experiment on Profiling Task with Impedance Controlled Manipulator Using Cutter Location Data, *Procs. of the IEEE International Conference on System, Man and Cybernetics (SMC '99)*, (1999) 848.
6. F. Nagata, K. Watanabe and K. Izumi, Profiling Control for Industrial Robots Using a Position Compensator Based on Cutter Location Data, *Journal of the Japan Society for Precision Engineering*, Vol. 66, No. 3, (2000) 473 (in Japanese).
7. K. Takahashi, S. Aoyagi and M. Takano, Study on a Fast Profiling Task of a Robot with Force Control Using Feedforward of Predicted Contact Position Data, *Procs. of the 4th Japan-France Congress & 2nd Asia-Europe Congress on Mechatronics*, Vol. 1, (1998) 398.

Human Friendly Mechatronics (ICMA 2000)
E. Arai, T. Arai and M. Takano (Editors)
© 2001 Elsevier Science B.V. All rights reserved.

Development of High Precision Mounting Robot with Fine Motion Mechanism: Design and Control of the Fine Motion Mechanism

A. Omari[a], A. Ming[a], S. Nakamura[b], S. Masuda[b], C. Kanamori[a] and M. Kajitani[a]

[a]University of Electro-Communications, Department of Mechanical Eng. and Intelligent Systems, 1-5-1 Chofugaoka, Chofu-City, Tokyo 182-8585, Japan

[b]Japan Radio Co. Ltd, Shimorenjaku 5-1-1, Mitaka-City, Tokyo, Japan

For the need of rapid and accurate mounting of a small electronic part on printed circuit board, SCARA type robot manipulator with a fine positioner, capable of high accurate positioning, has been considered. This paper describes basic concept of the system, design and control of developed X-Y axes fine positioner. The developed fine positioner consists of piezo-electric actuator and displacement magnifying mechanism with elastic hinge. The displacement magnifying system is designed by simulation based on finite element method. Asymmetry compensation of piezo-electric actuator, position feedback controller, feedforward controller and disturbance observer for the position loop form our proposed control system. Experimental results demonstrate significant improvements in endpoint accuracy and settling time achieved by the novel configuration of the coarse-fine robotic system.

1. INTRODUCTION

In recent years, portable electronic equipment has undergone rapid increases in density of surface mounting and production in diversified types. Therefore, it has become necessary to develop highly versatile mounting systems capable of loading ultra-miniature electronic components of non-standard shapes which cannot be mounted with high-speed chip mounters, and accurately mounting ball grid array (BGA) IC chips onto printed circuit board (PCB). The use of precision positioning mechanism such as the semiconductor manufacturing system is conceivable, but these systems lack versatility and are rather expensive. Meanwhile, classical industrial robots are generally insufficient to meet the target mounting accuracy and speed. Previous works have shown the improvement achieved by a two-stage approach, i.e. a specific fine positioner (FP) mounted at the tip of a conventional robot [1]. However, no practical robotic system is available, our work is thus motivated [2]. A SCARA type robot, featuring advantageous productivity and versatile, is used as the coarse positioning device (CP) and a specially developed piezo-electric based fine motion device [2] carried on the end-effector of the CP is used as the FP. The present work intends to reduce the cycle time of the coarse-fine system and improve its endpoint accuracy through the compensation for a transient and residual vibrations of the CP and also for its position errors. This paper introduces basic concept of the system

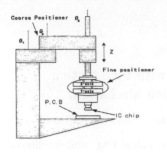

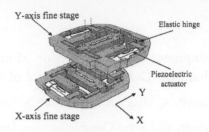

Figure 1. Coarse-fine manipulator system.   Figure 2. Two-axes fine motion.

briefly and describes design and control of the fine motion mechanism in detail.

2. COARSE-FINE ROBOTIC SYSTEM

The high precision mounting robot was developed with an arm fitted with an X-axis fine motion and a Y-axis fine motion mechanism below the X-axis stage. Figure 1 shows an overall view of the coarse-fine robotic system, and the developed two-axes FP is shown on Fig 2. The performance specification of the CP and design specifications of FP and system are set as in Table 1.

Table 1
Specifications of coarse-fine mechanism

	Coarse-fine positioner	Coarse positioner	Fine positioner
Repeatability	$\pm 1\mu m$	$\pm 15\mu m$	$\pm 1\mu m$
Stroke			$\pm 50\mu m$

3. DESIGN OF THE FINE MOTION MECHANISM

To satisfy the specification of table 1, piezo-electric actuator featuring excellent resolution and fast response has been selected. In general, the piezo-electric device enables precision positioning in a few nanometers, but has a small stroke. To cope with this situation, a displacement magnifying system with two-step lever structure, as shown on Fig. 3, that uses an arc notching single-side hinge and a leaf spring single-side hinge was developed. Finite element method (FEM) was employed to select the structural parameters and simulate its performance. To improve straightness of the system, symmetric structure is considered and one piezo-electric actuator is used on each side. Both actuators are driven by the same voltage, and then elongate with the same amount. Moreover, To achieve precise positioning, sensors with good resolution are required. Strain gage sensors are appropriate for this task at reasonable price and volume. It is set to the hinge part,

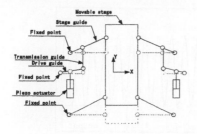

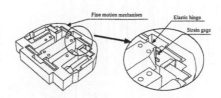

Figure 3. Fine motion mechanism model.   Figure 4. Location of strain gages.

as shown on Fig. 4, to measure the biggest distortion that occurs when the stage moves. The performance of the developed FP using a simple PI controller are listed in Table 2.

Table 2
Performance of the fine motion mechanism

	Mass	Resolution	Repeatability	Stroke
X-axis FP	$0.8Kg$	$\pm0.3\mu m$	$\pm0.18\mu m$	$139.8\mu m$
Y-axis FP	$0.8Kg$	$\pm0.3\mu m$	$\pm0.19\mu m$	$145\mu m$

4. CONTROL OF THE FINE MECHANISM IN A COARSE-FINE SYSTEM

Repeatability of the FP was evaluated using simple PI controller. However, when attached to the CP, it is necessary to consider restraint, such as the residual vibration and influence from the CP, to design a control system for the FP. A lot of work [3] is done about the cooperation control system in which both CP and FP models are considered. In this study, the control system of the FP is designed, as it satisfies the requirement of the coarse-fine system with no knowledge about the model of the CP, in the presupposition that the error of the CP is within the range of the FP. As for the actual system, the basic specification of the control system is shown below.

- Steady state error $E_{ss} <$ 0.1 μm: to assure an accuracy of the coarse-fine system < 1 μm;

- Phase margin PM > 40 deg: to assure a maximum overshoot < 25%, limited by the range of the FP;

- Bandwidth > 100 Hz: to compensate transient and residual vibrations of our CP in the range of 0 $\sim$ 50 Hz.

The designed controller, shown on Fig. 9, is composed of four elements: open-loop compensator, feedback controller, feedforward controller, and disturbance observer. Each element is explained below in detail.

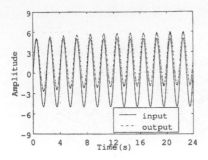
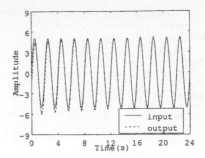

Figure 5. Open-loop response to sine wave. Figure 6. Open-loop response to sine wave.
(Without compensation for asymmetry)   (With compensation for asymmetry)

4.1. Improvement of open-loop characteristics

Since piezo-electric actuator is accompanied by the hysteresis, it has non-linear behavior. Figure 5 shows an open loop response of the FP to sine wave (amplitude $5\mu m$, frequency 0.5 Hz), in which the asymmetry characteristic modeled by Eq. 1 is of bigger effect compared to the hysteresis non-linearity and is to be compensated for.

$$S : y = ax + b \tag{1}$$

where x is an input signal, y is an output signal and a and b are constants determine by the data in Fig. 5. The input voltage signal v of the piezo-electric actuator is given by Eq. 2

$$v = \frac{1}{a}x - \frac{1}{b} \tag{2}$$

Using this asymmetry compensation, the response, shown by Fig. 6, was substantially improved.

4.2. The feedback controller

Lag-lead-phase compensation, shown below, is designed as a feedback compensator.

$$G_c(s) = K_c \left(\frac{s + \frac{1}{T_1}}{s + \frac{\beta}{T_1}} \right) \left(\frac{s + \frac{1}{T_2}}{s + \frac{1}{\beta T_2}} \right) \qquad \beta > 1 \tag{3}$$

where Kc is a direct current gain, and β is a number fixing position of zeros and poles. Bilinear transformation method [4] was used to derive the discrete form found out from Eq. 3.

4.3. The feedforward controller

Using the feedback controller, a bandwidth of more than 100 Hz was achieved. However, in the range of 0-50 Hz of frequencies the output does not track the input exactly. To solve this problem, a feedforward element is added. This leads to flatter gain and phase curves over a wide range of frequencies. The feedforward compensation is designed to be unit gain.

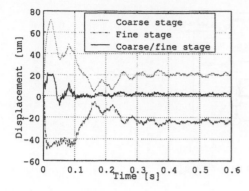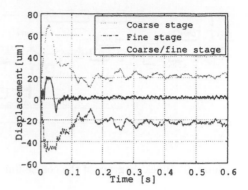

Figure 7. Positioning result by (FB+FF) controller.

Figure 8. Positioning result by (FB+FF) controller+DO.

4.4. Disturbance observer

The CP and the FP are mechanically connected. Therefore the vibrations of the CP affect the dynamic characteristics of the FP. To compensate for that influence, a disturbance observer [5] is used. It estimates the disturbance signal d and adjusts the control signal to compensate for it. By using the inverse transfer function of the plant, the estimate of d can be computed. A digital filter $Q(z)$ is used to shape the disturbance rejection characteristics.

5. EXPERIMENTAL RESULT OF THE COARSE-FINE POSITIONING

Using the designed controller, experiments were done on one axis positioning by the coarse-fine system. The setup and condition of the experiment are shown in Fig. 9. The SCARA type robot moves to the goal position and the FP begins to move when the CP reaches its movable range of 50 μ m. Capacitive sensors have been placed to measure the position of the CP, the position of the FP, and the position of the coarse + fine system. However, the position data of the CP is taken to show its displacement and is not used in the control of the system. The control voltage is sent to the actuator through the DA board and the voltage amplifier. The output displacement signal is feedback to the computer through AD board. As an example, the experimental result using the controller which consists of feedback + feedforward element (FB+FF) and the controller which is composed of feedback + feedforward element + disturbance observer (FB+FF+DO) is shown in Fig. 7 and Fig. 8. The FP moves in the opposite direction to the movement of the CP as shown in any figure. Therefore, the coarse-fine system demonstrates significant improvement for accuracy and settling time. Also, the Fig. 8 with DO shows better positioning performance of the coarse-fine system compared with Fig. 7.

338

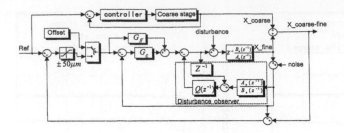

Figure 9. Block diagram of the coarse-fine control strategy.

6. Conclusion

A coarse-fine robot system for the purpose of precise mounting electronic components, has been described. The development of two-axis FP and its experimental evaluation were presented. Following results are obtained.

- A suitable structure of FP for this purpose has been designed by CAE based on FEM;

- Experimental results have shown that the design specifications are sufficiently satisfied. A stroke of 100 μm was achieved, and repeatability less than 0.2 μm was achieved for the two axes as well;

- Using only the control of the FP, transient and residual vibration of the CP has been compensated successfully. As the result of that, high speed and precision positioning can be realized by the developed coarse-fine mounting robot system.

REFERENCES

1. For example, A. Sharon, N.Hogan, and D.E.Hardt, The Macro/Micro Manipulator: An Improved Architecture for Robot, Robotics and Computer-Integrated Manufacturing, 10, 3, (1993), 209.
2. S.Masuda, S.Abe, N.Sakamoto, M.Kajitani, A.Ming and C.Kanamori, Research on Fine Motion Mechanism for End-effector of Assembly Robot(1st Report) - Characteristics of one Axis fine motion mechanism-, Proc. of the Autumn Conference of JSPE, Kobe University, Japan, p598, (1-3 October 1997) (in Japanese)
3. For example, A.Shimokohbe, K.Sato and F.Shinano, Dynamics and control of ultra-precision two-stage positioning system, Proc. Of the 9th Int. Precision Engineering Seminar, (1997) 495.
4. Katsuhiko Ogata, Modern Control Engineering. Upper Saddle River, NJ: Prentice-Hall,(1997).
5. C.M.Mok, M.Saitoh, and R.Bickel, Hard disk Drive Implementation of a New optimal Q-Filter Design for a Disturbance Observer, Proc. for Conf. on Motion and Vibration Control, Hitachi City, Japan, (1997).

Human Friendly Mechatronics (ICMA 2000)
E. Arai, T. Arai and M. Takano (Editors)
© 2001 Elsevier Science B.V. All rights reserved.

Development of Three Dimensional Measuring System for Making Precise Patterns on Curved Surface Using Laser Drawing

Choong Sik PARK*, Koichi IWATA **, Tatsuo ARAI***, Yoshikazu MURATA***
Shigeto OMORI*, Tetsuya HAMAMOTO*

*Regional COE Promotion,Osaka Science and Technology Center
2-7-1, Ayumino, Izumi City, Osaka, 594-1157, Japan
** College of Engineering, Osaka Prefecture University
1-1, Gakuen-cho, Sakai, Osaka 599-8531 Japan
***Graduate School of Engineering Science, Osaka University
1-3 Machikaneyama-cho, Toyonaka City, Osaka ,560-8531, Japan

Abstract

A three dimensional measuring system for making precise patterns on curved surface by laser exposure is developed. A interferometer which has been used for two dimensional measurement is improved to measure the position and orientation of the moving stage in three dimensional space. The moving stage with several corner cube arrays can reflect laser beams in the large moving range for interference of the reference laser beam. Processing interference images in the three different directions can realize the three dimensional measurement. Singular value decomposition method is applied to design of parameters for isotropic measuring resolutions. Moreover, error analysis is performed to obtain tolerances of various mechanical errors.

1. INTRODUCTION

Our project purpose is development of optical systems for processing great volumes of information at ultra-high speed. Diffractive optical elements with precise patterns on their surfaces are the key device of our studies. Making precise patterns on the three dimensional curved surface, is expected to produce more valuable diffractive elements with high performances and excellent functions, such as combination of diffraction and deflection, which are called three dimensional diffractive optical elements[1]. Recent developments in microfabrication technology including laser beam and electron beam lithography have several advantages, such as precise patterning and flexible fabrication. However, there are no instruments for maiking precise patterns on the three dimensional curved surface. Then, we are developing the laser drawing system for three dimensional diffractive optical elements[2]. The overview of the laser drawing system is shown in Figure 1. The system is based on the laser drawing system for flat surface of the element. Some improvements are made for precise patterning on the curved surface[3]. Especially, in order to expose the element correctly corresponding to the pattern, it is important to measure position and orientation of the element in the three dimensional space.

The interferometer is widely used for high accuracy measurement. When the measured object is moved over the scan range of the laser beam in the three dimensional space, it is impossible to measure. At present, the laser interferometer with tracking is only developed to measure the position and orientation of the target in the three dimensional space by micro meters order[4]. However, the measuring system has the complicated and large mechanism for tracking.

In this paper, we describe the method of three dimensional measuring, design of the measuring system using singular value decomposition and error analysis to obtain tolerances of various mechanical errors.

2. MEASUREMENT PRINCIPLE

Figure 2 shows the system of the measuring the position and the orientation of the moving stage in the three dimensional space. This principle is based on the laser interferometer. A laser beam is divided into two laser beams by the beam splitter. One beam goes to the reference mirror and another beam to the mirror attached to the measuring object. Tow laser beams reflected by each mirror interfere with each other. When the object moves in the direction of the laser beam, the difference of the light path between two laser beams changes, so that we can get the interference fringes according to the change of the distance by the detector such as CCD camera. The corner cube as shown in Figure 3 is employed in the interferometer which has been used. However, when the moving range of the object is large, the laser beam cannot irradiate the corner cube so that the measurement is impossible. In our measurement method, the corner cube array is employed in which many small corner cubes are arranged in order as shown in Figure 4. This allows us to get the reflected beam in the large moving range. Figure 5 shows the proposed measurement principle. The position vector $\mathbf{r}_{mk}$ (m=1,2,3, $1 \leq k \geq$ n)describes the position of the corner cube in the coordinate system placed on the center of the moving stage. The rotation matrix $\mathbf{R}$ describes the orientation of the stage in the global fixed coordinate assumed. The position vector $\mathbf{p}$ describes the relative position of the coordinate system set on the moving stage and the global fixed coordinate system. The unit vector $\mathbf{S}_m$ describes the measuring direction. The scalar value L_{mk} is the moving distance of the peak of a corner cube in the corner cube array. Assuming that $\mathbf{S}_m$ and $\mathbf{r}_{mk}$ are known, L_{mk} is defined as follows.

$$L_{mk} = \mathbf{S}_m \ (\mathbf{p} + (\mathbf{R}-\mathbf{E}) \ \mathbf{r}_{mk}) \tag{1}$$

where $\mathbf{E}$ is the unit matrix.

Analysis of the interference image of one corner cube gives the integer part of the number of interference fringes $\mathbf{F}_{mk}$, and its fraction part ε_{mk}. $\mathbf{L}_{mk}$ is expressed as following.

$$L_{mk} = (F_{mk} + \varepsilon_{mk}) \ \lambda /2 \tag{2}$$

where λ is the wavelength of the laser beam.

Measurements in more than three different directions can determine the six unknown values which are the components of the position and the orientation of the stage . However, it is difficult to solve equation (1), because the relationship of $\mathbf{L}_{mk}$ of the orientation including in $\mathbf{R}$ is non linear. Then, least square method is employed to obtain the approximate values.

3. DESIGN OF MEASURERING SYSTEM

3.1 Design Method

The motion of the moving stage should be precisely controlled in less than μ m order during drawing patterns. In the differential movements of the stage, the relationship of diffrencial displacements of the measurement value δL_{mk} of the position δp and the orientation $\delta \theta$ is derived by differentiating equation (1) relation to position and orientaion.

$$\delta L_{mk} = [S_m^T \quad (r_{mk} \times S_m)^T][\delta p^T \quad \delta \theta^T]^T \tag{3}$$

Applying equation (2) to all the measurement points obtains the following equation.

$$\delta L = J_x \quad \delta X \tag{4}$$

where

$$\delta L = [\delta L, \quad \cdots, \delta L_{1n}, \quad \cdots, \delta L_{mn}]^T \in R^{mn \times 1}$$

$$J_x = \begin{bmatrix} S_1^T & (r_{11} \times S_1)^T \\ \vdots & \vdots \\ S_1^T & (r_{1n} \times S_1)^T \\ \vdots & \vdots \\ S_m^T & (r_{mn} \times S_m)^T \end{bmatrix} \in R^{mn \times 6}$$

$$\delta X = [\delta p \quad \delta \theta]^T \in R^{6 \times 1}$$

J_x is the Jacobian matrix which expresses the linear transformation from the differential displacements of the stage to the differential measurement values.
Multiplying inverse matrix J_x^{-1} in equation (4) derives the following equation.

$$\delta X = J_x^{-1} \delta L \tag{5}$$

J_x^{-1} can be interpreted as the measurement resolutions. J_x^{-1} depends on the directions of laser beams S_m and the locations of corner cubes r_{mk}. The measurement system should be designed to be isotropic measurement resolutions in any directions. Singular value decomposition technique [4,5,6] is employed to search the optimum values of S_m and r_{mk}. The condition number is the ratio of the maximum singular and the minimum singular value. In other word, the number defines the maximum value in the ratio δL of δX. Therefore, the optimum values of S_{mk} and r_{mk} minimize the condition number.

3.2 Design Results

The condition number is calculated in two kinds of layout of corner cubes which are located at same intervals in one side direction minus related to the axis set on the moving stage and located at same intervals in both directions as shown in Figure 6. The condition number related to number of intervals of corner cube measured is shown in Figure 7. Number of measuring samples is fewer, the condition number is lower and changes smaller related to the interval change. The layout of the both sides gives the lower condition number than the layout of one side(No figure in this paper). The design parameters are chosen as the layout of both sides, number of intervals = 2, number of samples in each corner cube array is 2. Under these parameters, the condition number in laser beam angle related to each laser beam is shown in Figure 8. Laser beams should be perpendicular to each other.

4. ERROR ANALYSIS

4.1 Analysis Method

The relationship of the measurement errors of the mechanism errors is derived by differentiating equation (1) respect to the mechanical parameters.

$$\delta L = J_m \, \delta m \tag{6}$$

where

$$J_m = \begin{bmatrix} (p^T + (R-E)r_{j_1})^T & S_1^T(R-E) & \cdots & 0 & 0 \\ \vdots & \vdots & \ddots & \vdots & \vdots \\ 0 & 0 & \cdots & (p^T + (R-E)r_{j_n})^T & S_m^T(R-E) \end{bmatrix} \in R^{2mn \times mn}$$

$$\delta m = [\delta S_1^T, \delta r_1^T, \cdots, S_{mn}^T, r_{mn}^T] \in R^{2mn \times 1}$$

J_m expresses the effect of the mechanical errors on the measurement errors. Assuming the errors of laser beam direction and corner cube positions can estimate the mechanical errors for the given design parameters in one location of the moving stage.

4.2 Analysis Results

Figure 9 shows the errors of the measured position of the stage affected by the errors of position errors of corner cubes.   Figure 10 shows the errors of the measured position of the stage affected by the direction errors of laser beams. The errors of measurment are expected to be less than $0.1 \ \mu$m since the line width of diffractive patterns is aimed less than 1μm. The tolerable parameter errors are found as $\delta r < 4.5 \mu$m and $\delta s < 0.01$ degrees.

5. CONCLUSION

Three dimensional measuring method using interferometer was proposed and the measuring system was designed for micro patterning on the curved surface of the optical element using laser exposure. The method is characterized to measure the moving object in the large range precisely in comparison with the other interferometer. Singular value decomposition technique obtained to optimal design parameters for isotropic measuring resolutions . Error analysis resulted that tolerable errors of design parameters were found.

ACKNOWLEDGEMENTS

This work was supported by Development of Basic Tera Optical Information Technologies, Osaka Prefecture Joint Research Project for Regional Intensive, Japan Science and Technology Corporation.

REFERENCES

1. Steve M. Ebstein, Optics Letters, 21, 18,(1996),1454
2. C.S.Park and T.Arai and et.al.,Proc,International Symposium on Robotics,(1999),737
3. Y.Murata and T.Arai and et.al.,Proc,International Symposium on Robotics,(1999),645
4. T. Arai, JRSJ,10,4,(1992),100, (In Japanese)
5. T.Oiwa and N.Kuri and et.al., JSPE, 65, 2,(1999), 288, (In Japanese)
6. M.Naruse and M.Ishikawa,KOGAKU,29.2,(2000),101,(In Japanese)

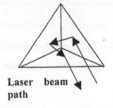

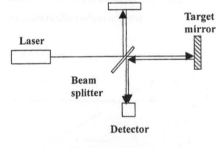

Figure 1 Developed laser drawing system

Figure 2   Measurement principle

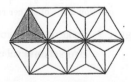

Figure 3   Corner cube

Figure 4   Corner cube array

344

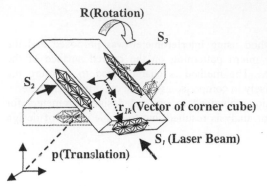

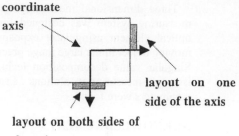

Figure 5   Measurement model

Figure 6   Corner cube layout

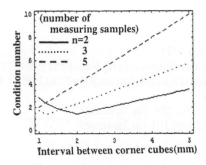

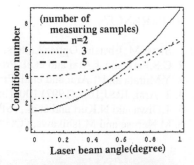

Figure 7   Relation between coner cube
position and condition number

Figure 8   Relation between laser beam direction
position and condition number

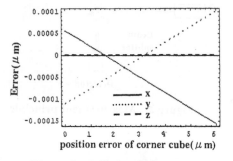

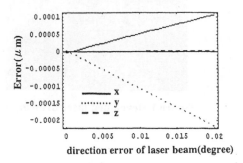

Figure 9   Influence of parameters errors on
errors of coner cube position

Figure 10   Influence of parameters errors on
errors of laser beam direction

Human Friendly Mechatronics (ICMA 2000)
E. Arai, T. Arai and M. Takano *(Editors)*
© 2001 Elsevier Science B.V. All rights reserved.

Autonomous Mobile Robot Navigation based on Wide Angle Vision Sensor with Fovea

Sohta SHIMIZU *, Yoshikazu SUEMATSU**, Tomoyuki MATSUBA**
Shunsuke ICHIKI**, Keiko SUMIDA** and Jian-ming YANG**

*   Department of Control Engineering, Tokai University
**   Department of Electronic-Mechanical Engineering, Nagoya University

In this paper it is presented that the author has examined Wide Angle Foveated Vision Sensor (WAFVS) with a special optical system designed by mimicking a human eye and has advanced it to a parallel processing system by plural computers with multi-task operation system (OS) under LAN network to make the most of its characteristics. WAFVS system carries out parallel image processing flexibly by plural computers with multi-task OS, based on timely task distributing. This system is applied to a navigation system for an autonomous mobile robot. As an example of the execution, experiments of navigation are carried out to obtain an image with sufficient resolution for face recognition, while a camera gazes at and tracks a human face by a proposed method characterized by WAFVS. In this method, central vision using color information, and peripheral vision using object motion information from gray scale images are combined cooperatively.

1. Introduction

Human eye has a 120 degrees wide-angle visual field. The visual acuity is the highest near a *fovea* in a central area of retina and gets lower rapidly as going to peripheral area. It is well-known that a human being behaves by processing such *foveated* visual information parallel and often cooperatively with high level *central vision* and low level *peripheral vision* [1]. Some researches on vision systems modeling after the human visual property have been reported [2]-[5]. For examples to have realized such vision by a single input device, the method to reconstruct a foveated image based on *log-polar mapping* by a computer [2], and the method to obtain it by using a *space variant* scan CCD [3] are well-known. In the authors' laboratory, a special super wide-angle lens (Wide Angle Foveated (WAF) lens) for robot vision was designed and produced to obtain a *wide-angle* foveated image for multi-purpose and multi-function by *optical* approach from a single CCD camera [4]. Compared with the above two methods, this method has an advantage to realize higher resolution and wide-angle visual field on the same time. In order to make the most of the characteristics, the authors have utilized WAF lens to a robot vision system with functions of image processing, camera motion control and so on, and have developed Wide Angle Foveated Vision Sensor (WAFVS) system [5].

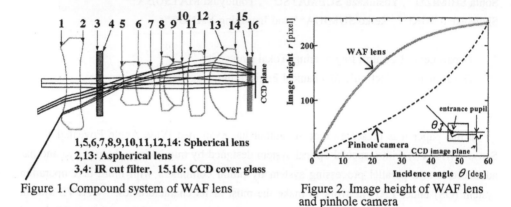

1,5,6,7,8,9,10,11,12,14: Spherical lens
2,13: Aspherical lens
3,4: IR cut filter, 15,16: CCD cover glass

Figure 1. Compound system of WAF lens

Figure 2. Image height of WAF lens and pinhole camera

2. Wide Angle Foveated Vision Sensor System

WAFVS system have been developed and advanced in the authors' laboratory[4][5]. The system has two CCD cameras for stereo vision and has *'fovea'* that is attentional area with high resolution in the visual field and has a function of rapid *camera motion control*. Special *super wide-angle* lens, named WAF lens(Fig.1), plays a major role in this system. This lens is attached to a commercially available CCD camera and *optically* realizes a wide-angle *foveated* image, that is 120 degrees wide visual field and local high resolution in the central area, *on the same time*. The input image can be utilized for some kinds of image processing with *multi-purpose* and *multi-function* because it has different levels of information, e.g. low resolution but wide visual field for *peripheral vision*, local but high resolution for *central vision* and so on. Concept of the different visions for robot vision is not only for a foveated image. However, it is clear that the former with a wide visual field has many advantages, when it is utilized *simultaneously* for *different levels* of image processing.

Figure 2 shows characteristics of image height r on CCD image plane versus incidence angle θ [rad] to WAF lens. For a comparison, characteristics of the *pinhole camera* with the same visual field and the same amount of information are also shown. These are represented as eq.(1) (WAF lens) and eq.(2) (pinhole camera), respectively. The inclination of each curve shows the image resolution along a radial direction of the visual field. It is noted that

WAF lens has higher resolution in the central area and, on the other hand, has lower resolution in the peripheral area, than the pinhole camera.

$$r = 596.200\theta^3 - 498.187\theta^2 + 147.395\theta \quad \text{[pixels]} \tag{1}$$

$$r = 247.300\frac{\tan\theta}{\sqrt{3}} \quad \text{[pixels]} \tag{2}$$

Further, *gaze control* (that is camera motion control for gaze) improves efficiency of utilizing the optical wide-angle foveated image. Camera motion control including gaze control is carried out by four stepping motors with neck rotation and two kinds of motion mimicking human *saccade* and *pursuit*. This time, *parallel image processing* function based on timely task distributing by plural network computers is added to WAFVS system, newly(Fig.3). This system has plural computers *with multi-task OS* under wireless and wired LAN network composed of *the main computer* and *support computers*, and carries out plural kinds of image processing distributively by network transmitting according to each processor stress to realize real-time processing. The main computer plays a role as a file server and has shared information among the computers or among plural tasks of image processing. Different kinds of image processing are carried out in combination with camera motion control parallel or selectively for various kinds of utilization.

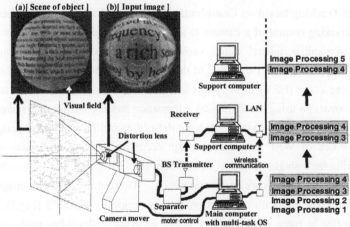

Figure 3. A scheme of WAFVS system and timely task distributing

Input images from WAF lens has the highest resolution in the central area than the other methods [2][3]. WAFVS system can make each algorithm of image processing simpler by obtaining a local image with high resolution based on gaze control, because generally, to use an image with high resolution makes it possible to realize the same precise of some information extracted from an image with low resolution by a simpler algorithm. Therefore utilization of WAFVS system can reduce calculation cost per an image. The authors think that combination between WAF lens and space variant CCD may be the best matching for wide-angle foveated vision, because this combination reduces visual information of the peripheral area further largely and get network transmitting cost lower.

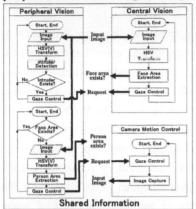

Figure 4. Flow chart of gaze and tracking

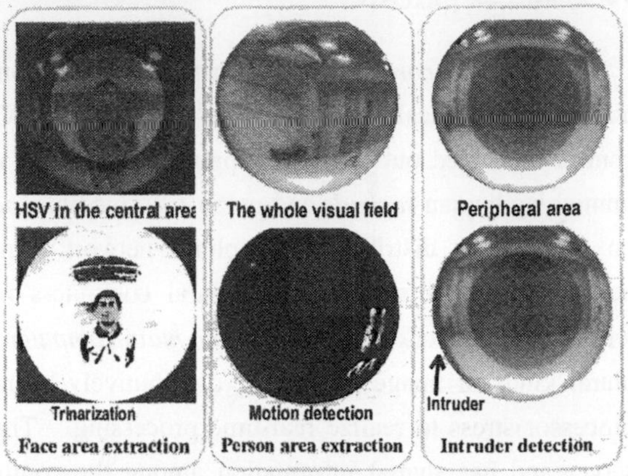

Figure 5 Three kinds of image processing for gaze and tracking

3. Gaze and Tracking based on Combination of Central Vision and Peripheral Vision

Gaze and tracking control of a camera to a human face based on *cooperative* combination of central vision and peripheral vision, each of which has a simple processing method, is proposed. Figure 4 shows a flow chart of the proposed method. In this method, *face area extraction* using *color* (HSV) information in the local central area of the visual field and *person area extraction* using object motion information based on difference images from gray scale images (V) in the whole visual field, are defined as central vision and peripheral vision respectively. Use of WAF lens realizes high accuracy tracking and wide-angle moving object capturing by a single camera. In addition, the object motion information is utilized for *intruder detection* as another peripheral vision. These three kinds of image processing are carried out repeatedly with different processing period by WAFVS (Fig.5). This paper does not describe in more detail each of image processing algorithm characterized by the resolution property of WAF lens, because of limited space. Gaze and tracking is executed under rules in the following.

(1) gaze at the extracted face area likely with pursuit, when it exists in the central area

(2) gaze at the extracted person area likely with saccade, when no face area exists

(3) gaze control is not carried out, when neither a face area nor a person area exists

(4) gaze at an intruder is carried out priorly likely with saccade, when it is detected

4. Autonomous Navigation based on Tracking a Human Face

4.1 Autonomous Mobile Robot Navigation System

Figure 6 shows a navigation system of a mobile robot with WAFVS. The robot is PWS vehicle-type with 2 DC servomotors. This system has the main computer and plural support computers under a wireless or wired LAN network and a BS wireless image transmitter to send images from WAFVS to one of support computers directly. The main computer on the robot plays a role as a moving file server, and the main computer and support computers carry out required tasks of image processing by timely task distributing. The support computers give back calculated results to the main computer on the robot by LAN network.

Similarly to WAFVS system, the computers of this system have multi-task OS and carry out parallel processing of plural tasks such as drive control of the mobile robot, communication between the computers, camera motion control, plural kinds of image processing, system management and so on by timely task distributing. This parallel processing is executed based on concept of *shared information* among the tasks. The shared information is shared on memory of the main computer (Fig.7).

Figure 6. Autonomous mobile robot navigation system   Figure 7. Shared information

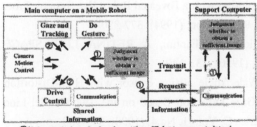

①If the resolution of a face image is sufficient, requests 'stop' and 'do gesture' are send.
②If no face area and no person area is extracted for a while, main computer on the the robot stops motions of the robot.

Figure 8. Relationship among plural tasks
for navigating a mobile robot

Figure 9. Scene of navigation experiments
to obtain sufficient image

4.2 Experiments

Figure 8 shows relationship among plural tasks for an experiment to obtain an image with sufficient resolution for personal identification by face recognition using the navigation system. While tracking a human face by the proposed method for WAFVS is carried out on a mobile robot, input images from WAFVS are *transmitted* to a support computer by a BS wireless transmitter and, according to processor stress of the main computer, a task of image processing is transmitted to the support computer by wireless LAN and is carried out on it, too. It judges whether to have obtained the sufficient image based on a size of the face area extracted from HSV information, and navigates the mobile robot automatically by sending back requests of approach till obtaining it. When the sufficient image is obtained, the robot does a gesture of eyes and neck movement repeatedly. Figure 9 shows a scene of the navigation experiments.

5. Conclusion

The results of this paper have been summarized as follows.

(1) WAFVS system, that is the parallel visual information processing system for wide-angle foveated images with camera motion control to carry out plural kinds of image processing parallel or selectively and cooperatively, has been examined and developed.

(2) WAFVS system is applied to a navigation system for an autonomous mobile robot. Experiments of navigation based on tracking a human face have been carried out to obtain an image with sufficient resolution for face recognition.

(3) A method to gaze at and track a human face based on combination of central vision and peripheral vision characterized by WAFVS system has been proposed.

REFERENCES

1. K.Hiwatashi, Human Information Engineering (in Japanese), (1979), 48

2. T.Baron, M.D.Levine and Y.Yeshurun, Exploring with a Foveated Robot Eye System, Proc. of 12th International Conference on Pattern Recognition, (1994), 377-380

3. G.Kreider, J.Spiegel, G.Sandini, P.Dario and F.Fantini, A RETINA LIKE SPACE VARIANT CCD SENSOR, SPIE, No.1242, (1990), 133-140

4. S.Shimizu, Y.Suematsu and S.Yahata, Wide-Angle Vision Sensor with High-Distortion Lens(Detection of Camera Location and Gaze Direction Based on the Two-Parallel-Line Algorithm), JSME International Journal, Series C, Vol.41, No.4, (1998), 893-900

5. S.Shimizu, Y.Suematsu and etc., Wide Angle Vision Sensor with Fovea(Navigation of Mobile Robot Based on Cooperation between Central Vision and Peripheral Vision)(in Japanese), Proc. of BRES99, (1999), 53-56

Human Friendly Mechatronics (ICMA 2000)
E. Arai, T. Arai and M. Takano *(Editors)*

351

Dead Reckoning of Multi-legged Robot —Error Analysis—

*Toru Masuda, Yasushi Mae, Tatsuo Arai, Kenji Inoue

Department of Systems and Human Science
Graduate School of Engineering Science, Osaka University
1-3 Machikaneyama, Toyonaka, Osaka, 560-8531, JAPAN

This paper proposes a method of dead reckoning of multi-legged robot. Multi-legged robots are required to do tasks on a rough terrain in an outdoor environment. It is important to estimate the self posture of the multi-legged robot for doing various tasks. Thus, dead reckoning of multi-legged robot is required for estimating posture. In order to validate dead reckoning of multi-legged robot, we analyze the error between joint parameter and the posture of the body of a multi-legged robot.

1. INTRODUCTION

Recently, we hope locomotive robots play an active part for rescue operation and construction and civil engineering in outdoor unknown environment (Fig.1). It is important to measure the posture of the robots for doing various tasks in such environments. A robot with external sensor such as camera and ultrasonic sensor can measure the self posture by observing landmarks in the known environment. In outdoor environment, however, it is difficult to detect landmarks stably. Besides the positions of the landmarks are not known previously. Thus, dead reckoning of locomotive robots in unknown outdoor environment is required.

As locomotion mechanisms of robots, wheeled and legged mechanisms have been studied Since most conventional robots use wheel mechanisms in planar environments dead reckoning of wheeled robots have been studied[1][2]. If rough terrain is anticipated as a work environment, however, wheeled robots have disadvantages in locomotion. Wheeled robots are not suitable for moving in a rough terrain which is not a planar environment, because the wheels must continuously contact the environment. Besides, in a rough terrain, not only position but also orientation of a robot must be measured. On the other hand, legged robots, especially multi-legged robots, are suitable for locomotion than wheeled robot, since multi-legged robots are superior to wheeled robots in moving and traverse avoidance ability.

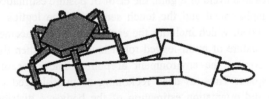

Fig.1 Image of Multi-legged robot

In this paper, we propose dead reckoning of multi-legged robots and analyze the effect of the error of the joint displacement of legs on the body posture. The body posture of a legged robot can be estimated by inverse kinematics if the joint displacement of legs is known. We assume the robot is equipped with touch sensors at every tip of legs. Thus, the current robot posture is estimated only from the joint displacement and the contact points of the legtips. This is called dead reckoning of multi-legged robot[3]. In a walking operation, the error of dead reckoning is accumulated, which includes the error of joint displacement and the error of touch sensors. The error will be analyzed for the purpose of raising the reliability of dead reckoning in rough terrain.

Furthermore, it is possible to model the unknown rough terrain environment, because the contact points of legs can be estimated. Therefore, the proposed method can be used as one of the environmental modeling techniques in rough terrain.

2. DEAD RECKONING OF MULTI LEGGED ROBOT

2.1. Procedure of dead reckoning by multi-legged robot

In walking, a multi-legged robot moves swing leg along a trajectory in a gait. If the joint displacement of legs is known when the legtips contact to the ground, the posture of the body can be estimated by inverse kinematics. Thus, the current posture of the body of the robot is estimated by iterating the following process.

1. Estimation of contact points of legtips based on forward kinematics.

2. Estimation of the body posture based on inverse kinematics.

We assume the mechanism parameters such as link length and link configuration are known previously. For posture estimation, the present displacement of each joint and the contact points to the ground of legtips must be known. If a touch sensor is at the tip of a leg, it can be judged whether swing leg contacts on the ground or not. We represent the relation between the contact point Pe and the joint parameter θ and the body posture P, Ω by equation (1).

$$Pe = F(\theta, P, \Omega)$$

(1)

When the body posture and the joint displacement are known, the contact point is obtained by the equation; forward kinematics. On the other hand, the contact points and the joint displacement are known, the body posture is obtained; inverse kinematics. Therefore, if the initial posture is known previously, the robot can estimate its posture when it moves on a rough terrain in unknown environment.

When a robot moves in a cycle of a gain, the error of posture estimation occurs due to the error of the joint displacement and the touch sensors of the legtips. The error of dead reckoning is accumulated, which includes the error of joint displacement and the error of touch sensors. As a feature of multi-legged robot locomotion, when the robot moves, the legtips of a robot contact the environment at points. The error of contact points is distributed in 3-D space. Thus, in dead reckoning of multi-legged robot, the errors of position estimation and orientation estimation of the body are distributed in 3-D space, respectively.

3. A MODEL OF MULTI-LEGGED ROBOT

A multi-legged robot for this study is shown in Fig.2, for instance. This multi-legged robot has been developed for rescue operation in rough terrain. The arrangement of the six legs is shown in Fig.3. The 6 legs are radically placed at every 60 degrees. Then the robot can move in omnidirection[4][5]. This robot can use legs as support legs and swing legs. This mechanism is suitable for locomotion in rough terrain. In this paper, the error of dead reckoning of this multi-legged robot is analyzed. In practice, it has no internal sensor currently. We assume each leg has a touch sensor at the tip and each joint displacement is obtained.

The robot alternately changes support leg and swing leg for locomotion. The flow of locomotion is as follows.

1. Trajectory generation and moving swing leg.
2. The confirmation of the grounding point of swing leg.
3. The changing of support leg and swing leg.
4. Adjusting the mass center to become a stable posture.   Return to 1.

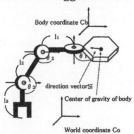

Fig.2 Multi-legged robot

Fig.3 Arrangement of six legs

4 WALKING ERROR ANALYSIS

4.1.Error of multi-legged robot

Fig.4 parameter of the leg

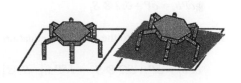

Fig.5 Image of Estimate field

The distribution of the error $(\delta X_i, \delta Y_i, \delta Z_i)$ between the actual contact point (X_i, Y_i, Z_i) and the estimated contact point (X_i', Y_i', Z_i') is assumed to be a uniform distribution. The suffix "i" indicates a leg. Since the robot only knows the (X_i', Y_i', Z_i'), the robot adjusts its body posture based on (X_i', Y_i', Z_i'). Thus the actual posture and the estimated posture are different. The position and orientation of the body is denoted by (X, Y, Z) and (A, B, C), respectively. The error between the actual posture and the estimated posture are denoted by $(\delta X, \delta Y, \delta Z, \delta A, \delta B, \delta C)$. This error occurs every time the robot moves. Therefore, the error of the posture estimation of the robot $(\delta X, \delta Y, \delta Z, \delta A, \delta B, \delta C)$ is accumulated in multi-legged dead reckoning.

4.2. Error analysis

We examine the effect of the error of the joint displacement $(\delta X, \delta Y, \delta Z, \delta A, \delta B, \delta C)$ on the posture of the body. In this analysis, we assume no mechanical error occur.

Brief figure and parameters of multi-legged robot are shown in Fig.1. The rotation angle of each joint is set with θ_1, θ_2 and θ_3 and each link length l_1, l_2 and l_3. And, central position coordinate of the body of the robot are set with $P=(Px, Py, Pz)$. The orientation vector is set with $\Omega = (\Omega x, \Omega y, \Omega z)$. The vector to the joint of the base of the leg from the center is set with $S_i = (S_{ix}, S_{iy}, S_{iz})$.

To begin with, we verify the error on a leg. The equation of the kinematics from the base of leg to thew legtip in the body coordinate Cb is differentiated. Then, the minuteness change of the leg tip can be described by (2).

$$\delta Pe_i = J_i \delta \theta_i.$$ (2)

The position error of the base of the leg δPb_i is represented by eq.(3) in the world coordinate system Co, where R_i represents the transformation matrix.

$$\delta Pb_i = -R_i \delta Pe_i = -R_i J_i \delta \theta_i$$ (3)

δPb_i is the error of the base of the leg.

The relation between Pb_i and the error of the position of the body P is represented by equation(4). In this equation, S_i denotes the vector from P to the ith joint at the base of the leg i.

$$\delta Pb_i = \delta P + \delta \Omega \otimes S_i$$ (4)

This is transformed to the following equation.

$$\delta Pb_i = \delta P + \delta \Omega \otimes S_i = I \delta P + \left[-S_i \otimes \right] \delta \Omega$$ (5)

$$= \left[I \ \left[-S_i \otimes \right] \right] \begin{bmatrix} \delta P \\ \delta \Omega \end{bmatrix}$$

$$-S_i \otimes = \begin{bmatrix} 0 & S_{iz} & -S_{iy} \\ -S_{iz} & 0 & S_{ix} \\ S_{iy} & -S_{ix} & 0 \end{bmatrix}, I = \begin{bmatrix} 1 & 0 & 0 \\ 0 & 1 & 0 \\ 0 & 0 & 1 \end{bmatrix}.$$

We assume a wave gait. Since three legs support the body in a wave gait, we obtain the following equation:

$$\delta P_B = A \begin{bmatrix} \delta P \\ \delta \Omega \end{bmatrix} \text{ ,where } A = \begin{bmatrix} I & [-S_1 \otimes] \\ I & [-S_2 \otimes] \\ I & [-S_3 \otimes] \end{bmatrix}, \quad \delta P_B = \begin{bmatrix} \delta Pb_1 \\ \delta Pb_2 \\ \delta Pb_3 \end{bmatrix}, \quad A \in R^{9 \times 6}. \quad (6)$$

The suffix i=1,2,3 represent the number of legs in fig.6 and fig.7.
This is solved using the pseudo inverse matrix.

$$\begin{bmatrix} \delta P \\ \delta \Omega \end{bmatrix} = \left(A^T A \right)^{-1} A^T \delta P_B \qquad (7)$$

Since δP_B is represented by the following equation,

$$\delta P_B = \begin{bmatrix} \delta Pb_1 \\ \delta Pb_2 \\ \delta Pb_3 \end{bmatrix} = \begin{bmatrix} -R_1 J_1 \delta \theta_1 \\ -R_2 J_2 \delta \theta_2 \\ -R_3 J_3 \delta \theta_3 \end{bmatrix} = \begin{bmatrix} -R_1 J_1 & 0 & 0 \\ 0 & -R_2 J_2 & 0 \\ 0 & 0 & -R_3 J_3 \end{bmatrix} \begin{bmatrix} \delta \theta_1 \\ \delta \theta_2 \\ \delta \theta_3 \end{bmatrix} \qquad (8)$$

equation (7) is represented as follows.

$$\begin{bmatrix} \delta P \\ \delta \Omega \end{bmatrix} = \left(A^T A \right)^{-1} A^T \begin{bmatrix} -R_1 J_1 & 0 & 0 \\ 0 & -R_2 J_2 & 0 \\ 0 & 0 & -R_3 J_3 \end{bmatrix} \begin{bmatrix} \delta \theta_1 \\ \delta \theta_2 \\ \delta \theta_3 \end{bmatrix}. \qquad (9)$$

This equation represents the relationship between error of the joint displacement θ and posture error P, Ω of the body of a multi-legged robot.

5 ANALYTICAL RESULT

In the model of the multi-legged robot described in chapter 3, a position of a legtip $P = (X, Y, Z)$ are represented by the following equations

$$X_i = \{l_1 - l_2 \cos\theta_2 + l_3 \cos(\theta_2 + \theta_3)\} \cos\theta_1. \qquad (10)$$
$$Y_i = \{l_1 - l_2 \cos\theta_2 + l_3 \cos(\theta_2 + \theta_3)\} \sin\theta_1. \qquad (11)$$
$$Z_i = l_2 \sin\theta_2 - l_3 \sin(\theta_2 + \theta_3). \qquad (12)$$

In the analysis, we set the length of links and the joint parameters are set as follows: $l_1 = 1, l_2 = 1, l_3 = 1, \ \theta_1 = 0°, \ \theta_2 = 180°, \ \theta_3 = 90°$. Then, Jacobian matrix in eq. (2) is represented by eq. (13).

$$J_i = \begin{bmatrix} 0 & 1 & 1 \\ 2 & 0 & 0 \\ 0 & 1 & 0 \end{bmatrix}. \qquad (13)$$

356

We analyze the error of the posture of the body based on eq. (9) in the above conditions. We analyze it using SVD(Singular Value Decomposition). The results of SVD are shown as follows.

Table 1 Translation error distribution

	ratio of the error
δPx	1.000
δPy	1.000
δPz	0.577

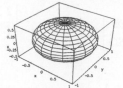

Fig.6 Translation error distribution

Table 2 Rotation error distribution

	ratio of the error
$\delta\Omega x$	0.816
$\delta\Omega y$	0.816
$\delta\Omega z$	1.154

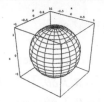

Fig.7 Rotation error distribution

Table 1 and **Table 2** show the distributions of the position and orientation error, respectively. The corresponding error ellipses are shown in **Fig.6** and **Fig.7**, respectively. From these results, we can see the vertical position error is small and the orientation error about Z axis is large comparatively. We can analyze the error for various postures in various gaits in the same way. For dead reckoning, for example, a multi-legged robot can choose a gait using its redundancy, which has a small error about horizontal direction.

6 CONCLUSION

We propose a dead reckoning of multi-legged robot and analyze its error. The error between the joint displacement of legs and the body posture is analyzed. The robot chooses its gait according to situations based on the analysis. In future, we can study for a new gait in which the posture estimation error is small based on the error analysis method.

References

[1]T Tsubouchi "A Review of Positioning Technologies for Autonomous Vehicles", Journal of the Japan Society for Precision Engineering, Vol.65,No.10,pp.1385-1388,1999.

[2]K Komoriya "Localization of a Mobile Robot", Journal of the Japan Society for Precision Engineering,, Vol.65,No.10,pp.1402-1406,1999.

[3]T Masuda, Y Mae, T Arai,K Inoue "Dead Reckoning of Multi Legged Robot", 2000 JSME Conference on Robotics and Mechatronics.

[4]Y Takahashi, T Arai, Y Mae, K Inoue, K Sakashita N Koyachi "Design of Robot with Limb Mechanism", Proceedings of the 17th Annual Conference of the Robotics Society of Japan, Vol.1,pp.349-350,1999.

[5]Y Takahashi, T Arai, Y Mae, K Inoue, N Koyachi" Design of Robot with Limb Mechanism", 2000 JSME Conference on Robotics and Mechatronics

Design and Manufacturing System

Human Friendly Mechatronics (ICMA 2000)
E. Arai, T. Arai and M. Takano *(Editors)*
© 2001 Elsevier Science B.V. All rights reserved.

USER FRIENDLY ENVIRONMENT FOR THE R&D OF CONTROLLERS IN HEAVY MACHINERY

T. Virvalo[a], M. Linjama[a], V. Aaltonen[b], M. Kivikoski[b]

[a]Institute of Hydraulics and Automation, Tampere University of Technology
P.O.Box 589, 33101 Tampere, Finland
[b]Institute of Electronics, Tampere University of Technology
P.O.Box 692, 33101 Tampere, Finland

ABSTRACT

Hardware-In-The-Loop (HIL) simulation is an effective way to speed-up and improve R&D of automation and control of heavy machines. In this paper a heavy large hydraulic press is presented. Its distributed control system is specified. The use of HIL in R&D is discussed. Requirements of HIL in this example case are presented and discussed. Examples of the use of HIL in a hydraulic press R&D is shown.

1. INTRODUCTION

Many real servo applications, especially hydraulic servo systems, are quite heavy, large, and often quite expensive. Tuning and testing, especially testing of new ideas, are time consuming and troublesome. The testing environment is often noisy, and different loading, malfunction, failure, and disturbance situations are difficult, expensive, and sometimes even dangerous to arrange in a preliminary testing stage.

During the last several years many advanced hydraulic control valves have come onto the market [1], [2], [3]. These valves include transducers, electronics, and software to realize advanced control and monitoring of valves via a CAN bus. Several transducers used in hydraulic control systems can also be connected into a CAN bus.

It is quite easy to conduct basic tests on the hardware and software of controllers, but it is often hard to test and tune the whole control system, especially when nonlinear controller solutions are used. The mechanical and servo system parts are most difficult to arrange for testing new electronic components and system ideas in as real a situation as possible.

One powerful and promising method of avoiding most of the above-mentioned problems is the so-called hardware-in-the-loop (HIL) method. The basic principle is that part of a real system is replaced with a real time simulation model. The HIL method has been used very successfully both in the industry and in universities [4]. Some very interesting HIL applications have been reported in the R&D of automotive electronic control units [5], [6] and [7]. The electronic transmission control unit has been tested against a simulated model of a car drive train utilizing the HIL idea [8]. Very high computing power has been required in some HIL applications [9]. A part of the mechanical system has been replaced with the HIL system in the research and development of a quite complicated hydrostatic power transmission system [10]. Other interesting HIL applications are also presented in [11].

The goal and purpose of this paper is to test the usability of the basic HIL idea when the mechanical and hydraulic servo system are replaced by the real time model and the distributed micro-controller based control system is real. The basic ideas and designed controllers have been tested with off-line simulations, but more realistic test systems are needed.

2. APPLICATION

The principle construction of a hydraulic press for heavy steel plates is depicted in Figure 1. Because the long upper tool is driven by four hydraulic cylinders, a difficult synchronizing problem arises. All four cylinders have to be synchronized with each other with the accuracy of 0.1-0.2 mm. There are three basic movement modes in the normal operation: an approach movement, a press movement, and a return movement. In addition there is one more important working phase; The release of pressure and other system tensions after pressing. The most difficult task in the design of a system is to avoid an undesirable situation where work occurs under one or two cylinders while the other cylinders are unloaded.

2.1 The controllers
The structuring and tuning of the basic controllers have been done in off-line simulations. Because the system damping is low and relative positioning accuracy is high the controller has to strongly increase effective damping. Because of the difficult task of synchronizing accuracy, the basic controller has to be a velocity controller. In practice two controllers are used; the velocity controller during the movement, and a position controller in final positioning. A user gives the end point of a movement directly to the position controller. The velocity controller generates a velocity profile for the movement. There is also a pressure controller that limits the maximum allowed force. All three different controllers are working simultaneously, as depicted in Figure 2.

2.2 The problem of impact
The impact phenomena are really problematic in the special case, when work occurs in the other end of the press. At the beginning of the pressing phase the tool approaches the plate at pressing velocity and impacts it. Lightly loaded cylinders continue their movement with very little disturbance, and heavily loaded cylinders almost stop. The phenomena are described in Figure 3.

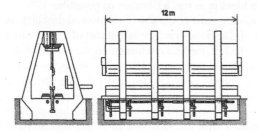

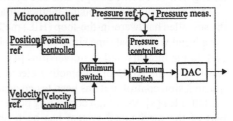

Figure 1 The principal construction of a hydraulic press.

Figure 2 Structure of controller in pressing mode.

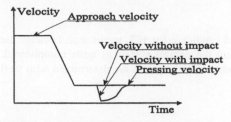

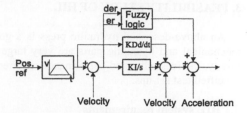

Figure 3 Velocities of different cylinders in impact.

Figure 4 Nonlinear velocity controller.

Because there is a follow-up error in the acceleration phase of each movement, especially in the impact case, it is a good practice to use a forward loop in these kinds of applications. One possible way to realize this kind of action is to use a nonlinear function in the velocity controller. An effective way is to use fuzzy logic [12], [13] as depicted in Figure 4.

2.3 Feedback signals

There are both velocity and position state controllers in control systems. The measurements of velocity and acceleration are a little bit problematic. One quite common way is to use high-resolution displacement measurement and then differentiate it twice. A 16 bit coded transducer is used in displacement measurement and an incremental transducer with resolution of 0.0001 mm is used for velocity and acceleration measurement purpose in all four cylinders. In addition a pressure transducer and 10 bit ADC are used in every cylinder.

2.4 Synchronizing principle

A traditional and proven synchronizing method is to choose one of the cylinders to act as a master and the others to act as slaves, Figure 5. The controllers of all cylinders are identical, with the same kind of movement profile generator in each controller. The master sends its position and velocity information to the slaves and their controllers make needed corrections.

2.5 Control system hardware

When each cylinder has its own controller and they are connected via a fast serial bus an easily expanding general solution can be achieved. The main controller in this kind of press application is typically CNC. Between the CNC and individual actuator controllers an interface unit is needed. It is called a machine controller, as depicted in Figure 6.

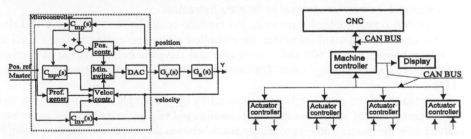

Figure 5 Master-slave synchronizing.

Figure 6 Structure of control system.

3. FEASIBILITY STUDY OF HIL

An above-described hydraulic press is a good candidate for HIL application because the mechanical and hydraulic parts are very large and the control system is quite complicated. Firstly, the feasibility and performance requirements of HIL system are determined with well specified test set-ups.

3.1 HIL system requirements
The basic requirements of the HIL approach were studied in [14] using a simple hydraulic servo system. The requirements can be summarized as follows:

Ideally, the HIL simulation model should give exactly the same response as the real system. This means that the simulation model must be able to simulate the dynamic behaviour as well as all I/Os of the real system.

Because hydraulic servo systems are strongly nonlinear and complex, the accurate model is usually computationally expensive. Thus, the computational power of HIL hardware must be high [14].

An incremental encoder is often used to measure the position of the servo system. The maximum pulse frequency might be very high in a fast high-resolution servo application. It is clear that such pulse frequencies cannot be generated from the simulation model because in the worst case it would lead to an extremely short simulation time step. Thus, a special hardware must be used to simulate the incremental encoder. Analog measurement signals can be generated from the model by DA-converters. Similarly, the analog output generated by the controller can be imported into the simulation model via an AD-converter [14].

The selected HIL hardware is produced by dSPACE GmbH [15]. The MathWorks' Simulink simulation tool is used in the modelling of the system [16]. Simulink, together with the MathWorks' Real-time Workshop [17] and dSPACE's Real-time Interface. [15] makes it possible to generate the whole HIL-model from the Simulink-model.

The controller module is based on the Motorola MC68HC16Z1 micro-controller, and contains a 16-bit CPU, external memory, serial communications interface, 12-bit DA-converter and two incremental encoder interfaces. The pulse counting is implemented with a HCTL-2016 counter manufactured by Hewlett-Packard.

3.2 Test system of HIL for press application
At this stage a simplified system is studied. Only two cylinders are assumed and the interactions between cylinders are assumed negligible. In principal the hydraulic system in the real time model is otherwise identical to the press hydraulics.

The control system contains three units: the machine controller, and the master and slave units. For communication with each other, the controllers are interconnected through a CAN bus that uses a transfer rate of 600 kb/s. The master and slave controllers consist of a micro-controller and a pulse counting boards. The structure of the controller is explained in more detail in [14].

The position of the master cylinder is controlled by the simple P-controller. The slave cylinder uses P-control but it also compares its position and velocity to the position and velocity received from the master, and tries to reduce errors between the master and the slave using PI-control.

The machine controller transfers all start, stop, and reset functions to the master and slave via

CAN bus. When the control is turned on, the machine controller transfers the reference position to the master and slave, and then both units calculate the reference trajectories. The master continuously transfers its position and velocity information to the slave.

4. RESULTS

The applicability of the HIL to this R&D is tested for the following conditions:
- ✔ the suitable complexity of the simulation model
- ✔ the interface functionality between simulation model and the hardware control system
- ✔ the operation of the synchronizing method via CAN bus
- ✔ possibilities to recognize the fast load change via CAN bus

The position responses of the master and slave are shown in Figure 7, when a force disturbance is applied on the slave. The responses are presented with and without the synchronizing information from the master; the position and velocity of the master. The position error between the master and slave is presented in Figure 8. These results show that the real time simulation works with the nonlinear models, including all the most important features of hydraulic servo drives. The results show also that the position and velocity information of the master can be transferred effectively via CAN bus from the master to the slave. The tests prove that the synchronizing principle works, and it can be realized with the help of the CAN bus.

5. CONCLUSIONS AND FUTURE RESEARCH

The results show that HIL is a good approach for testing distributed control systems. It is easy to generate different loading conditions in the simulation model and therefore the controller hardware can be tested in all necessary situations, some of which might be even dangerous in the real system. It can be expected that the HIL approach remarkably accelerates the testing of control systems.

The research will continue. The simulation model should be improved, and each controller also requires a lot of R&D. The reliability and performance of the distributed control system, including e.g. the analysing of delays, timing and drifts of the controllers clock in the CAN bus system require a lot of R&D, too.

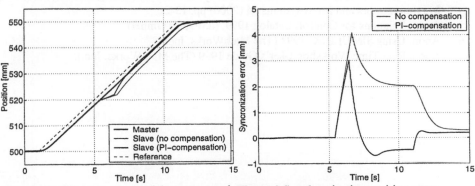

Figure 7 Position responses of the master and slave

Figure 8 Synchronisation position error between the master and slave.

6. REFERENCES

1. Lenz, W., Developments in high performance proportional valves with CANopen fieldbus interface, Proceedings of the Sixth Scandinavian International Conference on Fluid Power, May 26-28, 1999, Tampere, Finland, pp. 1177–1190

2. Gorzyza, P. and Haack, S., Proportional valves with bus interface for industrial hydraulics – strategy, development and application examples, Proceedings of the Sixth Scandinavian International Conference on Fluid Power, May 26-28, 1999, Tampere, Finland, pp. 1191–1202

3. Anon, Electonic-Hydraulic, Canbus Control System. (Ultronics Ltd, England 1999)

4. Feng, H., Toerngren, M and Eriksson, B., Experiences using dSPACE rapid prototyping tools for real-time control applications. DSP Scandinavian'97 conference, Stockholm, Sweden, June 3-4, 1997

5. Bechberger, P., Toyota controls under one roof. dSPACE News, Winter 1997/98, 2 p.

6. Knobloch, H., Maiwald, S., Schuette, H. and Chucholowski,C., Engine HIL simulation at Audi. dSPACE News, 1997/98, 2 p.

7. Suzuki, M and Inaba, Y., Vehicle dynamics models for HIL simulation. dSPACE News, 1998, 2 p.

8. Schlegel, D., Automatic gearbox HIL simulation at BMW, dSPACE News, 1997, 1 p.

9. Keller. TH., Scheiben, E. and Terweisch, P., Digital real-time hardware-in-the-loop simulation for rail vehicles: A case study. EPE'97, Tronheim, Norway, Sept. 8, 1997, 6 p.

10. Sannelius, M. and Palmberg J-O., Hardware-in-the-loop simulation of a hydrostatic transmission with sequence-controlled motors. Third JHSP Symbosium, Yokohama, Japan, 1997, pp. 301-306.

11. Maclay, D., Simulation gets into the loop. IEE Review, may 1997, pp. 109-112.

12. Zhao, T., Van Der Wal, A. J. & Virvalo, T., A Fuzzy Nonlinear State Controller for a Hydraulic Position Servo. A-Journal, Benelux Quarterly Journal on Automatic Control 36, 3, 1995, pp. 67-71.

13. Zhao, T. and Virvalo, T., Fuzzy Control of Hydraulic Position Servo with Unknown load. 2nd IEEE International Conference on Fuzzy Systems, San Francisco, USA, 1993.

14. Linjama, M., Virvalo, T., Gustafsson, J., Lintula, J., Aaltonen, V. and Kivikoski, M., Hardware-in-the-loop environment for servo system controller design, tuning and testing, Microprocessors and Microsystems 24 (2000) pp. 13-21.

15. Anon. Solutions for Control, Catalog 1999. (dSPACE GmbH).

16. Anon. Using SIMULINK. 1997 (The MathWorks, Inc.).

17. Anon. Real-Time Workshop User's Guide 1999 (The MathWorks, Inc.).

Human Friendly Mechatronics (ICMA 2000)
E. Arai, T. Arai and M. Takano *(Editors)*

365

User oriented definition of product requirements within mechatronic engineering

M. Gerst[a], H. Gierhardt[b] and T. Braun[c]

[a, b, c] Institute for Product Development, Technische Universität München
Boltzmannstraße 15, 85748 Garching, Germany

In this paper the application of a modeling technique from software engineering to mechatronic engineering is presented. The Unified Modeling Language (UML) is used to model users´ needs and relate them to product requirements. Within a bus seat development project a Use Case Model and an Analyses Model were elaborated. Advantages and Disadvantages of the presented procedure are shown.

1 Introduction

Through the integration of mechanical, electronic and software components in mechatronic systems the efficiency of products can be considerably increased. As consequence of that products are often equipped with features which have no real use for the user. Especially in this cases it is often not defined, which use cases should be realized with the product.

Capturing and formulating users needs, which represent the interaction between user and product is a challenge not considered in actual development strategies (e.g. VDI 2221). These development strategies are mostly oriented to the technical realization of a product and do not relate the technical solution to the use cases of a product. The interaction between user and product is not represented.

The only tool to analyze requirements for a new product within VDI 2221 are checklists, that enable to detect a large number of requirements by given criteria. This procedure has the risk that the needs of potential users may not be captured. Also quality-oriented approaches such as the Quality Function Deployment (QFD) and its variants do not clarify how determination and analysis of user requirements can be achieved.

One reason for the described problem is the fact that, within the product development process various not coherent models to represent product information are used. This problem is essential for the development of mechatronic products, because methods of modeling from different disciplines are utilized. There is no continuous modeling language which networks the information generated within the process. The correlation of information generated within the process is not represented. This means that the integration of results from one phase of the development process to the next phase is not transparent. This causes a vagueness in product information, which is especially for geographically distributed development processes a problem to solve [Steuer, 1999].

The goal of our work is to point out a way to capture and formulate the use cases of a product as well as to translate them into a product optimized for use. As a first step this paper presents the modeling of use cases and their translation into user requirements. The results have been elaborated within several industrial development projects.[⊗]

[⊗] Some results shown in this paper were generated in discussion with Dr. R. Stricker from BMW Group Munich.

2 Theoretical approach

Basis for our theoretical approach are methods and techniques from software engineering. These methods are applied to product development in mechanical engineering. So we interpret mechatronics also as interdisciplinary use of methods and techniques in order to develop new mechatronic products, which represent an integrated view of mechanics, electronics and software technology.

An object oriented strategy by using the Unified Modeling Language (UML) [Jacobson et al., 1999a] and the Unified Software Development Process (USDP) [Jacobson et al., 1999b] shows new opportunities for the development of new products as well as for organizing and controlling the development process.

According to [Jacobson et al. 1999b] the Unified Software Development Process is use-case-driven, architecture-centric, iterative and supports object oriented techniques.

In regard to this "use-case-driven" means that the interaction with the user is the central issue when developing a new product. An interaction of this sort is a use case. The use cases capture functional requirements[⊕] and can be gathered within a Use Case Model.

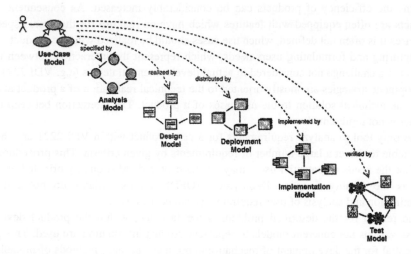

Figure 1: Submodels of the product model within the USPD [Jacobson et al., 1999b, p10]

The UML is an integral part of the USDP. The USDP uses the UML for modeling the results from the different stages of the process. These results can be seen as an integrated product model (see Figure 1), which represents the artifacts elaborated in the process and consists of the following submodels (the UML gives a language to build up the different models):

Use Case Model: The Use Case Model describes what the system does for each type of user. A Use Case Model consists of use cases (which are structured in use case packages),

---

[⊕] Within this paper we define function as the use of a product by an user. Here the function does not describe the technical operation.

actors and the relationship among use cases and between use cases and actors. This is represented and visualized by use case diagrams. The Use Case Model also shows the relationship between the Use Case Model and the Analysis Model represented by Collaboration Packages. The use case packages describe the different requirements for a system from the external view (user´s view). An Actor is everyone who is interacting with the product. That means external (e.g. consumers) as well as internal (e.g. manufacturing) customers are the actors that have certain use cases.

Analysis Model: The Analysis Model specifies the Use Case Model, that means the users´ needs are translated into product requirements (analysis of development task). The Analysis Model consists of the Collaboration Packages and the Use Case Realization Packages. The Collaboration Packages describe the internal and functional view on a product. They group the use cases from Use Case Model according to this internal and functional view. The Use Case Realization Packages realize the functional pools$^{\varnothing}$ represented by Collaboration Packages. They consist of so called analysis classes. The Analysis Model transforms the product behavior described by use cases into analysis classes.

Design Model: The design model is an object model that describes the physical realization of the use cases (synthesis of solution).

Deployment Model: The deployment model is an object model, that describes the physical distribution of the system in terms of how functionality is distributed among components.

Implementation Model: The implementation model describes a model, which enables the real use of a product. It shows the implementation of the use cases.

Test Model: The test model is a collection of test cases, test procedure and test components, which describe how specific aspects of products have to be tested. So the test model is basis for verifying the use cases.

Within our paper we present the Use Case Model and the basics of the Analysis Model. We do not present the different phases and workflows of the USDP.

3 Application

Within a development project in collaboration with industry we have applied the theoretical background to design and development work for a bus seat. Though a bus seat is not a mechatronic product, it is a very good sample for user product interaction and also a very good object to apply software engineering methods to the development of mechanical products.

The actors in the Use Case Model for the bus seat development are:

- Entrepreneur with various sub actors,
- Bus Manufacturer with two sub actors,
- Bus Carrier with two sub actors,
- Passenger,
- External Controllers.

$^{\varnothing}$ The functional pools of a product represent its main characteristics (cost, weight, space, technical functions, reliability, etc.)

We defined 5 different Use Case Packages related to the actors. The relationship between use cases and actors is an association. For the Analysis Model we were able to define 12 different Collaboration Packages (see Figure 1).

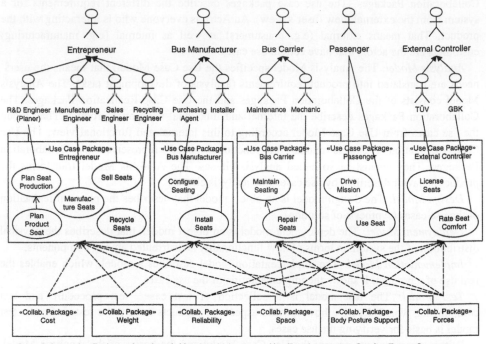

Other Collaboration Packages (not shown): Materials, Maintenance, Handling, Appearance, Comfort-Extras, Stowage

Figure 1: Use Case Model and Collaboration Packages

In Figure 2 the detailed Use Case Package "passenger" is shown. According to UML the relationship between different use cases in our sample is defined as generalization or dependency. A generalization is a relationship between a general thing (superclass or parent) and a more specific kind of that thing (subclass or child).

The generalization is visualized as a solid directed line with an large open arrowhead pointing to the parent. A dependency is a using relationship that states that a change in specification of one thing may effect another thing that uses it (but not necessarily the reverse). The dependency is visualized as a dashed directed line pointed to the thing being depended on.

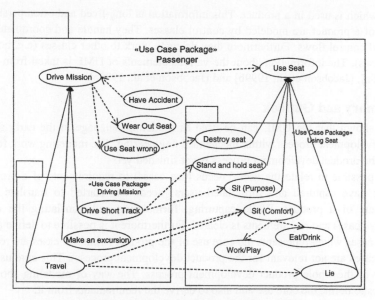

Figure 2: Use Case Package "Passenger"

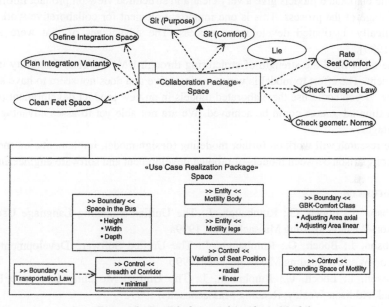

Figure 3: Detailed part of Analysis Model

In Figure 3 a detailed part of the Analysis Model is shown. Different use cases from the Use Case Model are gathered in the Collaboration Package "Space". The Use Case Realization Package "Space" contains entity, boundary and control classes. Boundary classes describe the product functionality across the system boundary. In an entity class information is posted which is used in a product. This information is long-lived and often persistent. The dynamics of a product are modeled by control classes. They handle and coordinate the main actions and control flows. Furthermore they delegate work to other classes (e.g. boundary or entity classes). The information about the various elements of UML is taken from [Jacobson et al., 1999a], [Jacobson et al., 1999b] and [Jacobson et al., 1999c].

4 Summary and Outlook

It was shown that the UML can be used as modeling language in the early stages of a product development process within mechanical engineering. The modeling work for the later stages of the product development process is not finished yet.

It was possible to relate the users´ needs (use cases) to requirements of a product. That means we have captured and formulated users´ needs and started to translate them into requirements of a product. While capturing, formulating and translating the correlation between use cases and requirements is cleared. Furthermore it is possible to define a reference set of use cases which represents the core use of the product (what use cases are relevant and what use cases are not relevant for the product development process). This core use becomes the center of the whole product development process. The way of modeling shown in this paper helps to define the core use and also to derive the requirements from it.

The coherence of product modeling within the development process could not be shown yet, but the elaborated models give a very clear and structured view on product information in an early stage of the process. This is one important element for collaborative work within a geographically distributed development process. The advantages of that were shown in [Steuer, 1999].

On the other hand the strength of structuring through UML restricts flexibility in product development. The effort to model is very time intensive and does not seem to have short term advantage. If you could use the elaborated models in an new product development process as pattern a time advantage could be achieved. We are not able yet to asses advantages versus disadvantages.

Future research will work on further modeling (design model, implementation model, etc.) and the comparison between models of product development and software engineering.

5 References

1. Jaccobson, I.; Booch, G.; Rumbaugh, J.: The Unified Modeling Language User Guide, Addison-Wesley, Reading Massachusetts, 1999a
2. Jaccobson, I.; Booch, G.; Rumbaugh, J.: The Unified Software Development Process, Addison-Wesley, Reading Massachusetts, 1999b
3. Jaccobson, I.; Booch, G.; Rumbaugh, J.: The Unified Modeling Language Reference Manual, Addison-Wesley, Reading Massachusetts, 1999c
4. Braun, T.: Einstieg in eine objektorientierte Entwicklungsvorgehensweise am Beispiel eines Bussitzes, not published Diploma Thesis, Institute for Product Development, Technische Universität München, 2000
5. Steuer, C.: Erstellung eines objektorientierten Anforderungsmodells für einen 1-Zyl. Versuchsmotor als Grundlage für ein 24h-Entwicklungsprojekt, , not published Diploma Thesis, Institute for Product Development, Technische Universität München, 1999

Human Friendly Mechatronics (ICMA 2000)
E. Arai, T. Arai and M. Takano *(Editors)*
© 2001 Elsevier Science B.V. All rights reserved.

A Friendly Open PC-Based Working Frame For Controlling Automatic Re-Configurable Assembly Cells

Jose L. Martinez Lastra and Prof. Reijo Tuokko

Institute of Production Engineering. Tampere University of Technology. P. O. Box 589. FIN-33101. Tampere, Finland

The paper deals with the design of a friendly working frame for controlling automatic assembly cells in the field of electronics production, specially in the final assembly tasks. The foundation of the system is an open architecture based in PC technology, standard communication systems and connectivity tools. The aim is to reconfigure manufacturing systems rapidly in response to changing needs and opportunities. The approach presented in this paper increases the level of "plug & assemble" due the open architecture control based in standard components and a friendly Windows environment.

KEYWORDS
Automation, Communication systems, Flexible system, Machine automation, Robotics

1. INTRODUCTION

The production of electrotechnical and electronics products has increased their volumes and the number of members within the same product family during the last years. Other trends include the reduction in the product life time and the worldwide presence of manufacturing close to markets and customers. To compote in this market environment involve the design of the product, the manufacturing system, and the distribution activities.

The need for flexibility is increasing and the design of factory automation system must allow the system to be easily reconfigured as requirements change. These manufacturing/assembly systems call for special flexible machine and line controllers, easy to integrate, maintain and reconfigure, decreasing the design cycle time and enabling continuos process environments.

The paper presents a design concept for creating flexible automatic assembly cells. The working environment is based in an open architecture control using standard hardware and software components allowing plug and assemble cells.

2. AN APPROACH TO OPEN ARCHITECTURE CONTROL FOR ROBOTICS

The current year has showed the starting of a promising discussion concerning the adoption of open architecture control for the robotics industry (1 / 2 / 3) following the never ended discussion in the PLC industry (4).

2.1 Hardware Platform

The main controller is based in an industrial configuration of Intel-based PC. The hardware subsystems for the motion control and the machine vision are attached to the main controller by a PCI card, the scanner card for the I/O network is using a standard ISA solt.

The selection of an industrial configuration for the PC main controller is due the possible vibration suffered by the robot cell (in our first approach the cartesian robot is built in modular aluminum profiles). The possible vibration may affect the hard disk life of the main controller.

The motion subsystem is a DSP-based card, the selection of this type of controller is based in a conservative approach for offloading the main controller CPU of mathematical calculations, for more details see (5)

The machine vision card capture the necessary data for locating the parts from pallet and find out the correct coordinates assisting the robot in pick and place movements, in our approach the machine vision system is not used as a servo-visual component.

2.2 Operating System

The main characteristics for operating systems OS of a robotic or intelligent machine are described in a critical technology roadmap by the US Department of Energy (6). The OS should: provide all the usual capabilities provided by an office OS; nearly "invisible" to users; support "plug and play"; assure the required reliability and stability.

Following these main guidelines is easy to understand the selection of a Microsoft Windows OS for certain applications. The selection approach of a Windows OS may differ within developers and any of the three generation may be choused: Windows OS plus a second real time OS, Windows OS plus a real time subsystem, or unaltered Windows OS. In our particular approach the selection is a second generation using a Windows NT OS powered by a real time subsystem called Hyperkernel.

The operating system allows the rapid integration of automatic assembly cell capabilities due their support for "plug and play" operations.

2.3 High Level Mechanical Language -HLML

The first programming languages were based in the G-codes and S-codes, using the earlier experiences from CNC programming. In the earlier eighties were born KAREL, RAIL, VAL, and ARLA. The advances in robot programming languages followed derivations from computer languages finding the main drawback no in the engineering staff but in the people on the factory floor.

The contemporary practice in language development is working toward very high-level representation of functionality in order to reduce the programming effort (7). Visual programming is a relative new programming technique having increasingly gained in importance (8). Two actual main techniques should be involved in the development of a robot programming language:

Rule Based Operation, defining all of the logic for a machine or process in the form of rules, often expressed in a natural language or English-like syntax.

Object-oriented programming of automation functions, reducing the complexity through encapsulation and achieving a direct parallel between the structure of the program and that of the machine or process (7)

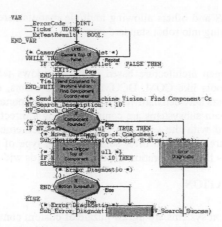

Figure 1. The HLML is an approach to visual programming. The ST created automatically in the background allowing the use of a standard IEC 61131-3 programming language.

The *High Level Mechanical Language* -HLML is a combination of visual programming and sensor integration techniques. It is based in visual flowchart control programming language developed by Nematron Corp. and widely used in large scale PC-based control projects at General Motors Powertrain Group. The HLML allows the user to program visually a robot application using two main visual components: process blocks and decisions elements. The process blocks are traditional logic actions like *Turn On*, *Start Counter*, etc. plus a new set of assembly actions like *Grasp Part*, *Peg in Hole*, etc. Since the traditional logic actions are standard process blocks included in the original flowchart language, the ones specific of assembly operations are custom developed and based in an auxiliary set of subflowcharts adding robot motion capacity and material handling. The decision elements include traditional *If-Then-Else*, *Do-Until*, etc. and have been implemented specific application decisions like *If-Component Founded*, *If Motion Done*, etc. allowing an easy integration of machine vision and motion control subsystems.

The visual programming HLML allows a friendly programming technique, it is an intuitive method, is self-documenting, modular and extensible. However, engineering staff is familiar with textual programming language. The original control editor incorporates an automatic translation to standard IEC-61131-3 Structured Text ST language. The result ST program may be used for software engineering proposes as an intermediate language. (Figure 1)

2.4 I/O Networks

The communication issues contains the communications between: main controller and sensors and actuators, and the main controller with other controllers in the same line or the line controller. The second one is explained in the fourth section of this paper and has a vital importance in the development of *plug & assemble* automation modules.

The communication between the main controller and the sensors and actuators is accomplished through the use of standard serial communication buses like DeviceNet,

Profibus DP, Interbus-S and others allowing multiple manufacturers to build I/O sensors and actuators that can plug into robot standard interfaces. (9)

2.5 Connectivity Tools

A system built in a open architecture based in PC Windows technology allows to use friendly connectivity tools like COM, DCOM, OPC and others, eliminating the need of complicated direct drivers between hardware and software components.

In our approach, the main subsystems are connected using a direct driver developed using the DDK tool provided with the real time extension Hyperkernel. Nevertheless, motion control cards makers are starting to offer solution of the type of OPC servers with their products, this will allow the incorporation of new control cards with less effort.

3. STANDARDIZATION

Conventional automation systems are jigsaw puzzle of custom components in conflict with current manufacturing needs. A key target for the authors is the equipment integration of multi-vendor assembly systems for electronic manufacturing. The research is accomplished within the initiative entitle *Specification of Open Control Systems for Light Assembly Applications -SpecOpen*, funded by the Finnish National Agency of Technology -TEKES and Finnish industry. (10)

Our approach to open architecture control for robotics is based in standard hardware components, communication systems and connectivity tools, this allow a complete standard controller where third party components can be added. On the other hand integration between machines from different suppliers must be accomplished through the use of a common standard interface. The SpecOpen initiative is working in the field of equipment integration based in the use of GEM/SECS or OPC-XML based interfaces. These guidelines will be added to our approach in the final integration of the assembly robotic cell and the other components of the production line.

4. PLUG AND ASSEMBLE

The United States National Research Council recently published a book on Visionary Manufacturing Challenges for 2020 (11). In it they outline six grand challenges. The grand challenge number 5 is to reconfigure manufacturing enterprises rapidly in response to changing needs and opportunities.

It is agreed that the system should be easily expanded and reconfigured as the company's needs and capabilities change in respond to market conditions.

The proposed working frame allows an easy reconfiguration of the robotic tasks due the particular open architecture used and the visual programming technique. Once, the guidelines according to the SpecOpen initiative will be implemented the reconfiguration of the system will allow the integration of third party modules in the same production line as well.

The interconnection issue also contains the communication necessary between machines. The connection of devices from different manufactures can consume up to 30 - 40% of the engineering necessary to build an integrated system. (3). The implementation of an open architecture will allow the integration of protocols between cell controllers like GEM/SECS and others allowing the desired *plug and assemble*

5. RELIABILITY AND STABILITY

The traditional and current role of the PC using Windows-based environment in the robotics industry has been and is relegated to interfacing tasks between the designer, developer or operator and the robotic system. The reasons giving by the major robot makers have been summarized by Rutledge in (3). In terms of reliability and stability the major concern is the real-time system that controls the operation of a high performance robot. The discussion usually concludes in the requirements of an operating system that can deterministically respond to input from sensors and maintain control and performance at the micro- or nano-second level. Normally, the last affirmation is that Microsoft Windows operating systems are not able to provide this feature following the experiences from the office automation.

Usually, the discussion on PC-based control systems lacks in veracity due two main assumptions. The first is the generalization of hard real time requirements for the totally of the automation applications without analyzing the process. Secondly, the people tend to assume the same PC-Windows-based environment that in the office without optimize the system for the new role, reducing the discussion to a set of very well known experiences suffered in the office world with PC Windows-based systems like the *Blue Screen of Death*. The result is a discussion on real time capacities of the operating systems without having in mind the other components of the final system.

The analysis should be made in terms of system performance and no in terms of the performance of single components, understanding components the ones presented in the second section of this paper.

Tuokko in (12) shows a collection of experiences using different commercial control application software combined with a set of standard communication systems. Lastra shows in (9) an approach for evaluating the system performance of control systems based in open architectures and test a collection of commercial solution based in PC Windows solutions using the developed metric and compare the results with two traditional PLC-based solutions. The results in both cases show differences between commercial solutions and in some cases the system performance reached by the PC Windows-based systems beat the one reached by the traditional proprietary PLC-based systems

The authors, actually are working in a new procedure for evaluating robotic operations based in the analysis of the cycle time and the variations of the cycle time for a giving sample time.

6. CONCLUSION AND SUMMARY

Due to the increasing competition in the manufacturing/assembly industry, assembly systems require rapid and flexible response to changing needs and opportunities . In order to solve this problem we have proposed an approach to open architecture control for controlling automatic re-configurable assembly cells. Furthermore we have extended the role of the PC Windows-based environment using it as programming, debugging, controlling and maintaining the robotic assembly cell enabling the use of an unique working frame and database.

The generic open architecture created offers an integrated and flexible platform for the development and/or modification of new/or existing automatic assembly cells.

7. REFERENCES

1. Fred Proctor. (ed.) Proceedings of the RIA / NIST Workshop on *Open Architecture Control for the Robotics Industry*. Orlando, Florida. 2000

2. Fred Proctor. (ed.) Proceedings of the RIA / NIST Workshop on *Open Architecture Control for Robotics - First wave objectives*. Ypsilanti, Michigan. 2000

3. Gary J. Rutledge. *The PC and its Influence on Robot Controllers*. Proceedings of the 2000 IEEE International Conference on Robotics & Automation. San Francisco, CA April 2000 pp. 717-721.

4. Reijo Tuokko. (ed.) Open Control Systems: State of the art and trends in Soft PLC and related technolgies. Proceedings of the International Symposium on Open Control Systems. Tampere, Finland. 1999

5. Jose L. Martinez Lastra and Reijo Tuokko, *Integration of PC-based Motion and Logic Control in High Performance Assembly Applications*, Scandinavian Symposium on Robotics. Oulu, Robotic Society, Finland, 1999, p. 82-87.

6. DOE, *Robotics and Intelligent Machines in the U. S. Department of Energy -A critical Technology Roadmap*. October 1998

7. Jose L. Martinez Lastra, PC-based Control Systems: the Controversy of the Programming Languages. In: *PC-based Automation Systems and Applications*, Report B, Helsinki University of Technology, 1999

8. Christoph Schroeder and Detlef Zuehlke. Integration of Sensor Technology in Visual Robot Programming Systems. Proceedings of the 30th International Symposium on Robotics. Tokyo. October 1999 pp. 297 - 304

9. Jose L. Martinez Lastra, *Evaluation of New Open Control Systems for Light Assembly Applications*, M.Sc. thesis, Tampere University of Technology. 2000, 116 p.

10. Reijo Tuokko, *Light Assembly Industry – LASSI, Technology Programme Final Report*, Helsinki, 2000, 107 p.

11. National Research Council, *Visionary Manufacturing Challenges for 2020*, National Academy Press, Washington DC. 1998

12. Reijo Tuokko, Ville Saarimäki, Niko Siltala, Petri Partanen and Jorma Vihinen, Experiences in Different Fieldbuses used together with PC-based Control Systems. In: *Fieldbus Technology. Systems Integration, Networking, and Engineering*, 1999 Spring Wien New York, p 164 – 169

Human Friendly Mechatronics (ICMA 2000)
E. Arai, T. Arai and M. Takano *(Editors)*

STRATEGIC MANAGEMENT OF ADAPTIVE, DISTRIBUTED, DEVELOPMENT OF A MECHATRONIC PRODUCT

Vesa Salminen [a] and Balan Pillai [b]

[a] *Massachusetts Institute of Technology, MIT*
Sloan School of Management, CIPD
Room E60-236, 77 Massachusetts Avenue
Cambridge, MA 02139, USA

[b] *Real Time Systems Inc.*
10627 Jones Street, Suite 301A
Fairfax, VA 22030, USA

Keywords:
Human cooperative, distributed product development, mechatronics, information management, knowledge management, nonlinear dynamics, fuzzy neural networks, fuzzy systems, embedded, adaptive control technology and sensing intelligence.

ABSTRACT

The manufacturing industries are forced to sustain with increased demand of competition and customized services that are adding value to the products. Embedded systems designers have adapted the web browser as a standard user interface technology for network-attached devices. This ubiquity has made possible a revolution in module management, improving the time to market and cost of sophisticated network interfaces for devices. In increment, they are innovative to be fast and preferred than the earlier ones. A unified framework for discrete-time efficiency of continuous nonlinear systems, based on neural network model, has not been built for the purpose of product development (PD) process. Therefore, it is necessary to build a tool system for adaptive integration in product development. A complex systems concept envelops unity and integration. This organizes otherwise the mystifying properties of diverse complex systems through the use of dynamic simulations.

Obviously, the product development with tools and skills is a complex system, which needs adaptive interpretations. This is, due to nonlinear activities. Therefore, the complex system is further re-engineered as modules. It is rather modeled to fit the man-machine interfaces. This is done through a black box approach. This black box is able to handle the dynamic feedbacks from the complex system trajectory. This, in turn, enhances the agent-network and identifies the goal for achievement. In this article we introduce a concept of adaptive, distributed, product development of mechatronic product.

1. INTRODUCTION

Increased competition and globalization are forcing manufacturing companies to develop new products better and faster. That is phenomenon that led companies to outsource more and more of its business, even Product Development (PD). This leads to a complex and distributed development process. The interactions require sharing and exchanging of not only information but also knowledge and methods with each other. The collaborative PD management is a complex task. To do better the complex system framework, then it has to be integrated with irrational and rational radicals. Common problems in design require that an engineer devise a control or decision algorithm that converts measurements of system and environment variables into signals that aid in system regulation. Design is made difficult by disturbances internal to the system and by noise at its output. The knowledge has increased rapidly. Systems interaction has become complex and subsystems share and exchange information with each other. The product architecture and modular product concept have become increasingly important to manage whole the business. To certain extent, it has become an impossible mission for an individual to manage the complexity in product development process.

The traditional value-chain based business is transformed to value networks. The trend seems to be towards dynamic markets in which the internal and external service providers will execute the development activities. The management will react according to value recognition and networking. It is obvious that Internet creates a new way of globally networking people to self-organized, complex, adaptive systems. The question is about what types of new-networked structures would emerge. Autonomous software agents will exploit in the future the information on the network. The reality is that product development of a complex system, with nonlinear activities, needs adaptive interpretations.

2. COMPLEX SYSTEMS AND NONLINEAR PARAMETERS

In our study we have used the theory of adaptive control. Any system comprises the varieties of interdependent and compound netting or parenting suffix, which we call as complex system or systems. A complex system, as we defined, has immense activities surrounded, which are not linear member of functions. It performs as linear state functions. This linearity shifts, when process non-stability appears, due to gray-areas in the netting. This gray-area evokes the process wandering. This phenomenon we call it as nonlinear dynamics. This nonlinear syndrome is seen in human organizations, man-machine-interfaces, partnership setups and breathings. These systems are physical cell bodies that acquire, store and utilize knowledge. The connection weight is adjusted to achieve a desired output. The process is embedded in such a manner it computes the derivative of the energy function with respect to the connection weight as an increment of the connection weight. This way, the derivative determines the rule of changing connection weights in order to minimize the descent of the participating energy function along the gradient. Therefore, we weigh here the feedback-error-learning and not linearizing the complex system of product development.

3. PRODUCT DEVELOPMENT PARADIGM

Contemporary aircraft are equipped with an abundance of automated controls and can fly great distances with little or no manual interventions. However, few of us would be willing to climb aboard an aircraft that didn't have a human pilot, because we understand that manual intervention is required to adjust the flight to changing conditions. In the real world of product development, the ability to manage a device or organization in the face of changing condition is often one of the most overlooked and underestimated aspects in business architecture. Many communication systems are deployed with essentially only autopilot capabilities and experience degradation of performance as the environment changes. An adequate management interface is therefore necessary to allow a device to be adjusted in response to changing conditions. One popular mechanism to manage complex system is to use hierarch to zoom in from a larger unit to component parts or zoom out to hide details. In a well-formed hierarchy, the larger unit is transparent: it simply groups, state machines together with no additional semantics. In a nontransparent hierarchy as employed, the larger units – the super-state have meaning, incorporating interactions with initial state, final state and deep and shallow histories. Therefore, the meaning of a state model is dependent on its context, which is akin to saying that a change in the outline changes the meaning of a product. The best practice [Pillai, 1999] shown that a complex product development scheme can model. This is done in decomposing like the model of an automobile into various subsystems including, e.g., a cruise control. The options are either re-engineering of an available system or starting new system development from scratch. The act of re-engineering– or better yet starting from the bottom and working up – almost always exposes problems that weren't visible in hierarchical state machines. This is a product development dilemma.

4. ADAPTIVE PRODUCT DEVELOPMENT PROCESS

When enterprises step by step move towards virtual enterprises also the product development process changes as adaptive process. Figure 1 describes the concept of Adaptive Product Development. There exist shorter and longer feedback loops during the executing of product development process. Shorter loops exist when we are simulating continuously over performance and process/knowledge models. The longer loop exists when we are simulating over available competences, resources, product models and available services.

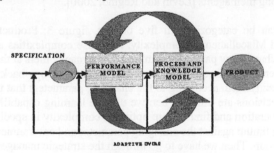

Figure 1. *The concept of Adaptive Product Development*

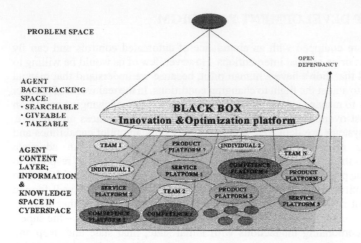

PROBLEM SPACE

OPEN
DEPENDANCY

AGENT
BACKTRACKING
SPACE:
· SEARCHABLE
· GIVEABLE
· TAKEABLE

BLACK BOX
· **Innovation &Optimization platform**

AGENT
CONTENT
LAYER;
INFORMATION
&
KNOWLEDGE
SPACE IN
CYBERSPACE

Figure 2. *Black Box and plane of possibilities used in Innovation and Optimization Processes*

We try to find gray areas in the plane of possibilities (figure 2) and in a black box dynamically simulate over all variables in the plane of possibilities in the distributed networked environment. The first basic law in product development is that there exist always deficit of resources with adequate knowledge. The knowledge and resources in networked environment has not been well benefited because of identifying, access and reliability problems. In the new generation of networked product development networked organizations use a common architecture to deliver independent pieces of value as priced services into adaptive product development process Product development is a continuous process of decision-making. First of all the main problem should be encapsulated and solved as an individual entity. That means a good knowledge of all the interrelations around the problem and possible solution. Network of all interrelating parameters is still too tight to understand. Using of nonlinear dynamics of interacting mechanisms helps us to find out gray areas and solve the problems in the space of that. Complex, adaptive system is fluctuating among three states: Static, Chaotic and Adaptive - depending on the prevailing environment and the nature of interactions among their agents [Levin and Regine, 2000].

The complexity can be categorized in five classes, figure 3: Product, Service, Process, Organizational and Miscellaneous Complexity. All these complexities are overlapped with each other: If you change any parameter it influences to many others at the same time. When optimizing the overall complexity we make short feedback loops back to the overlapping complexities, defuzzify them according to the changing parameters that traces the gray areas from which the decisions are drawn. Then we use the learning capability of the system by neural network generation and find a more optimum complexity in specific time scale. If we want to have the optimum against the changing parameter and environment, we have to make a longer feedback loop. Then we have to go through the strategic management cycle.

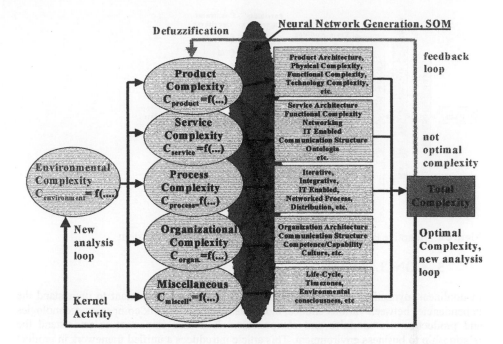

Figure 3. Categorization of complexities and feedback loop to reduce overall complexity.

Figure 4 describes the decision support system, which is used in adaptive product development process [Salminen and Pillai, 2000]. Applied engineering can be expressed in the form of possibility algorithms whereas solution engineering is expressed in the form of constraint algorithms. Both these algorithms are dynamic and form a modular rule base to be used in decision-making. The basic idea is to employ a nontransparent hierarchy in which the larger units – the super-state have meaning, incorporating interactions with initial state, final state and deep and shallow histories. A complex system is further re-engineered as modules. It is rather modeled to fit the man-machine interfaces. This is done through a black box. This black box is able to handle the dynamic feedbacks from the complex system trajectory. This, in turn, enhances the agent-network and identifies the goal for achievement. By using this type of state model means that a change in the outline changes the meaning of a product. A complex product development scheme can thus model in decomposing way. The management of existing nonlinearities happens in phased way. There are continuous condition monitoring activities and correcting operations influencing on the model. The feedback loop can influence on various things, also on the algorithm base. This type of open, adaptive, product development system is cost and time effective. By using dynamic product, networking and competence model we do not have gray areas any more to tackle with. Product grows up to an innovative, dynamic model, which seamlessly adapts on changes in our environment.

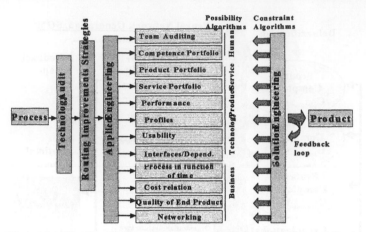

Figure 4. Adaptive Product Development and Decision Support System

5. CONCLUSIONS

In nonlinear, dynamic, systems as product development it is important to understand the dependencies between various entities; individuals, teams, network companies, technologies and product/service modules and components. It is also important to understand the relationship to business environment. This article introduces a unified framework in product development for discrete-time efficiency of continuous nonlinear systems, based on neural network model. A complex system concept envelops unity and integration. This organizes the otherwise mystified properties of diverse complex systems through the use of dynamic simulations. In virtual enterprise the product development process changes as dynamic, adaptive, process. The structure is available but invisible for the users of the service environment of adaptive product development. In this article has been introduced a new approach for the concept of adaptive product development. It is based on finding of gray areas in the plane of possibilities and in a black box dynamically simulate over all variables in the distributed networked environment.

REFERENCES

Levin, R., Regine, B., 2000, *The Soul at Work- Listen...Respond...Let Go-Embracing Complexity Science for Business Success*. Simon& Schuster, New York, USA, 2000.

Pillai, B., 1999, *Adaptation of Neural Network and Fuzzy Logic for the Wet-end-Control and Process Management of Paper or Board Machines – A Tool-making Approach, published in Acta Polytechnica Scandinavia, Mechanical Engineering Series, Me 138, Espoo 1999, Finland.*

Salminen V., Pillai Balan, *Non-linear Dynamics of Interacting Mechanisms in Distributed Product Development. International Conference on Complex Systems (ICCS), May 21-26, 2000, Nashua, NH, USA.*

Human Friendly Mechatronics (ICMA 2000)
E. Arai, T. Arai and M. Takano *(Editors)*

Development of Adaptive Production System to Market Uncertainty
- Autonomous Mobile Robot System -

Mineo HANAI[a], Hitoshi HIBI[b], Tatsumi NAKASAI[c], Kazuaki KAWAMURA[d] and Hiroyuki TERADA[c]

[a] Production Promotion Center, DENSO CORP. Kariya-shi, Aichi 448-8661, JAPAN

[b] Industrial Systems Eng. Dept. 3, DENSO CORP. Anjo-shi, Aichi 446-8507, JAPAN

[c] Production Eng. Dept., DENSO CORP. Kariya-shi, Aichi 448-8661, JAPAN

[d] Engine Electrical Systems Mfg. Dept. 1, DENSO CORP. Anjo-shi, Aichi 446-8511, JAPAN

It has been difficult with an automated production system to achieve compatibility between productivity and adaptability in the area where production volume and period are uncertain. Recently we have developed a new automated production system, APS(Adaptive Production System), by studying how a manual-operation system responds to uncertainties, and applied it to an automotive starter assembly line. The number of machines in the system can be easily increased or decreased according to the production volume fluctuation. The highlighted topic is integration of human-social functions such as autonomy, cooperation, and succession into a conventional automated line. As a result, we have realized an epoch-making automated production system which can change its production capacity between 4,700 and 61,000 units/month through a rapid transformation of system structure, while the automation ratio of this system is equivalent to that of the conventional transfer line whose production volume and period are certain.

1. Introduction

The manufacturing industry is exposed to severe global competitions and its business environment is changing rapidly. It is very difficult to forecast a market demand which was rather stable and predictable in the past. The industry is now required to adapt itself to this environmental change. Therefore, it is necessary to develop a new production system which can handle fluctuations quickly and flexibly in production volume and product life.

In recent examples of new production systems, we can see practical applications of human-based production systems, so-called cell production method in which attention is

384

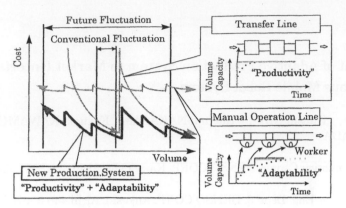

Fig.1 Direction of New Production System Development

given to human flexibility.[1] This production system, however, still cannot economically compete with automated production systems in the case that production volume and period do not fluctuate so much by chance.

In this paper, a newly conceived automated production system is introduced. The system, which can handle production fluctuation quickly and flexibly with economical efficiency, has been developed and applied to an automotive starter assembly line.

We will refer to a requirement for the new production system in chapter 2, the developed system in chapter 3, examples of developed technology for the system in chapter 4, results of this development in chapter 5, and conclusions in chapter 6 respectively.

2. Requirement for new production system

Let us pay attention to the adaptability of conventional production systems in terms of economical efficiency and flexibility to the production volume fluctuation first, in order to clarify a requirement for the new production system. Fig. 1 indicates the advantage and disadvantage of a conventional manual operation line and automated transfer line in terms of efficiency and flexibility, and shows the direction of the new production system development.

The transfer line can realize high productivity through optimizing the production system structure in the specified range of production volume. The cost, however, being very sensitive to the production volume fluctuation, will vary largely in the future volume fluctuation range.

On the other hand, the manual operation line has good adaptability to the demand fluctuation since production volume can be easily changed by increasing or decreasing the number of operators. Although its cost doesn't change largely against the volume fluctuation, it will not compete with the transfer line in terms of economical efficiency in the specified range of production volume.

From the above, we conclude that the requirement for the new production system is integration of high productivity, which is the advantage of the transfer line and adaptability to production volume fluctuation which is the advantage of the manual operation line.

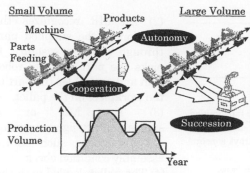

Fig.2 Basic Concept of APS

Fig.3 APS for Starter Assembly Line

3. Developed system(APS)

3.1 Concept

Fig. 2 shows a basic concept of the Adaptive Production System (APS), which fulfills requirements for high productivity and adaptability. This concept was developed based on an analysis of the manual operation line.

APS is an automated production system which keeps the system structure in optimum condition by adding or subtracting machines/system components according to the production volume fluctuation. In the development of APS, we pursued the integration of human-social functions such as autonomy, cooperation, and succession into the conventional automated line. The features of APS are as follows;

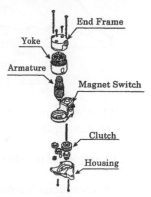

Fig.4 Automotive Starter

(1) Autonomy Function : Each robot can autonomously decide its next action, move to the corresponding process, and perform various kinds of operations.

(2) Cooperation Function : The robots can cooperate with each other and achieve efficient operations, i.e., the robots collaborate mutually, and each robot's job load can be normalized and its efficiency is maximized.

(3) Succession Function : A robot can transplant its operational expertise and know-how to other robots quickly when the production system structure needs to be modified.

3.2 Practical application to starter assembly line

We applied this concept to the automotive starter assembly line as Fig. 3. Fig. 4 shows components of the starter. Those 17 components, such as the clutch, magnet switch, armature, and housing, are assembled in the line.

This APS consists of a conveyor, Mobile Robot(MR)s, general parts-feeding units, and dedicated assembly units. The system is run by operators.

Following the basics of the APS;

(1) A pallet, loaded with a work on it, is transferred with the conveyor and positioned at an assembly process/station. Assembly parts are supplied toward the pallet with the general parts-feeding unit.

386

Table 1
Specification of Mobile Robot

Assembly	Number of Controlled Arm Axis	6
	Arm Length	900mm
	Arm Lifting Load	5kg
Traveling	Movable Load	80kg
	Velocity	40m/min
	Guidance	Magnetic tape guided
Recognition	Parts	Vision system
	Operator	Infrared rays sensors

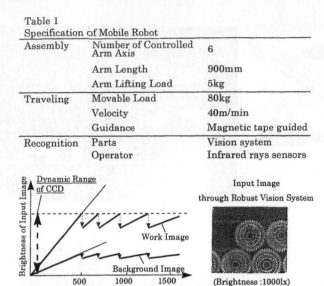

Fig.5　Robust Vision System

(2) An MR recognizes the position of the supplied part through a vision system. It picks the part with its robot arm and assembles the part to the workpiece on the pallet.

(3) The MR or a supplementary jig/unit carries out inspections such as an assembly condition check.

(4) The pallet, loaded with the workpiece on it, is transferred to the following process with the conveyor.

(5) The MR decides its next action such as assembly operation at the same process or moving to the preceding/following process, and takes the necessary action.

When machine trouble occurs, an MR informs the operator of the abnormality. Then the operator quickly fixes the trouble in cooperation with the MR.

4. Examples of developed technology

4.1 Autonomous function of Mobile Robot

The core structural component of the APS is an MR which possesses assembly, traveling, and recognition functions. The specifications of MR is in Table 1.

The MR has an arm with 6 axes, 900mm length and 5kg lifting load so that it carries out various assembly operations. It also has a 80kg movable load capacity and travels up to 40m/min guided by magnetic guide tape. It freely travels among processes/stations carrying jigs necessary for assembly operations. In addition, it has recognition functions; a vision system accurately recognizes the position of the assembly parts, and infrared-ray sensors system detects approaching humans for the MR's automatic stop. Owing to these functions, the MR is capable of carrying out various operations autonomously.

Let us refer to the detailed features of the vision system. Because it is a prerequisite for MR to travel to any place, the system shall perform well even in open surroundings inside the plant. Therefore a robust vision system was developed. The system measures the ambient brightness with an illuminometer and controls its camera aperture to get the suitable brightness for the sensor. As a result, the dynamic range of the sensing system can be expanded as shown in Fig. 5.

This development enables the MR to perform various operations with visual recognition regardless of the ambient brightness change caused by place or time in open surroundings.

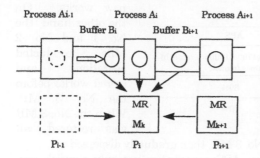

Process A$_{i-1}$ Process A$_i$ Process A$_{i+1}$
Buffer B$_i$ Buffer B$_{i+1}$

Fig.6 Autonomous Cooperation Model between MRs

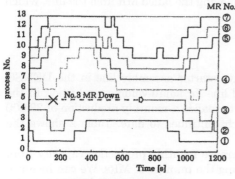

Fig.7 Behaviors of MRs

4.2 Cooperation function between Mobile Robots

Each MR has a cooperation function and complements the other MRs to carry out operations efficiently. The MR realizes this cooperation by autonomously deciding its action based on very simple criteria as follows.

Fig. 6 indicates the basic idea of cooperation between the MRs. This idea is based on a cooperation behavior in the manual operation line where operators help each other. The fundamental sets of information required for cooperation are presence of work in the current process/station, in the preceding buffer, in the following buffer, and presence of the other MRs in the preceding and the following processes.

The model in Fig. 6 can be expressed with the following equation:

$$Mk(Pi)=f(Ai, Bi, Bi+1, Pi-1, Pi+1)$$

where,

Mk stands for k-th MR destination pattern ("to stay", "to move to the preceding process", "to move to the following process"),

Pi for presence of MR in the i-th process,

Ai for presence of work for the i-th process, and

Bi for presence of work in the buffer between $(i-1)$-th and I-th processes.

f is a function or algorithm which determines the destination of the MR.

According to the combination pattern of 0, 1 of Ai, $\cdots$, Pi+1 based on the system status, the MR will make a selection among three actions(-1, 0, 1) of "to stay and operate", "to move to the preceding process" or "to move to the following process" respectively.

We have selected an algorithm, which maximizes system efficiency, among several algorithms with comparatively simple control rules by evaluating the MR's dynamic behaviors through computer simulation.

We will explain the effectiveness of this control method with a simulation results in which one MR fails and the others perform cooperative operations.

Fig. 7 demonstrates the behaviors of seven MRs in case that the No. 3 MR stops for a certain period of time at the 5th process. Note that works flow from process No. 1 to No.13.

When the No. 3 MR stopped, the No. 4 through No. 7 MRs moved downwards all together in accordance with the flow of existing works as there was no more downward flow of new work after process No.5. Then, when no work existed after process No.5, they moved upwards to process No.6 and waited for the restoration of the No. 3 MR or

388

Table 2
Development Results

	Conventional TR	APS
Production Capacity	Fixed	Adaptive(4700-61000pcs./month) every4700pcs./month
Automatic Ratio	80%	80%
Machine ConversionRatio	10%	70%

new works. On the other hand, the No. 1 and No. 2 MRs conducted operations and stored works before the No. 3 MR. When the No. 3 MR was restored, all the MRs gathered and operated around the No. 3 MR, then gradually disporsed.

Through this strategic operational system, APS can realize not only a quick and flexible response in increasing or decreasing the number of the MRs, but also system robustness through continuous operation by removing the failed MR from the line, which was unseen in conventional automated lines.

5. Results

This APS was first applied to the automotive starter assembly line in the DENSO Anjo Plant and started operation as an actual line for small-volume production in the spring of 1997. In the spring of 1998, it grew to realize a monthly production of 33,000 units by increasing the number of MRs, which continues to date. Table 2 shows the development results of this system.

This system made it possible to increase/decrease monthly production capacity from 4,700 up to 61,000 units by increasing/decreasing the number of MRs. We can increase the capacity by 4,700 units with one MR. The automation ratio is 80%, which is the same level as conventional transfer lines. Meanwhile, the machine conversion ratio can drastically be improved from 10% or less for the conventional line to 70% due to the usage of general-purpose equipment like the MRs.

6. Conclusion

Facing the 21st century where demand forecasts for products will become increasingly difficult, we have worked out a concept of the new production system, the Adaptive Production System(APS). This system realizes compatibility between adaptability and productivity in the area where the production volume and period are uncertain and automation was formerly deemed difficult. And we proved it with the automotive starter assembly line. We will continue our positive activities towards further APS application through the active utilization of the MR and further technology development.

Reference

1)T.Shinohara :Sensation of Conveyer-less production, Cell Production of one-man closed, NIKKEI MECHANICAL 1995.7.24 no.459 20-38 (in Japanese)
2)M.Hanai :Development of Adaptive Production System(APS) to Market Uncertainty, Int. J. Japan Soc. Prec. Eng., Vol. 33, No.1(Mar. 1999)

Human Friendly Mechatronics (ICMA 2000)
E. Arai, T. Arai and M. Takano *(Editors)*
© 2001 Elsevier Science B.V. All rights reserved.

In-process visualization of machining state with sensor-based simulation to support the recognition ability of operators

Koji Teramoto[a] and Masahiko Onosato

[a]Department of Computer Controlled Mechanical Systems, Graduate School of Engineering, Osaka University,
2-1 Yamadaoka, Suita, Osaka, 565-0871, Japan

This paper deals with an in-process visualization method of machining state to support the state recognition of human operators in machining. In order to realize the recognition support, we propose an accurate state estimation technique which associates sensed data with the machining simulation. A framework of Sensor-based Simulation (SBS) is presented. Furthermore, a classification and procedures of SBS are also discussed. The effectiveness of the proposed method is demonstrated by a prototype system.

1. INTRODUCTION

The agile and high quality parts fabrication becomes more important in the competitive circumstance of manufacturing. This trend requires machining systems to be more flexible, precise and productive. The requirement enhances the role of skilled technicians which enable the proficient machining[1][2][3].

In the other hand, it is usually said that longterm practice and a huge amount of trial are necessary to become the skilled technicians in the conventional manner. Furthermore, it becomes more difficult to master the skill by following two reasons. One comes from a trend of quick change of the products and the technological environment. This trend makes the machining operation more complicated and changeable. Then, the required levels of operational skill become higher than ones used to be. The other reason comes from the separation of the machining process and the operators. This is the result of the consciousness to secure the safety and comfortableness of the operators. However, the separation prevents to master the operational skill because it becomes hard to observe the actual machining processes.

Then, a support system which enhances the operators' skill in task execution is eagerly required for the operators. The objective of this research is to develop a method which supports the operators by amplifying the human abilities based on the simulation, sensing data acquisition, information technology and so on.

2. RECOGNITION SUPPORT BY IN-PROCESS VISUALIZATION

Skilled technicians can recognize the machining state from their own perception such as listening the machining sounds, watching the color and shape of swarf, and so on. From these sensing actions, they can estimate the various information like tool wear, cutting temperature, workpiece deformation, machining load, and so on. If operators can not recognize the machining state, they can not adjust the process correctly. So, it can be said the recognition ability acts the fundamental role of other abilities such as realtime adaptation and storing experience. Then, supporting the recognition ability of operators is expected to increase the proficiency of the operator-machine collaborative system.

We introduce the in-process visualization of the machining state to support the recognition ability. It is well known that visual information acts the effective roles for the human cognition. Furthermore, augmented reality (AR) technology becomes widely applicable to the various diagnostic applications[4] and the technology provides the effective presentation method of visual information.

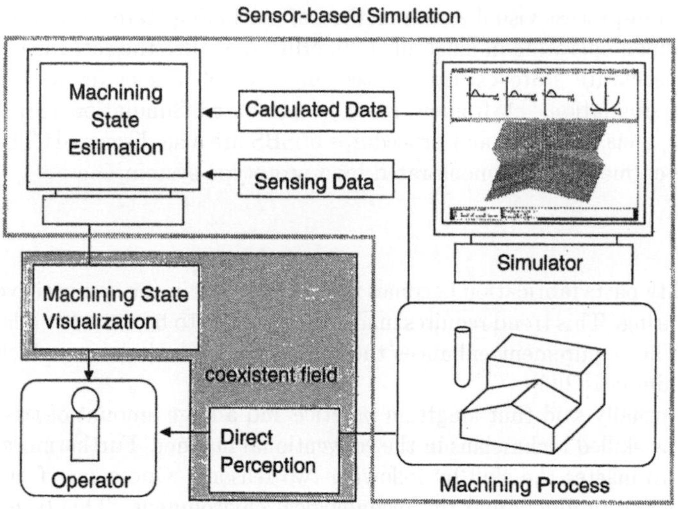

Figure 1. In-process visualization of machining states

Figure 1 illustrates a schematic diagram of the in-process visualization of the machining states. The method proposed in this paper is expected to have following three advantages.

- By visualizing the various information like workpiece temperature/strain, cutting force and so on, operators can recognize the machining process from various aspects.
- The coexistence of the real phenomenon and visualized information helps operators to organize their own sensation and indicated information. The organization is expected to reinforce their experience.
- When an in-process support method is established, it becomes possible to realize the task-wide support for operators by combining the pre-process[5] and post-process support technologies[6].

In order to achieve the advantages above-mentioned, it is very important to estimate the accurate machining states under the dispersion of phenomena. The accuracy of estimation is the key of appropriate support execution. In this research, a Sensor-based

Simulation(SBS) is proposed as an estimation method which associates sensing data with the machining simulation to improve the accuracy.

3. BASIC FRAMEWORK OF SBS

Two approaches to estimate the machining state have been researched for the different objectives. One is a simulation oriented approach (SOA) which aims to predict the machining state in advance. The other is a measurement oriented approach (MOA) which aims to detect the fault or to supervise the machining process. These two approaches are independently developed. The SOA can estimate the global machining state, but it is hard to adapt the situational variation. In the other hand, the MOA can sense the situational variation, but it is hard to evaluate the global state. So, we propose a Sensor-based Simulation(SBS) method to realize the global and adaptive state estimation.

The basic concept of SBS is adjusting the global estimation by locally sensed data. The adjusting techniques depend on the type of variation.

Since the simulation is considered as the repetition of the state transitions. The possible factors which make the simulation inaccurate are initial state dispersion, process model mismatch and unpredictable phenomena including the accumulation of the repetitious errors. In order to compensate these factors, there are three possible adaptation methods respectively; (1)preliminary state calibration, (2)realtime adaptation by enhancing the adaptive simulation technique[6][7], (3)state adjustment based on the data comparison. These strategies are shown in Figure 2.

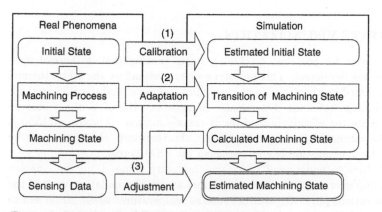

Figure 2. Framework of Sensor-based Simulation

Then, there are three types of SBS procedures. They are (1)condition adjustment, (2)model modification and (3)result transformation. Each of them has the trade-off relation between accuracy and time consumption. The model modification is the most flexible and time consume method. The calibration approach has the limit of adaptation because it only reflects the initial variation. Then, we investigate about the result transformation by constructing the transform function based on the measured data. The approach is based on following assumptions; (I)the result of simulation is qualitative correct, (II)the trend of difference between simulation and real phenomena does not vary

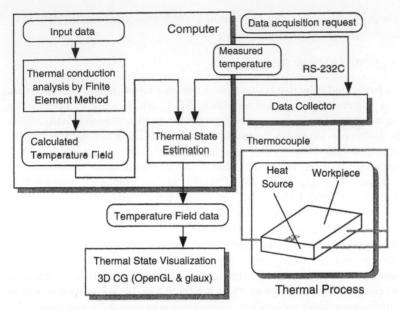

Figure 3. Configuration of Thermal State Visualization system

widely.

4. THERMAL STATE VISUALIZATION

As a prototype system for proposed machining states visualization, a thermal state visualization system has developed. This prototype system visualizes the static thermal state of a workpiece based on the 3D computer graphics technique and the SBS method. It is true real machining states are dynamically changeable. However, the prototype system aims to evaluate the effectiveness of result transformation. A static problem is easy to compare the estimated fields and real fields. Furthermore, the system can be easily enhanced to the dynamic problem by replacing the FEM kernel.

One of the difficulties to understand the machining state comes from the complex behavior of the machining errors. The machining errors are an accumulative result of the spatially distributed various physical quantities such as temperature, stress, strain and so on. Especially, thermal deformation of workpiece is a serious problem in the high-precision machining, because cutting with no coolant becomes more important as an environmental conscious machining method.

In the conventional machining research, there are many researches for predicting the thermal deformation of workpiece. Most of them are investigated based on FEM or FDM. In order to simulate the machining process as a thermal conduction process, model parameters should be determined in advance. In spite of some parameters, like a heat transfer coefficient and distribution of machining heat, are very sensitive to the machining environment, most of conventional researches have not deal with the variation of parameters. So, we adapt the SBS method to the thermal state estimation as the in-process and accurate state estimation method.

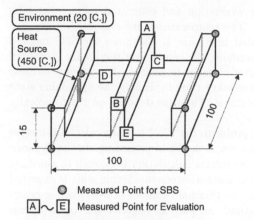

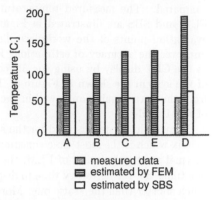

Measured Point for SBS

A~E Measured Point for Evaluation

Figure 4. Example problem

□ measured data
▤ estimated by FEM
☐ estimated by SBS

Figure 5. Temperatures at evaluation points [C.]

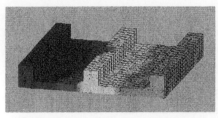

(a) Calculated Temperature Field by FEM   (b) Estimated Temperature Field by SBS

Figure 6. Visualized results of thermal state

Figure 3 illustrates the configuration of the prototype system developed. The prototype system consists of four major modules. They are a data collector, a simulator, a state estimator, and a visualizer. These modules are organized in order to implement the SBS method. The system acquires temperature of predetermined points by using thermocouples when a data acquisition request message is send to the data collector. A minimum interval of data collection is one second. At the same time, a thermal conduction analysis based on FEM is executed. A predetermined condition is used for FEM analysis. Based on the measured data and workpiece shape, a transformation function is constructed so as to reduce the difference between measured data and corresponding analysis results.

In this case, transformation function is generated as a weighted summation of the ratio between measured value and calculated value. Weights of the ratio are calculated every translation of data. The weight parameters are determined based on the distances between data position and sensors.

Figure 4 shows an example problem for an evaluation of the result transformation. The temperature field is estimated by SBS after a heat source is attached on the workpiece. The material of workpiece is aluminum alloy. A FEM model of the problem has 2079 nodes and 7800 tetrahedron elements. A result translation function is constructed based on the eight sensed data. A result transformation procedure is executed within 10 seconds by Sun Ultra 80 Workstation.

In order to evaluate the SBS method, estimated temperature by SBS and FEM are compared. The measured temperature for evaluation and estimated temperatures by FEM and SBS are illustrated in Figure 5. The temperatures are evaluated at the five evaluation points of the workpiece illustrated in Figure 4. It shows the SBS method improves the accuracy of estimation. Furthermore, the calculation results are visualized to the CRT display by using the 3D computer graphics technique. The graphic outputs of the system are shown in Figure 6. The necessary time to display the machining state from the data acquisition is within the 60 seconds. The time depends on the complexity of FEM models.

Because this system is used for the static problem, the FEM module calculates inverse matrix which is the most time consuming process of the FEM procedure. In the dynamic thermal state calculation of FEM, the inverse matrix calculation is enough to calculate one time. So, the necessary time to display the instantaneous machining state is expected much shorter than the static one. Moreover, the thermal behavior of workpiece is rather slow phenomenon. Then, the proposed method can play an effective role of in-process support of the operators.

5. CONCLUSION

In order to support the recognition ability of operators, an in-process visualization method of the machining state is proposed. The method is based on the Sensor-based Simulation, which associates sensed data with the machining simulation. The framework of SBS method is introduced for the accurate machining state estimation. The association procedure by result transformation is described. The effectiveness of result transformation is demonstrated by developing a prototype system of thermal state visualization.

REFERENCES

1. Kidd, P. T., On the Design of Skill Supporting Computer-Aided Technologies, IFAC Automated Systems Based on Human Skill, (1992) 35-41
2. Schulz, H. and Spath, D., Integration of Operator's Experience into NC Manufacturing, Ann. of the CIRP, 46/1, (1997) 415-418
3. Massberg, W., The Competitive Edge – The Role of Human in Production, Ann. of the CIRP, 46/2, (1997)653-662
4. Azuma, R. T., A Survey of Augmented Reality, Teleoperators and Virtual Environment, 6/4, (1997) 355-385
5. Teramoto, K., Onosato, M., and Iwata, K., Coordinative generation of machining and fixturing plans by a modularized problem solver, Ann. of the CIRP, 47/1, (1998) 437-440
6. Teramoto, K., Iwata, K., and Hirai, S., Development of Learning System with Process Model Selection for Control of Ball-end Milling, 7th Int. Conf. on Production/Precision Engineering, (1994) 461-466
7. Matsumura, T., Obikawa, T., Shirakashi, T. and Usui, E., An Adaptive Prediction of Machining Processes, Proc. of Int. Conf. Mach. Tech., (1993) 390-395

Human Friendly Mechatronics (ICMA 2000)
E. Arai, T. Arai and M. Takano *(Editors)*

Mediator-based modeling of factory workers and their motions in the framework of Info-Ergonomics

Hiroshi Arisawa[a]   Sayaka Imai[b]

[a]Department of Electrical and Computer Engineering, Faculty of Engineering
Yokohama National University, Japan

[b]Department of Computer Science, Faculty of Engineering
Gunma University

This paper presents modeling methodology of human workers motions in factory works for the purpose of precise evaluation of optimum movement. Multimedia Database plays an important role to describe, store and retrieve spatio-temporal information associated to human works.
Also, we consider it as a possible way for unifying all the data by a new concept "mediator". Using this we introduce a method for constructing a scene database based on an approximate polygon model that reflects the individual shape and meanings of an object.

1 INTRODUCTION

As a result of the significant advancement in computer technology many new applications of image and graphical data are brought to life. Those include the fields of Image processing and retrieval, Engineering Simulation, Tele-robotics, and so on. Recently some new applications supporting the designing process of a virtual factory and simulating its work are gaining popularity in areas like process planning and work scheduling. We think that human centered factors are usually overlooked or underestimated. Human labor has an important role in the manufacturing process. It is examined from the viewpoint of industrial engineering, ergonomics, etc. and results are used in product planning, working analysis, and work environment design. As a promising solution to the above problems we have proposed the use of Real World Database(RWDB) and Info–Ergonomics simulation [1]. RWDB is able to capture various types of data, namely video images, 3D graphic models, their motions and spatio-temporal events. The data of all types are unified and stored in a Multimedia Database(MMDB). We offer a model for analyzing employee's work evaluation and using the results in the manufacturing process design. Info–Ergonomics simulation in human–machine cooperation simulation would result in an integrated data model for manufacturing process applications. This paper presents an extended Info–Ergonomics system and model of factoryworkers which includes human body structure and work motions. For this purpose, we introduce "Mediator" to describe object's and motion's characteristics with small amount of data.

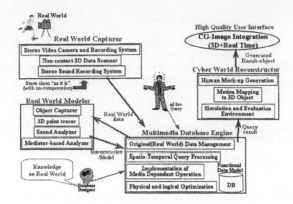

Figure 1: Conceptual architecture of RWDB

2 REAL WORLD DATABASE

The objective of the RWDB is to provide a total environment of capturing, modeling and storing physical or logical objects in the real world. Everything in the real world could be modelled through it and any type of data could be accumulated. For this purpose, RWDB must involve at least 4 components listed below. Conceptual architecture of RWDB is shown in Figure.1.

- Real World Capturer (RWC)

  The objective of RWC is to capture the external form of objects in the real world. There exists various types of 3D or spatial information depending on capturing devices. The simplest one is a set of video cameras. We can get a sequence of frames from, for example, a left-eye camera and a right one simultaneously. Another type of input device is "3D Scanner" by which we can get a perfect surface (polygon) model for static objects. The practical solution is to get above two kind of informations from the real world and to combine two models into one in the database level.

- Real World Modeler (RWM)

  RWM is a set of tools each of which analyzes original frame images and generate a new information. For example, the Outline Chaser[3] catches the outline of an object in a certain video frame, and then trace the outline in preceding and successive frames. Many algorithms are investigated and evaluated for range image generation in Image Processing area [4]. All the results of analysis are stored into database preserving the correspondences to the original images.

- Multimedia Database (MMDB)

  MMDB is a database which treats a variety of data types such as full texts, graphic drawing, bitmap images and image sequences. The features of such data are quite different from the conventional DBMS's ones, because some of them are continuous and might occupy much more space than traditional data types. As to data model, in order to integrate all types of data, introduction of simple primitives to describe real world entities and the associations between them are essential. Moreover, the query language of multimedia data handling must involve various types of media-dependent operations for retrieving and displaying. Especially, in RWDB, the result of a query creates a new "Cyberspace". A query may retrieve a certain unit work and project it to another human worker. The author proposed a total data model to describe the 2D or 3D data, and also presented query language MMQL for flexible retrieval [4].

- Cyber World Reconstructor (CWR)

  The result of database consists of various types of data such as frame sequence and 3D graphics data. In order to visualize the query result, RWDB should provide a "player" of result world. CWR is, in this sense, an integrated system of 3D computer graphics and 3D video presentation. Unfortunately, the modeling method of objects in the field of 3D graphics and VR systems are quite different from the DB approach because the former is focusing on natural and smooth presentation of surfaces of objects and their motions, whereas the latter makes deep considerations on semantic aspects os these objects.

3 INFO-ERGONOMICS MODELING

Based on the above RWDB Concept, we concentrate ourselves into modeling on human bodies and motions especially for factory workers. We offered Info-Ergonomics for storing Human Working and reconstructing it in CG Simulation.

3.1 Info-Ergonomics

Info-Ergonomics is one of the research priorities in the integrated computer-aided Manufacturing Process Design. Recently modeling the machines in the factory, creating virtual machines by using CG, simulating their work and evaluating it are coming into use gradually [5]. But modeling human beings and creating "virtual employee" by CG still has very limited use, because of the human body's high complexity and the limits imposed by the computer techniques. We focus on human body modeling along with the machines, and evaluating employee's working in the factory from human-machine co-operation and employee's point of view, so we propose Info-Ergonomics as the framework for CG simulation and evaluation of virtual employees' work.

The purpose of Info-Ergonomics can be defined as follows:

- Pursuing comfortable environment for employees (factory workers).
- Simulation and evaluation of employee's action with simplified human body.
- Modeling of skilled employees and storing their skills into Multimedia Database.

Info-Ergonomics provides the designers with modeling, evaluating, and visualizing tools for designing the optimized working environment for the factory employees.

As the whole informations are schematized and integrated based on the RWDB concept, the user can extract arbitral part of them and recognize multimedia data adapting to requrements of application.

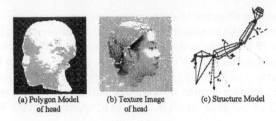

(a) Polygon Model
of head

(b) Texture Image
of head

(c) Structure Model

Figure 2: Human Model

The use of CG has significant advantages. For instance, Info-Ergonomics-based simulations allow more precise evaluation than real measurement because virtual employees can be made to do any more (even those causing pain, dangerous ones, etc) and, in addition, we can perform the evaluation itself at a much more detailed level visually.

And then, to replay the skilled employees' work in the Multimedia Database with CG, and to study from all angles is visual aids for beginners. In addition, we can keep safety and work space for employees, and design comfortable work environment.

3.2 The Human Body and Motion Modeling

In order to store and retrieve Human motion data, we need an integrated datamodel which can describe all kind of objects and materials about the human body, for use in all applications. Creating realistic model of the human body requires considerable amount of data because of object's complexity. Therefore from ergonomical point of view we must re-modeling the human body with small number of primitives. We call it "Simplified Human Body", which involves, for example, simplified head, arms, body and legs, connected by small number of joints each other.

The human body and motions can be described as follows.

1. Polygon Model

   Human body's figure can be described by using Polygon and Texture Models. An example of polygon model and texture image are shown in Figure.2(a)(b).

2. Structure Model

   Human body consists of more than 200 bones, which are connected so that the body can move using the power of the muscles [6]. But mechanism is too complex to model it completely. We focus on movable joints and select 24 of them in order to create a simplified model of the human body, Human Skeleton Model. Each child component of Human Skeleton Model has its own coordinate system, the origin of which is the connecting point (i.e. joint) to the parent component. An example of Human Skeleton Model's connections between parts are shown in Figure.2(c).

3. Motion Model

   For a certain employee's work, the action of each component of human body should be traced and modelled under the restriction of inter-joint structure. This Dynamic

Model is defined on the Human Skeleton Model discussed above. Each joint move along with time.

3.3 Mediator Concept

In order to handle a large number of entities like body–parts in databases, standarization or schematization of the description of such graphic objects are essential. For this purpose, we have proposed "Mediator" concept to describe object's characteristics with small amount of data. The mediator is an approximated model based on the meaning of the objects, with semi-automatic fitting on the sampled data from real world.

- Figure/Structure Mediator

  For example, as to figure of human head, we use a cylinder with 10 layer, polygonal column[2]. Model of a human head may have the additional information that the third layer from top is the height of eyes. Also, it is possible to supply position information that in the parameter of certain sections which vertex represents the left and right eye or the mouth. This way it becomes possible to supply general information about a certain type of parts, and integrate the individual parts data into a meaningful model representing each instance.

  We can assume most of objects consist of several "solid" parts and "joints" connecting parts each other. For example, a human body consists of head, neck, shoulder, chest, right upper arm, left upper arm, and so on, and there are specific joints between those parts with the specific DOF(Degree of freedom). Structure Mediator is defined as joint values – range of motion and other constraint between parts – for each person. Structure Mediator can be regarded as an extension of shape (figure) mediator which can describe not only "simple" "part" objects but "complex" object like human body. The above ideas are shown in Figure.3.

- Motion Mediator

  We have introduced "medator" to describe the shape and structure of parts of objects in abstract and uniform way. Now, we consider to extend this idea to motion description.

  Motion Mediator is defined as a sequence of intermediate postures within a motion which is done by a specific person. The intermediate posture must be preliminary assigned to each typical motion. (See Figure.4) The above intermediate postures can be detected by analyzing video images easily. Similar to the shape mediator, Motion mediator can represent the characteristics of a specific motion of a specific person, with a small amount of data. Therefore, from a set of motion mediators, the database user can calculate the average of the maximum height of right knee in a walking motion. Especialy for "work" motions in factory workers, detection of characteristic values from actual motions of workers is very important because we can pursue the optimum design of works in the human-machine co-operative environment. The authors group is now developing a formal description of generic motion and a motion evaluation system in the framework of RWDB.

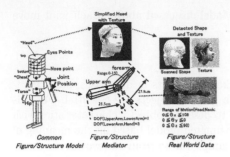

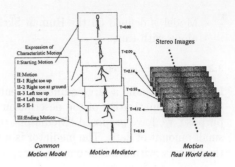

Figure 3: Common Model, Mediator and Real World Data about a figure and structure

Figure 4: Common Model, Mediator and Real World Data about a motion

4 CONCLUSION

In this paper we described a modeling techique of Human Body and its motion in the framework of Info-Ergonomics. The basic theory, Real World Database is also discussed. The Mediator concept is introduced that refers to the meaning of the actual data and therefore it can play the link between the acquired data from real world and its meaning. Many problems are left for future. Among them, we believe that the mediator can use not only "describe" but also "analyze" 3D objects and their behavior. This way it becomes possible to compute the spatial relation of the objects with fairly good precision and to use this cyberspace as the result of the queries. Another open problem is detection and schematization of higher-level semantics. To achieve them, much more "semantical" model based analysis might be required.

References

[1] S.Imai, K.Salev, T.Tomii, H.Arisawa: "Modelling of Working Processes and Working Simulation based on Info-Ergonomicas and Real World Database Concept", Proc. of 1998 Japan-U.S.A Symposium on Flexible Automation, pp.147-154, July, 1998

[2] Takashi Tomii, Szabolcs Varga, Sayaka Imai and Hiroshi Arisawa: "Design of Video Scene Databases with Mapping to Virtual CG Space", Proceedings of the 1999 IEEE International Conference on Multimedia Computing & Systems ICMCS'99, pp.741-746, Florence, Italy, 7-11 June, 1999.

[3] Michihiko Hayashi, Takashi Tomii, Hiroshi Arisawa: "A Form Acquisition of Subjects in Image Databases", IPSJ SIG Notes Vol.97, No.7, 97-DBS-111-13, 1997

[4] H. Arisawa, T. Tomii, H.Yui, H. Ishikawa: "Data Model and Architecture of Multimedia Database for Engineering Applications," IEICE Trans. Inf. & Syst. ,Vol.E78-D No.11, November,1995

[5] Kageyuu Noro: "Illustrated Ergonomics", JIS, 1990

[6] R. Nakamura, H. Saitoh: "Foundmental Kinematics", Ishiyakushuppan Cop,1995

[7] H.Arisawa, T.Tomii, K.Salev : "Design of Multimedia Database and a Query Language for Video Image Data", International Conference on Multimedia Conputing and Systems, IEEE, June 1996

Human Friendly Mechatronics (IMCA 2000)
E. Arai, T. Arai and M. Takano *(Editors)*

Author Index

Printed and bound by CPI Group (UK) Ltd, Croydon, CR0 4YY

03/10/2024

01040330-0009